Advances in Photoelectric Tracking Systems

Advances in Photoelectric Tracking Systems

Editors

Yao Mao
Yutang Wang

MDPI

Basel • Beijing • Wuhan • Barcelona • Belgrade • Novi Sad • Cluj • Manchester

Editors

Yao Mao
Institute of Optics and Electronics
University of Chinese Academy
Chengdu
China

Yutang Wang
Changchun Institute of Optics, Fine Mechanics and Physics
Chinese Academy of Sciences
Changchun
China

Editorial Office
MDPI
St. Alban-Anlage 66
4052 Basel, Switzerland

This is a reprint of articles from the Special Issue published online in the open access journal *Photonics* (ISSN 2304-6732) (available at: https://www.mdpi.com/journal/photonics/special_issues/TDMAJT6AKO).

For citation purposes, cite each article independently as indicated on the article page online and as indicated below:

Lastname, A.A.; Lastname, B.B. Article Title. *Journal Name* **Year**, *Volume Number*, Page Range.

ISBN 978-3-7258-0966-0 (Hbk)
ISBN 978-3-7258-0965-3 (PDF)
doi.org/10.3390/books978-3-7258-0965-3

Contents

photonics

MDPI

Editorial

Special Issue on Advances in Photoelectric Tracking Systems: An Overview

Jiuqiang Deng [1,2,3,4], Qiliang Bao [1,2,3,4], Yutang Wang [4,5,6,*] and Yao Mao [1,2,3,4,*]

1 National Key Laboratory of Optical Field Manipulation Science and Technology, Chinese Academy of Sciences, Chengdu 610209, China
2 Key Laboratory of Optical Engineering, Chinese Academy of Sciences, Chengdu 610209, China
3 Institute of Optics and Electronics, Chinese Academy of Sciences, Chengdu 610209, China
4 University of Chinese Academy of Sciences, Beijing 101408, China
5 The Key Laboratory of Airborne Optical Imaging and Measurement, Chinese Academy of Sciences, Changchun 130033, China
6 Changchun Institute of Optics, Fine Mechanics and Physics, Chinese Academy of Sciences, Changchun 130033, China
* Correspondence: ytwang@ciomp.ac.cn (Y.W.); maoyao@ioe.ac.cn (Y.M.)

1. Introduction

The photoelectric tracking system, which consists of optical, mechanical, electrical, and computer components, is in charge of tracking targets at extremely long distances. Tracking accuracy and stability accuracy are two of the system's most crucial performance metrics. Currently, quantum communication, free space laser communication, aerospace, space observation, and other fields make extensive use of the photoelectric tracking system [1,2]. Quantum and laser communication systems, which are based on the photoelectric tracking system, are able to create high-speed data transmission channels with great flexibility that are resistant to electromagnetic interference, and have low power consumption and huge capacity [3,4].

Tracking accuracy and stabilization accuracy are two crucial performance metrics of photoelectric tracking systems, and they are influenced by a variety of factors, such as mechanical attributes, sensor capabilities, drive characteristics, control algorithms, and more [5,6]. The tracking accuracy indicates how quickly the device can follow a moving target. The photoelectric tracking system often uses a composite axis design, which consists of a coarse tracking axis and a fine tracking axis [7]. The coarse tracking axis has a relatively low bandwidth and an obvious tracking error, due to the frame's large inertia. A tip-tilted mirror is the controlled object of the fine tracking axis, which has a high resonance frequency and low inertia. As a result, the fine tracking axis has a high bandwidth and small tracking error [8]. The coarse tracking axis' error is corrected by the fine tracking axis, so as to further increase the tracking accuracy of photoelectric tracking systems [9]. The ability of the system to suppress various disturbances is referred to as stabilization accuracy. The photoelectric tracking system is subject to external disturbances, including vibrations from the ground, carrier platforms, wind, and other sources [10]. Furthermore, friction, parameter perturbation, torque variation, and other unknown factors, collectively referred to as internal disturbances, are present in the photoelectric tracking system [11,12]. In photoelectric tracking systems, there are two types of disturbance suppression characteristics: passive and active. The vibration isolation structure, such as the spring, provides passive disturbance suppression, while the actuator's driving control provides active disturbance suppression [13,14]. The photoelectric tracking system is dependent on the image data for target recognition and tracking; hence, image signal processing is critical to the system. The photoelectric tracking system's image signal processing presents significant challenges in complicated settings with factors including object occlusion, ultra-long distance, inadequate brightness, and complex backgrounds [15].

Citation: Deng, J.; Bao, Q.; Wang, Y.; Mao, Y. Special Issue on Advances in Photoelectric Tracking Systems: An Overview. *Photonics* **2024**, *11*, 314. https://doi.org/10.3390/photonics11040314

Received: 21 February 2024
Revised: 20 March 2024
Accepted: 25 March 2024
Published: 28 March 2024

The purpose of this Special Issue, "Advances in Photoelectric Tracking Systems", is to present the most recent findings and creative solutions in the field. The following are the original research papers and review paper on optical and mechanical design, image processing, and motion control planning for photoelectric tracking systems.

2. An Overview of the Published Articles

The composite axis structure, comprising a coarse tracking loop driven by a rack and a fine tracking loop driven by the tip-tilt mirror, is typically used by photoelectric tracking systems. To increase tracking accuracy, the coarse tracking loop's residual tracking errors are corrected by the fine tracking loop. Consequently, designing and optimizing the tip-tilt mirror is crucial. A multi-objective optimization technique, based on a two-parameter coupling performance study, is presented by Contribution 1. The balanced solution of numerous goal functions is attained by introducing physical involvement, and the tip-tilt mirror's overall performance is enhanced. In order to address the issues with the high cross-linking value and poor capacity to rectify high-order wavefront aberration of piezoelectric deformable mirrors, a novel single-piezoelectric-piece deformable mirror driving structure was suggested in Contribution 2. By laser-cutting piezoelectric ceramics, the structure lowers the cross-linking value of deformable mirrors and lessens the coupling between driving units. X-Y polynomials are the basis for a large field of view and low-distortion three-mirror anastigmat system that is proposed for coarse-tracking optical imaging systems in Contribution 3. When compared to the conventional spherical or aspherical three-mirror anastigmat system, the suggested system uses a free-form surface, which increases the system's design flexibility and enhances its capacity to balance aberrations. To enhance the precision of measuring infrared radiation properties in the aviation domain, a double swing solid angle reflector-based infrared Fourier transform imaging spectrometer is devised in Contribution 4. The suggested hood can reduce external stray light point source transmission to an order of 10^{-4}.

One of the research focuses of the photoelectric tracking system is image processing technologies. Referencing partition processing, Jiang, C et al. suggest an enhanced spatial image non-uniformity correction technique to address the issue of uneven skylight background and detector noise in high-resolution imaging of ground-based photoelectric tracking systems in Contribution 5. This technique satisfies the need for real-time processing of high-resolution photographs under prolonged exposure conditions, and efficiently eliminates the uneven background of spatial images. In Contribution 6, an adversarial network with an error correction function is proposed as a joint input generation method to address reconstruction errors in compressed imaging using a high-resolution digital micromirror array. For a higher peak signal-to-noise ratio and improved visualization, the network models the deterioration of image quality brought on by alignment errors. Using the super-resolution technique to improve the optical system's image performance has significant practical implications. Super-resolution reconstruction is achieved through the application of kernel regression technology in Contribution 7, while micro-displacement information is obtained using a 2D scanning galvanometer. The technique enhances the 16-frame image's optical resolution by 39% in a lab setting. Furthermore, the low-altitude bird flock targets' detection ability is strongly impacted by the clutter reduction effect of ground objects. A beam-constraint-based collision detection approach is presented to increase ray tracing collision detection efficiency in Contribution 8.

The photoelectric tracking system is driven by the motion control unit to track and capture the moving target, based on the tracking error of the image processing system's output. The acquisition of moving targets is, hence, the basis for tracking in the photoelectric tracking system. A novel approach to target acquisition issues in airborne radar, laser communication, and other fields, is offered by Contribution 9, which suggests an enhanced scanning acquisition technique based on hexagonal spiral scanning. Liu, C et al. suggest a linear quadratic regulator optimal control approach based on the disturbance observer and reference model controller to enhance the dynamic properties of the photoelectric tracking system in Contribution 10. This technique increases the photoelectric tracking system's rising time, settling time, overshoot, and ability to suppress disturbances. Furthermore, Singh, J et al. suggest a joint transceiver equalization methodology for space division multiplexed multiple-input multiple-output optical fiber communication in Contribution 11. This technique is important for enhancing the spectral efficiency of critical dispersion-impaired transmission links by mitigating the dispersion impairments caused by 1.4 km of multi-mode fiber. To address the issues of sluggish convergence and the low precision of tracking errors in photoelectric tracking systems, a nonlinear differentiator based on the Softsign excitation function is suggested by Contribution 12. The backward step control approach is then used with this solution to address the complexity explosion problem.

3. Conclusions

In summary, photoelectric tracking systems are a typical example of an interdisciplinary research object. Researchers primarily focus on optical mechanical hardware, control algorithms, and image processing algorithms, all of which work together to foster the invention of photographic tracking systems. The growth of laser communication, quantum communication, astronomical observation, and other sectors is being quickly boosted by photoelectric tracking systems. As we come to the end of this Special Issue, we hope that readers will gain an improved comprehension of photoelectric tracking systems, and that this will inspire additional scholars to enter the field to further the world through exceptional research.

Funding: This project was sponsored by the Special Research Assistant Program, Chinese Academy of Sciences, China (No. E329691C21), Natural Science Foundation of Sichuan Province for Youths, China (No. 24NSFSC3777), and the Key Research Program of Frontier Sciences, Chinese Academy of Sciences, China (No. ZDBS-LY-JSC044).

Acknowledgments: The guest editors would like to express sincere thanks to all of the authors and reviewers for their contributions to this Special Issue as well as their appreciation to the *Photonics* editors for their outstanding support.

Conflicts of Interest: The authors declare no conflicts of interest.

List of Contributions

1. Dong, Q.; Wang, Q.; Wang, C.; Luan, Y.; Wang, X.; Wang, X. Multiobjective optimization of SiC mirror based on dual-parameter coupling. *Photonics* **2023**, *10*, 171. https://doi.org/10.3390/photonics10020171.
2. Li, B.; Tian, D. Comparative analysis of the effect of cutting piezoelectric ceramics on optically oriented compensation Capability. *Photonics* **2023**, *10*, 1136. https://doi.org/10.3390/photonics10101136.
3. Jia, B.; Li, Y.; Lv, Q.; Jin, F.; Tian, C. Design of a large field of view and low-distortion off-axis optical system based on a free-form surface. *Photonics* **2023**, *10*, 596. https://doi.org/10.3390/photonics10050506.
4. Ben, C.; Shen, H.; Yu, X.; Meng, L.; Cheng, H.; Jia, P. Stray light analysis and suppression for an infrared fourier imaging spectrometer. *Photonics* **2024**, *11*, 173. https://doi.org/10.3390/photonics11020173.
5. Jiang, C.; Chen, T.; Lu, C.; Wu, Z.; Liu, C.; Shao, M.; Cao, J. Automatic inhomogeneous background correction for spatial target detection image based on partition processing. *Photonics* **2023**, *10*, 433. https://doi.org/10.3390/photonics10040433.
6. Xu, M.; Wang, C.; Shi, H.; Fu, Q.; Li, Y.; Dong, L.; Jiang, H. Deep compressed super-resolution imaging with DMD alignment error correction. *Photonics* **2023**, *10*, 581. https://doi.org/10.3390/photonics10050581.
7. Ma, T.; Liang, C.; Han, Y.; Yuan, F.; Meng, L.; Xu, Y.; Shen, H.; Liu, Y. Super-resolution imaging enhancement through a 2D scanning galvanometer: Algorithm Formulation and Application in Aerial Optoelectronic Systems. *Photonics* **2023**, *10*, 113. https://doi.org/10.3390/photonics10111203.

8. Su, Z.; Sang, L.; Hao, J.; Han, B.; Wang, Y.; Ge, P. Research on ground object echo simulation of avian lidar. *Photonics* **2024**, *11*, 153. https://doi.org/10.3390/photonics11020153.
9. Jia, B.; Jin, F.; Lv, Q.; Li, Y. Improved target laser capture technology for hexagonal honeycomb scanning. *Photonics* **2023**, *10*, 541. https://doi.org/10.3390/photonics10050541.
10. Liu, C.; Mao, Y.; Qiu, X. Disturbance-observer-based LQR tracking control for electro-optical system. *Photonics* **2023**, *10*, 900. https://doi.org/10.3390/photonics10080900.
11. Singh, J.; Ahrens, A. Joint-transceiver equalization technique over a 1.4 km multi-mode fiber using optical MIMO technique in IM/DD systems. *Photonics* **2023**, *10*, 696. https://doi.org/10.3390/photonics10060696.
12. Li, J.; Zhuang, S.; Wang, H.; Deng, J.; Mao, Y. Design of backstepping control based on a softsign linear-nonlinear tracking differentiator for an electro-optical tracking system. *Photonics* **2024**, *11*, 156. https://doi.org/10.3390/photonics11020156.

References

1. Yin, J.; Cao, Y.; Li, Y.H.; Liao, S.K.; Zhang, L.; Ren, J.G.; Cai, W.Q.; Liu, W.Y.; Li, B.; Dai, H.; et al. Satellite-based entanglement distribution over 1200 kilometers. *Science* **2017**, *356*, 1140–1144. [CrossRef] [PubMed]
2. Liao, S.K.; Cai, W.Q.; Liu, W.Y.; Zhang, L.; Li, Y.; Ren, J.G.; Yin, J.; Shen, Q.; Cao, Y.; Li, Z.P.; et al. Satellite-to-ground quantum key distribution. *Nature* **2017**, *549*, 43–47. [CrossRef] [PubMed]
3. Li, Z.; Mao, Y.; Qi, B.; Zhou, X.; Liu, Q.; Zhou, Q. Research on control technology of single detection based on position correction in quantum optical communication. *Opto-Electron. Eng.* **2022**, *49*, 210311.
4. Luo, Y.; Liu, K.; Yang, F.; Wen, X.; Huang, Y.; Guo, S.; Ren, G.; Li, T. Observation and compensation control of sliding mode compound layered interference for the fast steering mirror system. *Opto-Electron. Eng.* **2023**, *50*, 220330.
5. Bharathi, M.; Bhatt, V.; Kumar, V.R.; Sharma, R.J.; Hemavathi, S.; Pant, B.; Arani, R.P.; Sathish, T.; Mohanavel, V. Developing a dual axis photoelectric tracking module using a multi quadrant photoelectric device. *Energy Rep.* **2022**, *8*, 1426–1439.
6. Lu, C.Y.; Cao, Y.; Peng, C.Z.; Pan, J.W. Micius quantum experiments in space. *Rev. Mod. Phys.* **2022**, *94*, 035001. [CrossRef]
7. Qi, B.; Chen, H.; Ren, G.; Huang, Y.; Ding, K.; Ma, J. ATP technology for 100-kilometer quantum entanglement distribution experiment. *Opt. Precis. Eng.* **2013**, *21*, 1628–1634.
8. Wu, Y.; Wen, L.; Tang, T. Line-of-sight stabilization enhancement with hybrid sensing in a piezoelectric mirror-based cubic stewart platform. *IEEE Trans. Aerosp. Electron. Syst.* **2024**, 1–11. [CrossRef]
9. Cao, K.; Hao, G.; Li, B.; Du, H.; Tan, L. Modified steady tracking strategy for composite axes system in free-space laser communication. *Opt. Eng.* **2023**, *62*, 078101. [CrossRef]
10. Glück, M.; Pott, J.U.; Sawodny, O. Piezo-actuated vibration disturbance mirror for investigating accelerometer-based tip-tilt reconstruction in large telescopes. *IFAC-PapersOnLine* **2016**, *49*, 361–366. [CrossRef]
11. Tang, T.; Niu, S.; Ma, J.; Qi, B.; Ren, G.; Huang, Y. A review on control methodologies of disturbance rejections in optical telescope. *Opto-Electron. Adv.* **2019**, *2*, 190011. [CrossRef]
12. Luo, Y.; Mao, Y.; Ren, W.; Huang, Y.; Deng, C.; Zhou, X. Multiple fusion based on the CCD and MEMS accelerometer for the low-cost multi-loop optoelectronic system control. *Sensors* **2018**, *18*, 2153. [CrossRef] [PubMed]
13. Tang, T.; Qi, B.; Yang, T. Youla–Kucera parameterization-based optimally closed-loop control for tip–tilt compensation. *IEEE Sens. J.* **2018**, *18*, 6154–6160. [CrossRef]
14. Bian, Q.; Wang, Y.; Ruan, Y.; Tang, T. Accelerometer-assisted disturbance feedforward control of an inertially stabilized platform. *IEEE Sens. J.* **2023**, *23*, 9880–9888. [CrossRef]
15. Yang, H.; Li, W. Performance measurement of photoelectric detection and target tracking algorithm. *Int. J. Smart Sens. Intell. Syst.* **2015**, *8*, 1554–1575. [CrossRef]

photonics

Article

Stray Light Analysis and Suppression for an Infrared Fourier Imaging Spectrometer

Chenzhao Ben [1,2], Honghai Shen [1], Xiao Yu [1], Lingtong Meng [1], Huishi Cheng [1,2] and Ping Jia [1,*]

1 Key Laboratory of Airborne Optical Imaging and Measurement, Changchun Institute of Optics, Fine Mechanics and Physics, Chinese Academy of Sciences, Changchun 130033, China; benchenzhao21@mails.ucas.ac.cn (C.B.); shenhh@ciomp.ac.cn (H.S.); yuxiao17@mails.ucas.ac.cn (X.Y.); zy1517111@buaa.edu.cn (L.M.); chenghuishi21@mails.ucas.ac.cn (H.C.)
2 Daheng College, University of Chinese Academy of Sciences, Beijing 100049, China
* Correspondence: jiap@ciomp.ac.cn

Abstract: To improve the accuracy of infrared radiation characteristics measurement in the aviation field, an infrared Fourier transform imaging spectrometer based on a double-swing solid angle reflector was designed. This imaging spectrometer operates in the 3–5 μm wavelength range and has a field of view of $1.7° \times 1.7°$. This article presents a comprehensive analysis of the system's stray light and also studies the impact of external stray light on the imaging quality, along with the influence of internal stray light on the interference effects and the spectral resolution. It also present the design of a hood that suppresses the point source transmittance of the external stray light down to the order of 10^{-4}. Based on this, we propose a method that incorporates the introduction of wedge and inclination angles. Additionally, a numerical range is provided for the addition of these angles on the beam splitter mirror and compensation plate. This ensures the effective suppression of any internal stray light. This study fills the gap in the knowledge about Fourier transform imaging spectrometers operating in the mid-infrared band for aviation applications, and proposes a suppression method suitable for interference systems, which is also suitable for Fourier transform imaging spectrometers based on other types of interferometers. This study broadens the application field of Fourier transform imaging spectrometers in stray light, and has great significance to promote the development of Fourier transform imaging spectrometer.

Keywords: Fourier transform imaging spectrometer; mid-infrared band; stray light analysis; interference fringe contrast; point source transmittance

Citation: Ben, C.; Shen, H.; Yu, X.; Meng, L.; Cheng, H.; Jia, P. Stray Light Analysis and Suppression for an Infrared Fourier Imaging Spectrometer. *Photonics* **2024**, *11*, 173. https://doi.org/10.3390/photonics11020173

Received: 11 January 2024
Revised: 8 February 2024
Accepted: 9 February 2024
Published: 12 February 2024

1. Introduction

At present, in modern aviation-based high-tech warfare, methods to enable the rapid identification of camouflage and false targets have become an important research field and a major application direction for military reconnaissance [1,2]. With the increasing demand for the application of imaging spectrometers in a variety of fields, the development of the Fourier transform imaging spectrometer [3–6] has also become a research hot spot. The fundamental task of the Fourier transform imaging spectrometer is to detect and acquire the image signal and the spectral function of a specific target [7,8]. During the target detection process, stray light, as a non-imaging beam that can be received by the detector [9], will cause degradation of the imaging quality of the optical system and will reduce the signal-to-noise ratio of the imaging spectrometer. To meet the technical requirements of the Fourier imaging spectrometer [10–13] and provide high performance, researchers are now paying increasing attention to the study of stray radiation.

As part of the complex development process for Fourier transform imaging spectrometers, the influence of stray light on these spectrometers and methods to suppress it have become an important research issue in several countries. Du et al. analyzed and calculated the stray light characteristics [14] of a spaceborne Fabry-Pérot interference imaging spectrometer operating in the visible band, and then suppressed the stray light in this system by adding a blocking stop to the system structure, thus reducing the stray radiation ratio effectively. Chen et al. from the Beijing Space Electromechanical Research Institute analyzed the effects of multiple reflections of stray light on the modulation regime [15,16] in a near-infrared band hyperspectral resolution Fourier transform spectrometer, and proposed the introduction of wedge angle and tilt angle suppression methods to separate the stray light spots from the normal light spots. Dussarrat et al. from the European Meteorological Satellite Organization analyzed the stray light present in the background of a Michelson interferometer in a far-infrared Fourier transform spectrometer [17] and proposed relevant correction methods for image processing. The research described above indicates that at this stage, there have been certain studies of stray light analysis for spaceborne visible, near infrared, and far infrared interference imaging spectrometers; various stray light suppression schemes have also been proposed for spectrometers with different structures, thus effectively reducing the effects of stray light on these optical systems. There have been few reports, however, on the analysis and suppression of the stray light in Fourier transform imaging spectrometers for use in the aviation field.

The purpose of this paper is to improve the imaging quality and spectral resolution of mid-infrared Fourier transform imaging spectrometers in the airborne aviation field, and analyze the influence of stray light on spectral resolution according to the potential situation in actual processing. In this paper, a new mid-infrared Fourier transform imaging spectrometer based on a double-oscillating solid angle reflector is proposed. To provide a comprehensive analysis of the impact of stray light on these systems, theoretical research is performed on both the stray light that occurs outside the field of view and the stray light within the system. By designing the internal and external hood and blocking rings for the Cassegrain telescope system, the external stray light is effectively suppressed. On this basis, the non-ideal transmission phenomenon inside the interference system is analyzed. The decrease in the contrast of interference fringes and spectral resolution caused by the reflected stray light produced by this phenomenon is studied. Finally, the scheme of introducing wedge angle and inclination angle is proposed. The theoretical research and analysis of this phenomenon is carried out, which can effectively suppress the influence of this phenomenon.

At present, the Beijing Space Electromechanical Research Institute has conducted stray light analysis and suppression on the visible to near-infrared band of the Fourier transform interferometer with high spectral resolution on Gaofen-5 [18], and finally obtained the impact of stray light on the contrast of interference fringes. In this paper, the stray light analysis and suppression of a Fourier imaging spectrometer in the mid-infrared band are carried out, and finally the influence of stray light on the contrast of interference fringes is further translated into the spectral resolution. Therefore, this study has a more direct application value for the spectral resolution index parameters of Fourier imaging spectrometer.

In this paper, the TracePro 7.4 software was used to trace the stray light path to study and analyze the stray light phenomenon. The designed model and the results will offer reference values for the development of spectrometers for military and civil applications. Moreover, the innovative design of the optical system and the analysis of stray light in this paper play a driving role in improving the imaging quality and spectral resolution of imaging spectrometers in military and civilian fields. The measurement accuracy of the radiation characteristics of the imaging spectrometer in the middle infrared band can be effectively improved.

2. Optical System

2.1. Solid Angle Mirror

The system proposed in this work uses a solid angle mirror rather than a plane mirror and omits the system for compensation of the plane mirror moving tilt [19] caused by vibrations.

A solid angle mirror is an optical instrument consisting of three plane mirrors. The solid angle mirror is based on the law of light reflection and the structure of the angle reflector, when the light is incident on the angle mirror's three mutually perpendicular reflecting surfaces, it will be reflected by the three plane mirrors and emitted in the opposite direction of the incident light. The structure diagram of the solid angle mirror is shown in Figure 1.

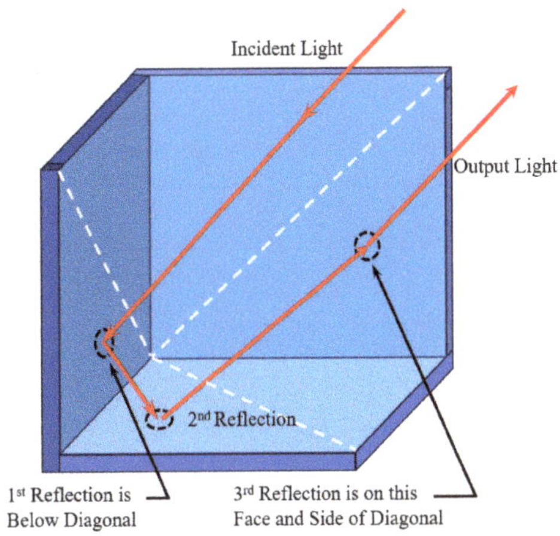

Figure 1. Structure diagram of solid angle mirror.

2.2. The Front Telescope System

The front telescope system uses a Cassegrain coaxial folding transmission structure [20] to achieve high imaging quality and a compact structure; this structure also avoids the difficulties involved in the fabrication and processing of large-aperture transmissive optical infrared materials required by the conventional transmission system. The Cassegrain telescopic system structure is generally a reflective system composed of a primary mirror and a secondary mirror. The structure diagram of the Cassegrain telescopic system is shown in Figure 2.

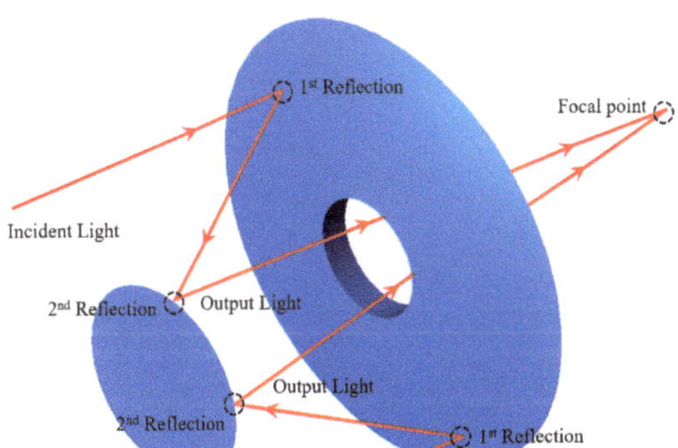

Figure 2. Structure diagram of Cassegrain telescopic system.

As shown in the Figure 2, first, the primary reflector reflects and focuses light onto a single point, creating a sharp image. The secondary mirror then reflects this image onto the focal point of the telescope. By precisely designing and manufacturing the shape and position of the two mirrors, it is possible to ensure that the light remains focused after passing through the mirrors and a high-quality image is formed. To achieve compact size, Cassegrain telescope structures are typically designed with folded optical paths. This design reduces the length and diameter of the telescope, making it easier to transport and install.

Compared with the traditional transmission system, the Cassegrain system has the advantages of having a large aperture, long focal length, and small size, and being multi-band, colorless and so on. Therefore, it is widely used in aerial optical systems. The characteristics of common telescopic systems are compared as shown in Table 1.

Table 1. Comparison of characteristics of common telescopic systems.

Type	Characteristic
Traditional transmission structure	Low installation difficulty, low cost
Off axis three reverse	The center is open, high installation difficulty, high cost
Cassegrain system	Large aperture, long focal length, small size, colorless

2.3. Overall Optical System

The interference system uses a swinging scanning structure [21] in place of the traditional translation and push sweep structure, which increases the optical path length and improves the spectral resolution; finally, the most commonly used silicon and germanium materials, which offer good stability and mature processing technology, are selected as the lens materials [22]. The proposed optical system consists of three parts: the front telescope system, the interference system, and the rear imaging system. A diagram of the optical system is shown in Figure 3.

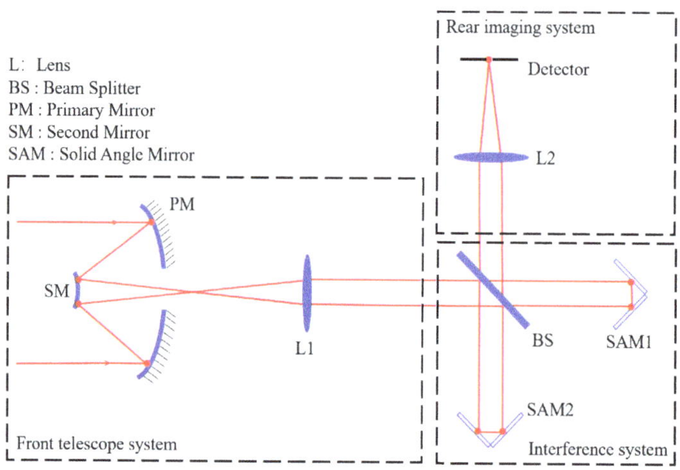

Figure 3. Optical structure of the mid-wave infrared Fourier imaging spectrometer.

The front telescope system can achieve five times the beam contraction performance, and can collimate the beam into the interference system to match with the pupil plane of the interferometer. Then, by swinging the solid angle reflectors SAM1 and SAM2, the optical path is changed such that the beam interferes at the beam splitting system. The rear imaging system is a secondary imaging system that realizes 100% matching between the exit pupil and the cold stop of the detector [23]; the interference beam is then focused on the detector through the rear imaging system and forms continuously varying interference fringes. The parameters of the proposed optical system are listed in Table 2.

Table 2. Parameters of the optical system for the medium-wave infrared Fourier imaging spectrometer.

Parameter	Value
Spectral range (μm)	3~5
Diameter (mm)	105
FOV (°)	1.7
F number	4
MTF@35 lp/mm	>0.5
Relative distortion	<1%
Energy concentration	>70%
Infrared detector pixel size (μm)	30
Number of infrared detector pixels	320 × 256

The mechanical assembly diagram for this system is shown in Figure 4. Germanium was selected as the material for the beam splitter mirror and the compensation plate, and the surfaces of the primary and secondary mirrors and the solid angle mirrors of the Cassegrain telescope system are plated with gold films to achieve high reflection effects.

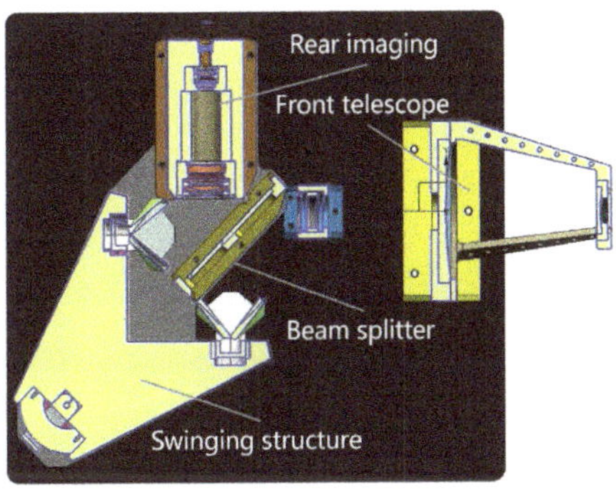

Figure 4. Mechanical structure of the mid-wave infrared Fourier imaging spectrometer.

3. Stray Light Sources and Evaluation Criteria

3.1. Sources of Stray Light

Based on the characteristics of infrared imaging spectrometers and the analysis method used for stray light in traditional optical systems, the following potential sources of stray light were determined [24]: (1) Stray light can be caused by light originating outside the system's field of view entering the optical system, which mainly comes from direct irradiation by sunlight, surface radiation, and atmospheric scattering. The stray light is formed after it reaches the image plane through the optical system, and this light is generally present in various imaging systems. (2) Imaging target stray light can originate within the field of view, i.e., imaging target light that reaches the image plane by passing through the system along an abnormal imaging path; this light is mainly generated from the imaging target light through residual reflection, scattering, and diffraction processes on the surfaces of the optical and structural elements.

3.2. Harm Caused by Stray Light

In the imaging optical system of an infrared Fourier imaging spectrometer, stray light will increase the noise that occurs on the image plane, and the stray light convergence point near the image plane will have a particularly serious impact on the imaging performance [25]. For the Michelson interferometer system, because the beam interference occurs inside the system, the quality of the interference effect becomes the main factor that affects the ability to recover the spectrum, and the interference fringe contrast [26] is one of the criteria used to evaluate the quality of the interference effect. When stray light is generated in the interference system through reflection processes, the interference phenomenon that occurs will be aliasing; the interference fringe contrast will then be reduced, and this will have a serious impact on the spectrum recovery capability.

3.3. Design of the Anti-Stray Light Structure

As part of the design of the anti-stray light structure, the external hood design must usually meet the following principles [27]: (1) non-imaging light should be prevented from reaching the image surface directly; (2) normal light from the edge field of view must not be blocked by the system hood; (3) strong stray light sources such as sunlight at angles greater than the avoidance angle may only enter the optical system after at least two or more scattering processes; (4) black coatings with high absorption rates should be used. The external hood design was thus completed according to the principles above. For a general Earth observation system, a first-level hood [28] can meet these requirements. The external hood used by the system proposed in this paper is shown in Figure 5.

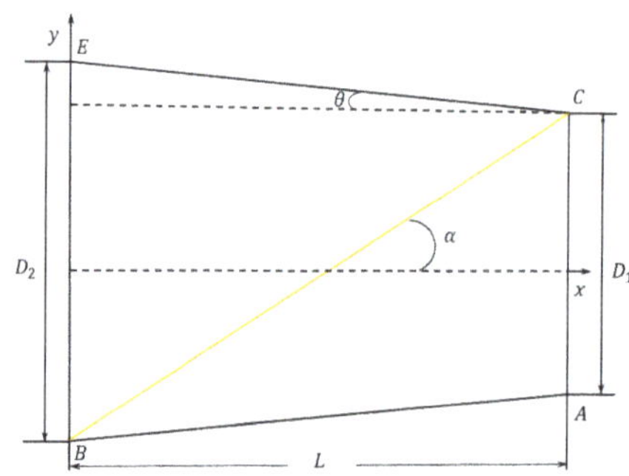

Figure 5. Schematic design for the external hood.

The Sun's avoidance angle α is known to be 30°, θ is the system's field of view angle and has a value of 0.8°, L is the hood length, D_1 is the diameter of the inner aperture of the hood, i.e., the diameter of the system's entrance pupil, and D_2 is the diameter of the outer aperture of the hood. The following relationships can be obtained from the geometric relationship shown in Figure 3:

$$D_2 = D_1 + 2L\tan\theta \quad (1)$$

$$D_1 + L\tan\theta = L\tan\alpha \quad (2)$$

When the system's field of view angle θ = 0.8°, the entrance pupil diameter D_1 = 105 mm, and the avoidance angle α = 30° are known, then values of D_2 = 110.20 mm and L = 186.37 mm can be obtained using Equations (1) and (2).

Aviation imaging systems require a high stray light suppression effect and it is necessary to add a blocking ring inside the hood for these systems. The blocking ring can prevent stray light from being caused by multiple reflections inside the hood. The blocking ring usually takes the form of a black coating with a high absorption rate. The hood structure [29], when combined with a cylindrical outer wall and a conical inner wall, can effectively reduce the number of blocking rings required and increase the depth of these blocking rings, and it is thus highly conducive to the elimination of stray light; therefore, the structure shown in Figure 6 was used for the design of the blocking rings.

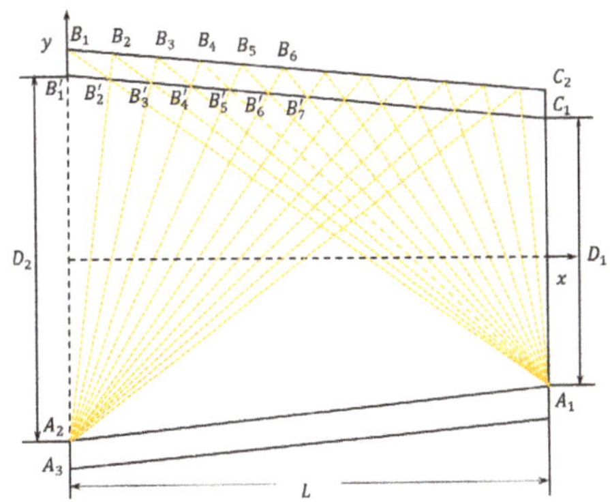

Figure 6. Schematic diagram for the design of the inner blocking rings of the external hood.

Connecting the upper vertex B_1 of the external hood and the upper inner wall B_1' of the cone gives the first stop ring position; A_1 at the main mirror of the lower inner wall of the cone is then connected with B_1 at the edge of the external hood, and the intersection of this line with the upper inner wall of the cone at point B_2' is the second stop ring position. Then, by connecting A_2, B_2', and the upper wall of the external hood, which intersects at B_2, and continuing in a similar manner, the positions of all of the blocking rings in the hood can be determined.

According to the design principle for the inner hood [30], the inner hood sizes for the primary mirror and the secondary mirror in this system will be determined using the focal length of the Cassegrain system, the focal length of the primary mirror, the positions of the stop and the back intercept, and other parameters. The design process is illustrated in Figure 7.

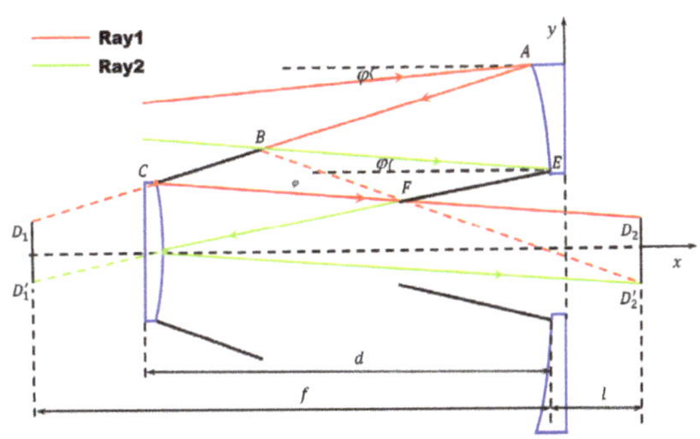

Figure 7. Schematic of the design of the internal hood.

As shown in Figure 7, ray 1 and ray 2 are the two edge rays, φ is the half-field angle, d is the distance between the primary mirror and the secondary mirror, f is the focal length, and l is the back intercept. The apex of the hood in the secondary mirror is the point of intersection of light rays 2 and 1 when reflected by the primary mirror, and the apex of the hood in the primary mirror is the point of the intersection of light ray 2, when reflected by the primary mirror, and light ray 1, when reflected by the primary mirror and the secondary mirror in turn.

The front aperture and the rear aperture of the hood in the primary mirror are 29.4 mm and 52.85 mm in diameter, respectively. The front aperture diameter of the secondary lens hood is 47.98 mm and the rear aperture diameter is 57.03 mm. ProE 5.0 software was used to model the hood. The stereogram designed for the external hood is shown in Figure 8a, and the stereogram for the inner hood is shown in Figure 8b.

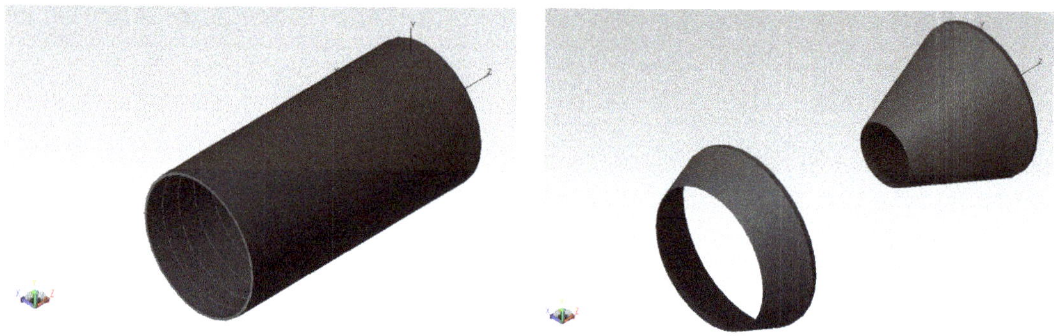

(a) (b)

Figure 8. Structural drawings of the hood. (**a**) External hood structure. (**b**) Inner hood structure.

3.4. Evaluation Criteria for Stray Light

The point source transmittance (PST) is a transfer function that is commonly used to describe the effects of stray light on optical systems [31]. The PST is equal to the total amount of stray light incident on the focal plane of an optical system divided by the total amount of light incident on the optical system that enters the pupil, and it can be calculated as follows:

$$PST = \frac{E_{SL}}{E_{INC}} \quad (3)$$

In Equation (3), E_{SL} is the irradiance of the stray light on the focal plane and E_{INC} is the irradiance when the light emitted by a point light source located at infinity (through collimation) is incident on the plane perpendicular to the source. If the irradiance of the light source is normalized to 1, then the irradiance on the image surface is the PST.

4. Stray Light Analysis

4.1. Stray Light Model Establishment

At present, the software packages with stray radiation analysis functionality generally include LightTools 8.6, TracePro 7.4 and ASAP 2019, and among these packages, TracePro 7.4 software is compatible with ProE 5.0, SolidWorks 2023, and other mechanical design software. At the same time, TracePro 7.4 is also compatible with ZEMAX OpticStudio 2023 and CODE V 11.5 optical design software, and has a user-friendly operation interface [32]. TracePro 7.4 is a powerful lighting and optical design simulation software that simulates and analyzes the propagation and behavior of all kinds of light, including stray light. In TracePro 7.4, users can evaluate and analyze the effects of stray light by simulating the propagation and interaction of light in an optical system. This includes the simulation and analysis of light sources at different angles, reflections, and refractions of optical surfaces, internal and external scattering, etc. The comparison table of the functions of different stray light analysis software is shown in Table 3.

Table 3. Comparison of functions of different stray light analysis software.

Function	TracePro	Light Tools	ASAP
Interface operation	Easy	Difficult	Difficult
Simulation accuracy	High	Higher	Higher
Operational speed	High	High	Higher
Modeling function	Comprehensive	Partial	Comprehensive

In this paper, TracePro software was used to simulate the optical surface and analyze the stray light, so that the source, degree and influence of stray light in the infrared Fourier imaging spectrometer designed in this paper can be understood, and different measures can be taken to reduce stray light according to different sources of stray light, and finally achieve the purpose of optimizing optical design.

4.2. Analysis of the System Stray Light

Reasonable settings and distributions of the surface properties of the different optical elements can provide better simulations of the actual situation. The specific parameters that were set in this process are listed in Table 4.

Table 4. Optical surface parameter settings.

Element	Transmittance	Absorptivity	Reflectance
Lens surface	0.99	0.01	0
Solid angle mirror surface	0	0.01	0.99
Mirror surfaces	0	0.01	0.99
Detector surface	0	0.99	0.01

Because the optical system is located far away from the Sun, the Sun can be regarded as a point light source, and reverse light tracing is required to locate the important surfaces. The specific light source setting parameters are listed in Table 5. The positioning of the mirror is based on the origin of the axes.

Table 5. Light source model settings.

Light Source	Solar Radiation
Light source type	Circular source
Light source size	Diameter 110 mm
Light source position	−200 mm
Number of rays	10 million
Ray accuracy	10^{-7}

As a result of the symmetry of the Cassegrain system, only the off-axis angle from 0° to 90° must be taken into account when analyzing the orientation of the sunlight. In this paper, the stray light analysis was performed on optical systems without optimization or with an added hood. The PST was analyzed for systems with off-axis angles ranging from 0° to 80°, and the PST of the solar radiation was calculated at various off-axis angles via ray tracing. The analysis results are shown in Figure 9.

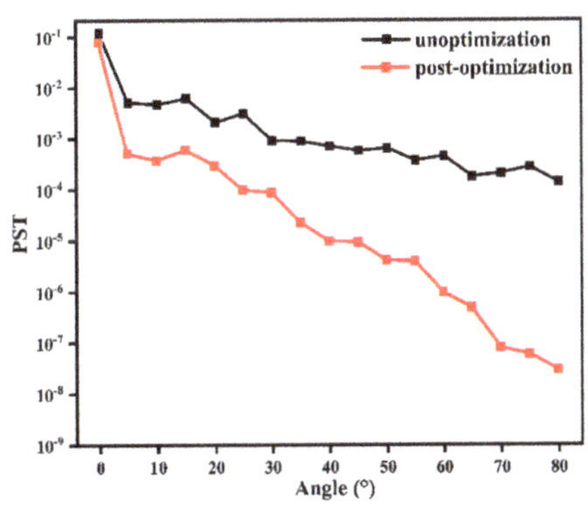

Figure 9. Designs of the PST analysis curves before and after use of the hood.

Through a simulation analysis, it was found that in the system without the hood, the PST is mainly distributed with values of the order of 10^{-4}, which will have a considerable impact on the imaging quality of the system. After the hood is set, the PST can be controlled to be within the order ranging from $10^{-4} \sim 10^{-7}$ at angles between 25° and 70°, whereas at angles above 70°, the PST is basically controlled to be below the order of 10^{-8}. The imaging effect of the system will be unaffected. The feasibility of the suppression scheme design proposed in this paper was verified by a data comparison process. The designs of the internal and external hoods and the blocking ring can suppress the influence of stray light on the system effectively to a certain extent.

5. Stray Light Interfering with the System

5.1. Establishment of an Interference Model

Because the core aspect of interference spectroscopy technology is the extraction of an interferogram, the interference effect becomes the key factor affecting the system's spectral recovery capability. When incident light with the wave number γ interferes, the interference signal light intensity [33] satisfies the following equation:

$$I(x) = \int_0^{\Omega} I_R(\gamma) + I_T(\gamma) + 2\sqrt{I_R(\gamma)I_T(\gamma)}\cos\,(2\pi\gamma x\cos\,\omega)d\Omega \tag{4}$$

In Equation (4), γ is the wave number of the light source, $I_R(\gamma)$ is the light intensity on the interference system's reflected light path, $I_T(\gamma)$ is the light intensity on the interference system's transmitted light path, x is the optical path difference, ω is the half angle of the light source's field of view, Ω is the solid angle of the light source's field of view, and $\Omega = 2\pi(1 - \cos\,\omega)$.

The contrast of the interference fringes has become the standard used to evaluate the quality of the interference effect, and the contrast M of the interference fringes is determined using the following equation:

$$M = \frac{I_{max} - I_{min}}{I_{max} + I_{min}} \tag{5}$$

In Equation (5), I_{max} and I_{min} are the maximum and minimum values of the light intensity $I(x)$, respectively.

In the Fourier transform imaging spectrometer [34], when incident light with the wave number γ interferes, Equation (4) can be substituted into Equation (5), and the equation for the interference fringe contrast is as follows:

$$M = \frac{2\sqrt{I_R(\gamma)I_T(\gamma)}}{I_R(\gamma) + I_T(\gamma)} \tag{6}$$

In an ideal Fourier transform imaging spectrometer, where it is assumed that the ideal device does not absorb light, if $I_R(\gamma) + I_T(\gamma) = 1$, then if and only if $I_R(\gamma) = I_T(\gamma)$, the interference fringe contrast is at a maximum. However, if $I_R(\gamma) \neq I_T(\gamma)$, the interference fringe contrast will be affected, and then $M < 1$.

5.2. Simulation of an Ideal Interferogram

In this work, a monochromatic light source was used to verify the interference phenomenon in the optical system. A rectangular light source was selected in this case and the incident wave was a plane wave. The reflectance of the solid angle mirror was set to have the ideal value of 100%, and the transmittance and reflectance of the back surface of the beam splitter were both set at 50%. Finally, the interference phenomenon of the 4.44 µm wavelength beam in the optical system was simulated using simulation analysis software, and the ideal interferogram obtained is as shown in Figure 10, where the contrast of the interference fringe $M = 1$.

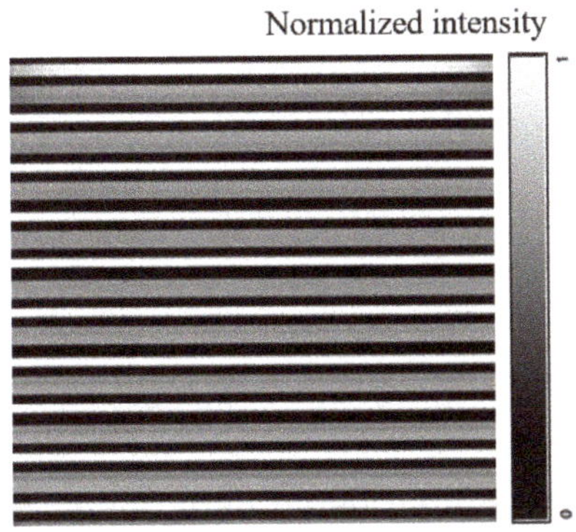

Figure 10. Ideal interferogram.

5.3. Michelson Interferometer Stray Light Analysis

Because the Fourier transform imaging spectrometer is based on the working principle of the Michelson interferometer [35], the corresponding spectral diagram is obtained by performing a Fourier transform on the interferogram obtained from the interference, and this means that the interference effect will affect the spectral recovery accuracy of the imaging spectrometer directly. However, the transmittance values of the beam splitter and the compensator in the interference system cannot reach their ideal 100% values during actual processing. Therefore, multiple reflections will occur between the beam splitter and the compensator, and the stray light formed by these reflections will enter the design field of view parallel to the normal light; this stray light cannot be filtered out by the cold stop, and will then affect the interference modulation system, leading to reduced spectral resolution. The stray light path obtained by this reflection process is as shown in Figure 11.

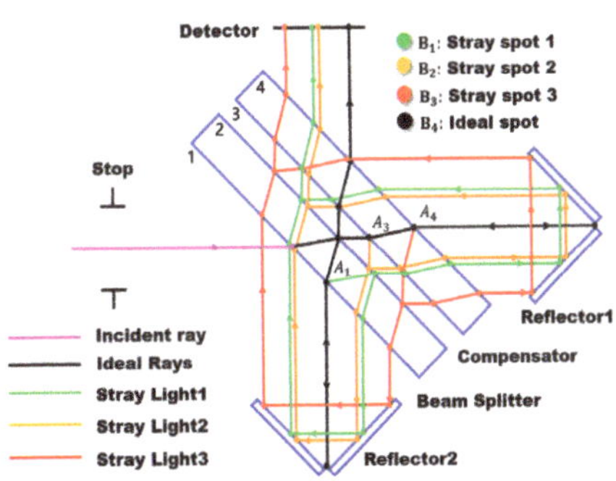

Figure 11. Schematic diagram of the stray light path.

As shown in Figure 11, when the transmittance values of the beam splitter and the compensator cannot reach their ideal values of 100%, the light that enters the interference system will be reflected at points A_1, A_3, and A_4 at surfaces 1, 3, and 4, respectively, thus forming new stray light beams that enter the detector, in which these beams cross with the light from different fields of view; this affects both the imaging quality of the system and the interference effect of the system. Assuming again that the ideal optical device does not absorb light, the reflectivity values of surface 1, surface 3, and surface 4 are given by 100% minus their transmittance. We therefore simulated the actual processing conditions. As shown in Table 6, the different transmittance values of beam splitter surface 1 and compensator surfaces 3 and 4 were set within the range from 95% to 100%. The other lenses in the system were set at the ideal transmittance of 100%.

Table 6. Transmittance setting table for different anti-reflection coatings on opposing sides 1, 3, and 4.

Transmittance	Reflectance
0.99	0.01
0.98	0.02
0.97	0.03
0.96	0.04
0.95	0.05

The optical path shown in Figure 11 was simulated and the corresponding interference graphs generated under the different transmittance conditions are shown in Figure 12.

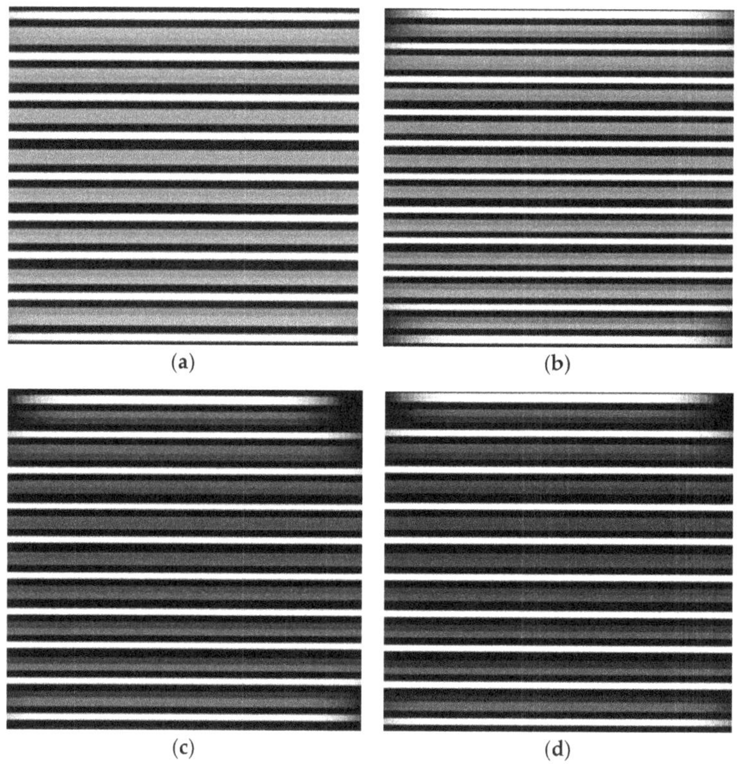

Figure 12. Interferograms generated by the optical system at different transmittance values. (**a**) T = 99%; (**b**) T = 98%; (**c**) T = 97%; and (**d**) T = 96%.

The maximum value of the light intensity I_{max} and the minimum value of the light intensity I_{min} corresponding to the interference fringes were obtained via a simulation analysis, and were then substituted into Equation (6) to perform the calculations. The calculated data were then sorted and used to draw a dot plot, as shown in Figure 13.

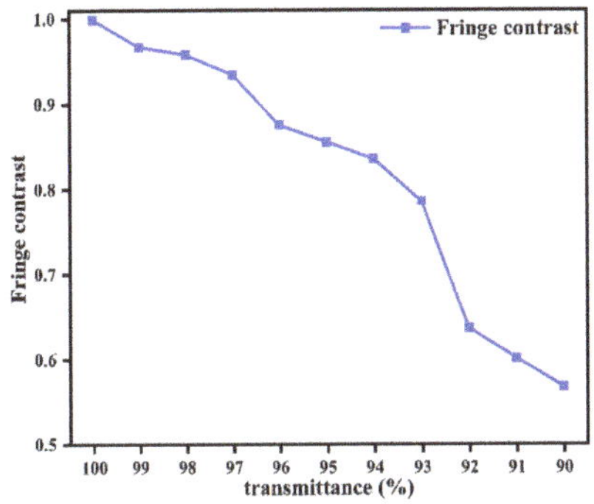

Figure 13. Curve of fringe contrast versus transmittance.

The analysis results show that when the transmittance decreases, the contrast of the interference fringes also decreases. This occurs because the stray light caused by the reflection from the beam splitter mirror and compensating mirror surfaces causes the interference beams to begin aliasing, and the contrast of the interference fringes then decreases. Spectral restoration was performed on the interferogram with the appropriate transmittance to obtain the corresponding spectrum at the central wavelength of 4.44 μm, as shown in Figure 14c. Other parameters remain unchanged. Under different transmittance conditions, the interferograms with wavelengths of 3.64 μm, 4.00 μm, and 5.00 μm are transformed by Fourier transform, and the restored spectral diagrams are shown in Figure 14a,b,d.

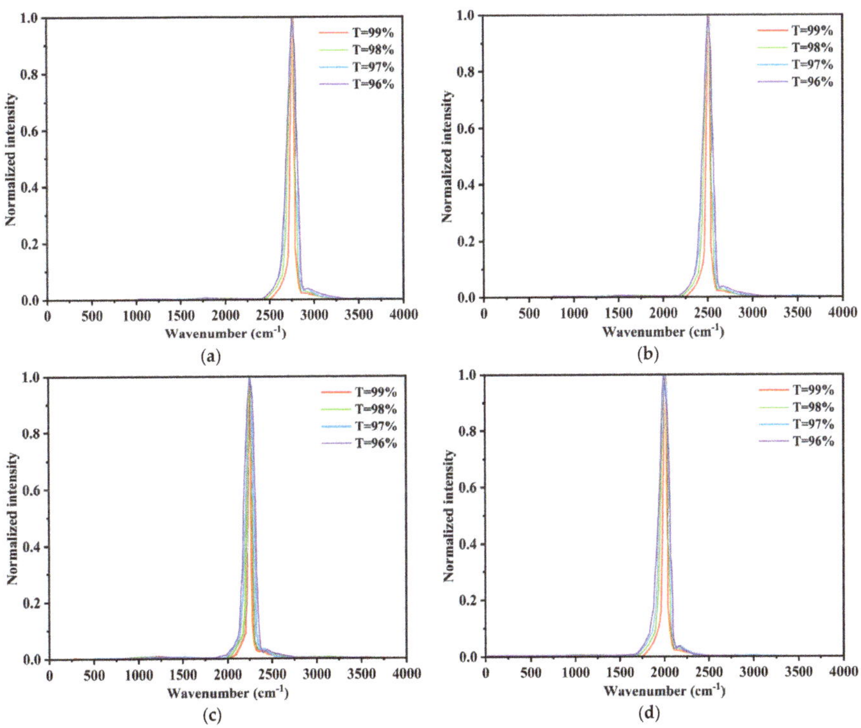

Figure 14. Recovery spectral curve of interferograms with different transmittance and wavelength: (**a**) 3.64 μm; (**b**) 4.00 μm; (**c**) 4.44 μm; and (**d**) 5.00 μm.

As shown in Figure 14a–d, the half-peak width of the reconstructed spectral line gradually increases with decreasing T, and the spectral resolution then gradually decreases. The analysis results show that as the transmittance decreases, the fringe contrast also decreases, and the spectral resolution decreases in turn.

In order to eliminate the influence of reflected stray light on the contrast of interference fringes and spectral resolution, it can be designed to separate the reflected stray light from the spot of the normal optical path. In this paper, wedge angle and inclination angle are introduced into the beam splitter and compensator to eliminate the influence of reflected stray light on the contrast and spectral resolution of interference fringes. The wedge angle and inclination angle of the beam splitter and compensator can make the reflected stray light path no longer parallel to the normal light path. When the angle between the reflected stray light path and the normal light path is greater than the field of view angle of the interferometer, the stray light can be separated from the normal light.

5.4. Increase the Wedge Angle and Inclination Angle to Eliminate Multiple Reflections of Stray Light Design

In order to eliminate the reflected stray light, the beam splitter and compensator increase the wedge angle, and the compensator increases the inclination angle. According to the requirements of the optical path equivalent post-parallel plate for transmitted and reflected light paths, the placement mode of the beam splitter and compensator after increasing the wedge angle and inclination angle is shown in Figure 15. The wedge angle θ_1 of the beam splitter is equal to the wedge Angle θ_2 of the compensator, and the inclination angle of the compensator is θ_3, α is the angle of the incident light, and n_{Ge} is the refractive index of germanium metal (in the infrared range, germanium has a refractive index of about 4.0).

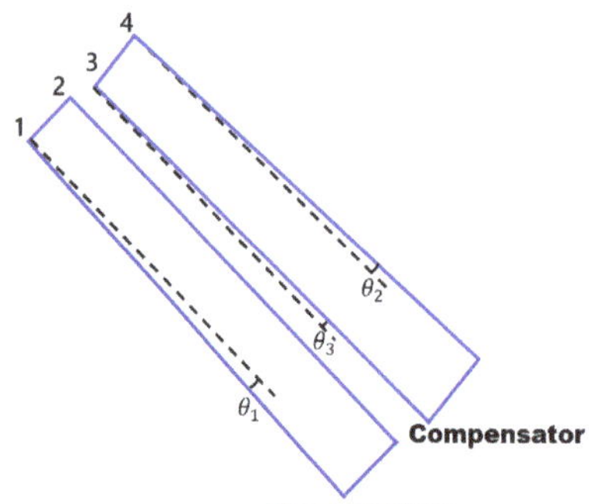

Figure 15. Diagram of placement of beam splitter and compensator.

In order to eliminate the influence of stray light reflected by surfaces 1, 3, and 4 on the interference system, the spot of this stray light should be deviated from the spot of the normal optical path.

The stray light path is shown by the green line in Figure 11. The relation between the angle δ_1, δ_2, and wedge angle of the stray light reflected by surface 1 and the normal optical path is shown in Equations (7) and (8):

$$\delta_1 = arcsin\left\{n_{Ge}\sin\left[arcsin\left(\frac{\sin\alpha}{n_{Ge}}\right) + 3\theta_1\right]\right\} - arcsin\left\{n_{Ge}\sin\left[arcsin\left(\frac{\sin\alpha}{n_{Ge}}\right) + \theta_1\right]\right\} \tag{7}$$

$$\delta_2 = arcsin\left\{n_{Ge}\sin\left[arcsin\left(\frac{\sin\alpha}{n_{Ge}}\right) - \theta_1\right]\right\} - arcsin\left\{n_{Ge}\sin\left[arcsin\left(\frac{\sin\alpha}{n_{Ge}}\right) + \theta_1\right]\right\} \tag{8}$$

The stray light path is shown by the yellow line in Figure 11. The relation between the angle δ_3, δ_4, and wedge angle of the stray light reflected by surface 3 and the normal optical path is shown in Equations (9) and (10):

$$\delta_3 = arcsin\left\{n_{Ge}\sin\left[arcsin\left(\frac{\sin(\alpha+\theta_3)}{n_{Ge}}\right)+\theta_3\right]\right\} - arcsin\left\{n_{Ge}\sin\left[arcsin\left(\frac{\sin(\alpha-\theta_3)}{n_{Ge}}\right)+\theta_3\right]\right\} \tag{9}$$

$$\delta_4 = arcsin\left\{n_{Ge}\sin\left[arcsin\left(\frac{\sin(\alpha-3\theta_3)}{n_{Ge}}\right)+\theta_3\right]\right\} - arcsin\left\{n_{Ge}\sin\left[arcsin\left(\frac{\sin(\alpha-\theta_3)}{n_{Ge}}\right)+\theta_3\right]\right\} \tag{10}$$

The stray light path is shown by the red line in Figure 11. The relation between the angle δ_3, δ_4, and wedge angle of the stray light reflected by surface 4 and the normal optical path is shown in Equations (11) and (12):

$$\delta_5 = arcsin\left\{n_{Ge}\sin\left[arcsin\left(\frac{\sin\alpha}{n_{Ge}}\right)+3\theta_2\right]\right\} - arcsin\left\{n_{Ge}\sin\left[arcsin\left(\frac{\sin\alpha}{n_{Ge}}\right)+\theta_2\right]\right\} \tag{11}$$

$$\delta_6 = arcsin\left\{n_{Ge}\sin\left[arcsin\left(\frac{\sin\alpha}{n_{Ge}}\right)-\theta_2\right]\right\} - arcsin\left\{n_{Ge}\sin\left[arcsin\left(\frac{\sin\alpha}{n_{Ge}}\right)+\theta_2\right]\right\} \tag{12}$$

In the reflected stray light, with the increase in the number of reflections, the angle between the stray light and the normal light path will increase, so as long as the angle between the reflected stray light and the normal light path is greater than the angle of the field of view, the higher order reflected stray light can also be separated from the normal light path. In order to separate the spot that reflects stray light from the spot of the normal light path, the following conditions must be met: $\delta_1 = \delta_5 > 2\omega$, $\delta_2 = \delta_6 > 2\omega$, $\delta_3 > 2\omega$, and $\delta_4 > 2\omega$, ω is the field angle of the interferometer. When $\alpha = 45\,^{\circ}$, $\omega = 0.8^{\circ}$ is taken, and the value range of wedge angle and inclination angle can be calculated according to Formulas (7)–(12): $\theta_1 = \theta_2 > 0.2\,^{\circ}$, $\theta_3 > 0.8\,^{\circ}$.

5.5. A Flow Chart of the Whole Simulation Work

In the whole simulation process, the optical and mechanical structure of the system is first designed, and then the physical model is established in TracePro 7.4 software, and the relevant parameters and conditions are set. Finally, the external stray light and the internal stray light of the interference system are analyzed by simulation, and the corresponding stray light suppression scheme is proposed. The simulation flow chart of the overall work is shown in Figure 16.

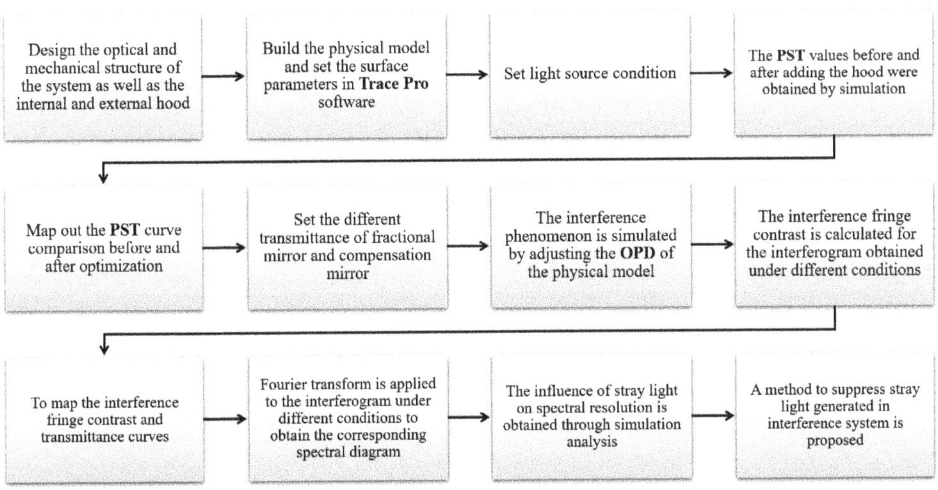

Figure 16. A flow chart of the whole simulation work.

6. Conclusions

In this paper, the stray radiation sources of a Fourier transform imaging spectrometer for use in the airborne mid-infrared band are discussed, and the external stray radiation and the internal stray light characteristics of the interference system are studied in depth. In this paper, the PST is used as an index to evaluate the suppression level of the stray solar radiation. An optical-mechanical simulation model and a light source simulation model were established using TracePro software, and ray tracing was performed. The image plane irradiance and the system of the stray solar radiation were calculated, the system hood was designed, and its shading effect was analyzed. In addition, the influence of stray light from internal reflections on the contrast of the interference fringes was simulated and analyzed, and corresponding improvement measures were proposed. The results showed that when the off-axis angle was greater than or equal to 25°, the stray solar radiation on the phase surface was very small, and the order of magnitude of the system transmittance was less than or equal to 10^{-4}. When the off-axis angle was greater than or equal to 70°, the solar stray radiation could be suppressed completely. For stray light caused by internal reflections from the interference system, the spectral resolution decreases as the magnitude of this stray light increases; a scheme for the addition of a wedge angle and an inclination angle to the beam splitter mirror and the compensating mirror is proposed in this work. It is deduced theoretically that when the wedge angle $\theta_1 = \theta_2 > 0.2°$ and when the inclination angle $\theta_3 > 0.8°$, stray light spots inside the interference system can be separated from the light spots obtained from the normal light path, thus reducing the influence of internally reflected stray light on the interference fringe contrast.

7. Discussion

In this study, we have undertaken a comprehensive examination of the stray light associated with a custom-designed mid-infrared Fourier transform imaging spectrometer. By implementing internal and external hoods in combination with a blocking ring for the Cassegrain telescope system, we mitigated stray light outside the field of view effectively, thus leading to enhanced imaging quality. Our analysis of the stray light within the interference system showed that as the transmittance of the beam splitting system decreases, the contrast of the interference fringes also diminishes, and this ultimately results in a reduction in the spectral resolution of the corresponding spectra. The root cause of this phenomenon lies in the fact that, during practical processing, the ideal transmittance of 100% for the optical lenses cannot be achieved. Consequently, reflected stray light appears on the non-ideal transmission surfaces; this stray light persists within the field of view and cannot be filtered out using the detector's cold diaphragm. This stray light introduces aliasing into the interference light, which then leads to the observed non-ideal transmission phenomenon. To address this issue, we used a method that involved the introduction of wedge and inclination angles to the beam splitter mirror and the compensation plate. This approach was successful in deflecting stray light out of the field of view of the cold diaphragm, thus effectively separating the stray light spot from the interference spot.

In terms of the limitations of the proposed methodology, the analysis of the stray light inside the interference system for the mid-infrared Fourier transform imaging spectrometer under study in this paper was limited to an analysis of the Michelson interferometer structure. Although beam splitting systems exist in most interferometers, based on the results of this research, the stray light analysis of infrared Fourier transform imaging spectrometers based on other interferometer structure types has not yet been verified, which will become the main direction of our future work, which has great significance for the comprehensiveness of the stray light analysis of Fourier transform imaging spectrometers.

In terms of the application field, the aim of this research is to solve the problems of performing quantitative measurements of mid-wave infrared radiation characteristics in the aviation field; therefore, the measurement range is limited to the mid-infrared spectral band from 3 to 5 μm, and the measurements of the visible light to near infrared band and the far infrared band are limited. We intend to continue to meet the challenges in the research into multispectral Fourier transform imaging spectrometers in our future work. The aim of this work is to improve the measurement accuracy, expand the application range, and reduce the limitations of the equipment.

Author Contributions: Conceptualization, methodology, software, formal analysis, investigation, writing—original draft preparation, visualization, C.B.; resources, L.M.; data curation, H.C.; writing—review and editing, X.Y.; supervision, H.S.; project administration, P.J. All authors have read and agreed to the published version of the manuscript.

Funding: This research was funded by the innovation project of the Changchun Institute of Optics, Fine Mechanics and Physics, Chinese Academy of Sciences (CXJJ-22S014).

Institutional Review Board Statement: Not applicable.

Informed Consent Statement: Not applicable.

Data Availability Statement: No new data were created or analyzed in this study. Data sharing is not applicable to this article.

Conflicts of Interest: The authors declare no conflicts of interest.

References

1. Shen, H.H.; Huang, M.; Li, J.Q.; Liu, J.H.; Dai, M.; Jia, P. Recent progress in aerial electro-optic payloads and their key technologies. *Chin. Opt.* **2012**, *5*, 20–29.
2. Lu, K.; Li, H.; Lin, L.; Zhao, R.; Liu, E.; Zhao, R. A Fast Star-Detection Algorithm under Stray-Light Interference. *Photonics* **2023**, *10*, 889. [CrossRef]
3. Jiang, H.; Niu, X. Stray Light Analysis and Suppression of the Visible to Terahertz Integrated Cloud Detection Optical System. *Sensors* **2023**, *23*, 4115. [CrossRef] [PubMed]
4. Tao, H.; Lv, J.; Liang, J.; Zhao, B.; Chen, Y.; Zheng, K.; Zhao, Y.; Wang, W.; Qin, Y.; Liu, G.; et al. Polarization Snapshot Imaging Spectrometer for Infrared Range. *Photonics* **2023**, *10*, 566. [CrossRef]
5. Wang, N.; Wu, J.H.; Meng, H.; Gao, J.B. Performance of a long-wave infrared Fourier Transform imaging spectrometer using a corner-cube Michelson interferometer and an uncooled microbolometer array. In Proceedings of the AOPC 2015: Optical and Optoelectronic Sensing and Imaging Technology, Beijing, China, 15 October 2015.
6. Revercomb, H.E.; Sromovsky, L.A.; Fry, P.M.; Best, F.A.; LaPorte, D.D. Demonstration of imaging Fourier Transform Spectrometer (FTS) performance for planetary and geostationary Earth observing. In Proceedings of the Hyperspectral Remote Sensing of the Land and Atmosphere, Sendai, Japan, 8 February 2001.
7. Soucy, M.A.; Chateauneuf, F.; Deutsch, C.; Etienne, N. ACE-FTS instrument detailed design. In Proceedings of the Earth Observing Systems VII, Seattle, WA, USA, 24 September 2002.
8. Matallah, N.; Sauer, H.; Goudail, F.; Fontanella, J.C.; Ferrec, Y.; Taboury, J.; Chavel, P. Design and first results of a Fourier Transform imaging spectrometer in the 3–5 μm range. In Proceedings of the Optical Design and Engineering IV, Marseille, France, 21 September 2011.
9. Hu, B.L. Review of the development of interferometric spectral imaging technology (invited). *Acta Photonica Sin.* **2022**, *51*, 751401.
10. Wu, C.K.; Zhang, W.; Hao, Y.Z. Design of a control system for a visible/near-infrared real-time imaging spectrometer. *Chin. Opt.* **2022**, *15*, 348–354.
11. Grotevent, M.J.; Yakunin, S.; Bachmann, D.; Romero, C.; de Aldana, J.R.V.; Madi, M.; Calame, M.; Kovalenko, M.V.; Shorubalko, I. Integrated photodetectors for compact Fourier-transform waveguide spectrometers. *Nat. Photonics* **2023**, *17*, 59–64. [CrossRef]
12. Afara, I.O.; Shaikh, R.; Nippolainen, E.; Querido, W.; Torniainen, J.; Sarin, J.K.; Kandel, S.; Pleshko, N.; Töyräs, J. Characterization of connective tissues using near-infrared spectroscopy and imaging. *Nat. Protoc.* **2021**, *16*, 1297–1329. [CrossRef]
13. Wang, H.; Chen, Q.; Ma, Z.; Yan, H.; Lin, S.; Xue, Y. Development and prospect of stray light suppression and evaluation technology(Invited). *Acta Photonica Sin.* **2022**, *51*, 0751406.
14. Du, S.S.; Wang, Y.M.; Du, G.J.; Wang, Y.-J. Stray light analysis of Fabry-Perot interference imaging spectrometer. *J. Appl. Opt.* **2009**, *30*, 246–251.
15. Chen, F.; Gao, C.; Xu, P.M. Analysis of stray light influence to modulation on fourier transform spectrometer. *Spectrosc. Spectr. Anal.* **2018**, *38*, 2966–2970.
16. Chen, F.; Gao, C.; Bai, J. Thickness matching design between splitter and compensator high spectral resolution Fourier transform spectrometer. *Spectrosc. Spectr. Anal.* **2020**, *40*, 3941–3945.
17. Dussarrat, P.; Deschamps, G.; Dehnavi, S. Impact of straylight in Michelson Fourier transform spectrometers. In Proceedings of the International Conference on Space Optics—ICSO 2022, Dubrovnik, Croatia, 12 July 2023.
18. Fan, B.; Chen, X.; Li, B.C.; Zhao, Y. Technical innovation of optical remote sensing payloads onboard GF-5 satellite. *Infrared Laser Eng.* **2017**, *46*, 0102002. [CrossRef]

19. Feng, M.C.; Xu, L.; Wang, Y.J. Analysis and simulation of moving mirror tilt in michelson interferometer. *Laser Optoelectron. Prog.* **2021**, *58*, 0122001. [CrossRef]
20. He, L.; Wang, R.H.; Hou, B.; Si, H.; Yang, K.; Hu, H. Athermalization design of a low-cost medium-wave infrared optical system based on Si/Ge material. *Infrared Technol.* **2023**, *45*, 527–533.
21. Byun, W.J.; Kim, K.S.; Kim, B.S.; Lee, Y.S.; Song, M.S.; Choi, H.D.; Cho, Y.H. Multiplexed Cassegrain Reflector Antenna for Simultaneous Generation of Three Orbital Angular Momentum (OAM) Modes. *Sci. Rep.* **2016**, *6*, 27339. [CrossRef] [PubMed]
22. Oishi, Y.; Sawada, Y.; Kamei, A.; Murakami, K.; Nakamura, R.; Matsunaga, T. Impact of Changes in Minimum Reflectance on Cloud Discrimination. *Remote Sens.* **2018**, *10*, 693. [CrossRef]
23. Zhu, J.L.; Zhang, P.; Li, G.; Wei, X.; Fei, C.; Gan, J.; Yuan, Y.; Ji, Y. Stray light analysis of low-level-light optical system with LightTools software. *J. Appl. Opt.* **2022**, *43*, 1061–1065.
24. Li, J.; Yang, Y.; Qu, X.; Jiang, C. Stray Light Analysis and Elimination of an Optical System Based on the Structural Optimization Design of an Airborne Camera. *Appl. Sci.* **2022**, *12*, 1935. [CrossRef]
25. Wei, L.; Yang, L.; Fan, Y.-P.; Cong, S.-S.; Wang, Y.-S. Research on Stray-Light Suppression Method for Large Off-Axis Three-Mirror Anastigmatic Space Camera. *Sensors* **2022**, *22*, 4772. [CrossRef]
26. Zhang, X.; Zhu, J.; Huang, L.; Zhang, Y.; Wang, H.; Li, H.; Guo, F.; Deng, J. Hyperspectral Channel-Modulated Static Birefringent Fourier Transform Imaging Spectropolarimeter with Zoomable Spectral Resolution. *Photonics* **2023**, *10*, 950. [CrossRef]
27. Li, Y.Y.; Liu, L.; Peng, Q.; Yang, J. Stray radiation analysis of cassegrain system. *Laser Infrared* **2019**, *49*, 987–991.
28. Jiang, S.W.; Xia, Z.T.; Sun, Y.X.; Wang, K. Optical Design and Stray-Light Analysis of Urban Night-Light Remote Sensing Imaging System. *Laser Optoelectron. Prog.* **2020**, *57*, 206–213.
29. Chen, X.; Hu, C.H.; Yan, C.X.; Kong, D.C. Analysis and suppression of space stray light of visible cameras with wide field of view. *Chin. Opt.* **2019**, *12*, 678–685. [CrossRef]
30. Li, J.; Meng, X.; Xu, D.; Song, H.; Wang, L.; Zhu, R. Near-infrared Fourier transform imaging spectrometer for remote sensing. In Proceedings of the International Symposium on Optoelectronic Technology and Application 2014: Imaging Spectroscopy; and Telescopes and Large Optics, Beijing, China, 18 November 2014.
31. Key, R.; Sander, S.; Eldering, A.; Blavier, J.F.; Bekker, D.; Manatt, K.; Rider, D. The Geostationary Fourier Transform Spetrometer. In Proceedings of the 2012 IEEE Aerospace Conference, Big Sky, MT, USA, 3–10 March 2012.
32. Liu, Y.N. Development of hyperspectral imaging remote sensing technology. *J. Remote Sens.* **2021**, *25*, 439–459. [CrossRef]
33. Dong, X.; Xu, P.M.; Hou, L.Z. Design and Implementation of Atmospheric Infrared Ultra-spectral Sounder. *Spacecr. Recovery Remote Sens.* **2018**, *39*, 29–37.
34. Wang, S.; Yetisen, A.K.; Wang, K.; Jakobi, M.; Koch, A.W. Dependence of the Michelson Interferometer-Based Membrane-Less Optical Microphone–Photoacoustic Spectroscopy Gas-Sensing Method on the Fundamental Parameters of a Photoacoustic Gas Cell. *Photonics* **2023**, *10*, 888. [CrossRef]
35. Louis, M.; Frédéric, G.; Stéphane, L.; Raphaël, D.; Marc-André, S. An interferometer for compact imaging spectrometer. In Proceedings of the Imaging Spectrometry XII, San Diego, CA, USA, 12 September 2007.

photonics

Article

Design of Backstepping Control Based on a Softsign Linear–Nonlinear Tracking Differentiator for an Electro-Optical Tracking System

Jiachen Li [1,2,3,4], Shanlin Zhuang [1,2,3,4], Haolin Wang [1,2,3,4], Jiuqiang Deng [1,2,3,4,*] and Yao Mao [1,2,3,4]

1 National Key Laboratory of Optical Field Manipulation Science and Technology, Chinese Academy of Sciences, Chengdu 610209, China; lijiachen22@mails.ucas.ac.cn (J.L.); zhuangshanlin22@mails.ucas.ac.cn (S.Z.); wanghaolin20@mails.ucas.ac.cn (H.W.); maoyao@ioe.ac.cn (Y.M.)
2 Key Laboratory of Optical Engineering, Chinese Academy of Sciences, Chengdu 610209, China
3 Institute of Optics and Electronics, Chinese Academy of Sciences, Chengdu 610209, China
4 University of Chinese Academy of Sciences, Beijing 100049, China
* Correspondence: jqdeng@ioe.ac.cn

Abstract: To address the problems of a low tracking accuracy and slow error convergence in high-order single-input, single-output electro-optical tracking systems, a backstepping control method based on a Softsign linear–nonlinear tracking differentiator is proposed. First, a linear–nonlinear tracking differentiator is designed in conjunction with the Softsign excitation function, using its output as an approximate replacement for the conventional differentiation process. Then, this is combined with backstepping control to eliminate the "explosion of complexity" problem in conventional backstepping procedures due to repeated derivation of virtual control quantities. This reduces the workload of parameter tuning, takes into account the rapidity and stability of signal convergence, and improves the trajectory tracking performance. This method can ensure the boundedness of the system signal. The effectiveness and superiority of this control method are verified through simulations and experiments.

Keywords: backstepping control; tracking differentiator; electro-optical tracking system; Softsign function; linear–nonlinear systems

Citation: Li, J.; Zhuang, S.; Wang, H.; Deng, J.; Mao, Y.; Design of Backstepping Control Based on a Softsign Linear–Nonlinear Tracking Differentiator for an Electro-Optical Tracking System. *Photonics* **2024**, *11*, 156. https://doi.org/10.3390/photonics11020156

Received: 21 December 2023
Revised: 23 January 2024
Accepted: 31 January 2024
Published: 6 February 2024

1. Introduction

An electro-optical tracking system (ETS) is a piece of complex and high-precision servo tracking equipment integrating optical, mechanical, and electronic technology. It has the characteristics of a fast response speed, a low tracking error, and strong disturbance resistance [1]. These devices are mainly used for real-time tracking and measurements of moving targets, and are widely employed in the fields of target observation, laser communication, quantum communication, aviation, aerospace, automated production, and other fields [2,3]. In the aerospace field, they are used for spacecraft guidance, positioning, attitude control, etc. [4]. In the field of automated production for automated processing, they are used for machine vision [5], target tracking and pointing, motion target trajectory measurements [6], etc. In the medical field, they are used for target tracking in surgical robots, etc.

For this type of system, some scholars have proposed PID control, sliding mode control, adaptive control, and neural network control methods, among others. These control methods have effectively improved the tracking performance of ETS control from different aspects. PID control is a simple and effective control method that meets the steady-state and dynamic performance requirements of a system by adjusting three parameters: the proportional, the integral, and the derivative. However, for electro-optical tracking systems with nonlinearity and uncertainty, the performance of PID control may be limited [7].

Sliding mode control uses a sliding surface and switching logic to achieve the desired system performance, exhibiting strong robustness and adaptability. However, it may suffer from chattering issues [8]. Fuzzy control is a control method based on fuzzy logic, capable of handling uncertainty and nonlinear problems. Yet, in high-dimensional and complex electro-optical tracking systems, the design of fuzzy rules may become complex and time-consuming [9]. Neural network control is an artificial-intelligence-based control method suitable for highly nonlinear and uncertain systems. However, training neural networks requires substantial amounts of data and time, and there may be some delay in real-time applications [10].

Furthermore, an electro-optical tracking system is limited by the sensor frequency and the performance of the driving mechanism, which leads to the problems of a low tracking accuracy and slow error convergence. To address this problem, backstepping control techniques [11–13] have been developed rapidly in recent years. Backstepping control is not only characterized by an ease of design and implementation but also by the ability to measure the state of the system in real time and adjust the inputs according to the difference between the target output and the actual output, thus achieving highly accurate tracking. It can also effectively suppress the instability of the system and thus ensure its stability. A backstepping design allows the control scalar functions and controllers to be systematic, structured, and jitter-free, making them widely available [14,15].

The backstepping control method suffers from the "complexity explosion" problem [16,17], which leads to a complex solution process and slow convergence.

In the process of backstepping control design, the quality of the position feedback signal, the speed feedback signal, etc., is key to improving the control accuracy, and it is largely influenced by signal filtering and signal differentiation. According to the position signal, the traditional differentiation method can calculate a series of differential signals such as the velocity and acceleration. There are multiple successive differentials in backstepping control, which can easily lead to the problem of a "complexity explosion" [16,17]. The traditional differentiation process can reduce the accuracy of the signal differentiation estimation when it is affected by signal noise, and, in turn, the control effect of the backstepping controller is affected.

To avoid a "complexity explosion" and to suppress signal noise, past studies usually used filters and various signal processing methods to preprocess the signal, such as dynamic surface control (DSC) [18–21] and command filtering (CF) control [22–24]. The main feature of DSC is the introduction of a first-order filter in the backstepping design process to replace the differential operator in each virtual controller design step; however, the DSC scheme does not consider the impact of filtering errors on the control system [25]. To solve this problem, researchers proposed a command-filtered [23,24,26] backstepping control method by introducing virtual input second-order filtering in each step of the conventional backstepping design process to replace the conventional differential process and by using an error compensation mechanism to overcome the shortcomings of DSC. However, the transient performance of the filtered signal of the conventional command filter in this scheme is poor and the performance of its differential process decreases when the frequency of the input signal increases.

To solve these problems, different tracking differentiators have been used to improve the transient performance of the filtered signal and the differentiation performance at higher frequencies [27–34]. These differentiators can complete the process of tracking and differentiating a real-time signal without relying on the controlled object model, and they use an integration process instead of the differentiation process used in the traditional numerical differentiation method [35] to avoid a "complexity explosion" while performing signal filtering. With in-depth research on tracking differentiator technology in recent years, various types of improved tracking differentiators, such as Linear Tracking Differentiators (LTDs) [27], High-Speed Tracking Differentiators (HSTDs) [28], a New Simple Linear Tracking Differentiator (T-D) [29], and High-Gain Tracking Differentiators (HGTDs) [30], have been gradually developed to improve the dynamic performance of differentiators.

However, the tracking function of these tracking differentiators is not able to simultaneously consider the rapidity and stability of signal convergence, which may cause the rapidity or stability of signal convergence to deteriorate when adjusting the convergence trend near the equilibrium point.

When a tracking differentiator has a relatively fast convergence speed, the speed of the change in state quantities near the equilibrium point increases simultaneously, which leads to convergence chattering and other problems, resulting in a decrease in the tracking accuracy of the differentiator. In this regard, Arctangent Tracking Differentiators (ATDs) in the form of an inverse tangent using an inverse tangent function [31], Modified Tracking Differentiators (MTDs) designed using a nonlinear odd-exponential continuous function that is stable at only one equilibrium point [32], New Nonlinear–Linear Tracking Differentiators (NTDs) using hyperbolic tangent (Tanh) functions [33], and Hyperbolic-Sine-Based Tracking Differentiators (HNTDs) [34] using hyperbolic sinusoidal functions have been proposed, which are based on a common improvement strategy: introducing both nonlinear and linear links into the differentiator design. Linear and nonlinear links exhibit different degrees of action when the state is far from or close to the equilibrium point, ensuring the rapidity and stability of differentiator convergence. However, the structural form of a tracking differentiator designed in this way is relatively complex, with more parameters, and the functions have the problem of a faster gradient disappearance, and so a more accurate approximate differentiation process may not be achieved.

In summary, in high-order electro-optical tracking systems characterized by nonlinearity and uncertainty, the performance of PID control may be limited [7], sliding mode control may encounter significant chattering issues [8], fuzzy control requires the design of complex fuzzy rules [9], and neural network control demands substantial additional data and time for training [10]. Therefore, the decision has been made to adopt backstepping control, which can circumvent these issues and provide effective control for an electro-optical tracking system. When solving the complexity explosion present in backstepping control, the existing dynamic surface will introduce filtering errors and reduce the tracking accuracy [21]; the transient performance of the filtered signal of the existing command filter is poor [24]; and existing tracking differentiators have a complex structure and more parameters and the gradient of the function they use disappears faster, so it is not possible to realise an approximate differentiation process with the proposed accuracy [34]. Thus, this paper proposes a linear–nonlinear tracking differentiator based on the Softsign excitation function (SL-NTD) for an electro-optical tracking system. The results show that the proposed control design ensures that all signals are in a bounded set and the tracking error converges to the desired neighborhood of the origin. Compared to all existing technology, the innovations and main contributions of the proposed control scheme can be summarized as follows:

1. A linear–nonlinear tracking differentiator is designed using the Softsign excitation function for the first time. Compared to the dynamic surface used in Liang and Qiu's work, the method proposed in this paper does not introduce filtering errors [21]; compared to the command filter used in Han and Yu's work, the method in this paper can simultaneously take into account the speed and stability of signal convergence [24]; and compared to the tracking differentiator used in Fan and Jing's work, the method in this paper has fewer parameters, the disappearance of the gradient is slowed down, and thus differentiation can be approximated more accurately [34].
2. This paper proposes introducing a Softsign tracking differentiator in each step of backstepping control for the first time, using an approximation of the output of the tracking differentiator instead of the traditional differentiation process in virtual control, which solves the problem of the "complexity explosion" in backstepping control.
3. The number of parameters in the whole backstepping control process is significantly reduced, improving parameter tuning in the scheme. Finally, the feasibility and superiority of the backstepping control method of the linear–nonlinear tracking differentiator based on the Softsign excitation function are verified via simulations and experiments.

The remainder of the paper is structured as follows. Section 2 presents the problem description and introduces the research objectives and methodology of the paper. Section 3 describes the backstepping control design process and the proof of its stability for a linear–nonlinear tracking differentiator based on the Softsign excitation function. Section 4 presents a specific control object example and compares the proposed method with other methods in simulations and experiments to solve the "complexity explosion" problem. The effectiveness and superiority of the proposed design method are verified in simulations and experiments. Section 5 provides the conclusions of the paper.

2. Problem Description

Consider the following electro-optical tracking system that can be expressed as:

$$\begin{cases} \dot{x}_1 = x_2 \\ \dot{x}_2 = x_3 \\ \cdots\cdots \\ \dot{x}_{n-1} = x_n \\ \dot{x}_n = u \\ y = x_1 \end{cases} \tag{1}$$

In Equation (1), $x = [x_1, x_2, \ldots, x_n]^T \in R^n$ is the state quantity, and it is measurable, u is the control rate, and y is the output.

The ultimate goal of this paper is to make the class of ETSs achieve high-accuracy trajectory tracking, where the root mean square error can be used as a reference for adjustment and measurements, as shown in Equation (2) below:

$$RMSE = \sqrt{\frac{1}{N}\sum(v_t - x_1)^2} \tag{2}$$

where v_t is the input standard trajectory and x_1 is the output actual trajectory. In order to improve the trajectory tracking accuracy, it is necessary to reduce the error between the actual trajectory and the standard trajectory; therefore, in this paper, a new backstepping controller is designed that achieves high-precision tracking while ensuring system stability by measuring the system state in real time, adjusting the input according to the difference between the target output and the actual output, and using the state feedback information of the system to offset the effects of perturbation.

The backstepping control method used in this paper is a commonly used control method [11,12], the basic idea of which is decomposing a complex system into subsystems that do not exceed the order of the system. As shown in Equation (1) above, the system can be divided into n subsystems, and then a Lyapunov function V_n and an intermediate virtual control quantity $x_{(i+1)d}$ can be designed, respectively, for each subsystem x_{i+1}. In each subsystem, $\dot{x}_i = x_{i+1}$ is virtual control, and appropriate virtual feedback $x_{i+1} = x_{(i+1)d}$, where $i = 1, 2, \ldots, n-1$, allows the previous state of the system to reach asymptotic stability and then "back up" the entire system until the entire controller design is completed. In this process, it is necessary to differentiate the virtual control quantity $x_{(i+1)d}$ used in each step, and after simplification, this is equivalent to multiple successive differentiations of the first virtual control quantity x_{1d}. This can lead to the phenomenon of a "complexity explosion". To solve the problem of the "complexity explosion ", this paper uses a Softsign tracking differentiator to replace the traditional differential derivation process.

3. SL-NTD-Based Backstepping Controller Design

3.1. Description of the Softsign Tracking Differentiator

In this paper, we propose a linear–nonlinear tracking differentiator based on the Softsign excitation function. This section focuses on the introduction of the Softsign excitation function; an image of the function is shown in Figure 1, and the mathematical equation is of the form shown in Equation (3):

$$Softsign(x) = \frac{x}{1+|x|} \tag{3}$$

where x is the independent variable of the Softsign excitation function, and the value of $Softsign(x)$ converges to 1 when the value of x is positive infinity and to -1 when the value of x is negative infinity.

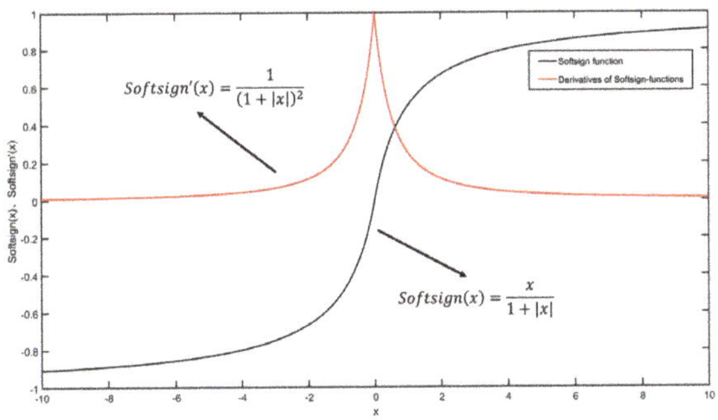

Figure 1. Softsign function image.

The input value of the Softsign function is small and its output value in the interval close to 0 presents non-linear characteristics, so it can eliminate or alleviate the problem of jittering to a certain extent.

The magnitude of the Softsign function and the coefficients of the independent variables can be extended to obtain the function $f = m * Softsign(nx)$, and changing the values of m and n can change the magnitude of the Softsign function and the rate of change. Changing the magnitude of m changes the range of function values; when n decreases to nearly 0, the output of the function will tend to be linear and the slope will decrease as well. This indicates that as n decreases, the function f becomes more linear, smoother, and more saturated. This characteristic can alleviate or even eliminate the impact of jitter and can also reduce the risk of overfitting in many cases, making the output more stable.

In addition, it can be seen from Figure 1 that the Softsign function has properties such as smoothness and continuity, and the saturation interval is small, giving it a slower decreasing derivative, which also indicates that it can learn more efficiently and can better solve the problem of gradient disappearance. Thus, it is suitable and convenient for use in the design of tracking differentiators.

In order to design a linear–nonlinear tracking differentiator based on the Softsign excitation function, the following theorem/lemma and related proofs need to be given first.

Theorem 1. *There is the following system:*

$$\begin{cases} \dot{k}_1(t) = k_2(t) \\ \dot{k}_2(t) = -a * [k_1(t) + Softsign(k_1(t))] - b * [k_2(t) + Softsign(k_2(t))] \end{cases} \tag{4}$$

If the parameters a and b in the above system are greater than 0, then the system is uniformly asymptotically stable at the origin $(0,0)$*. This means that for any initial condition, the system state will converge to the origin as time tends to infinity and the rate of convergence is consistent for all initial conditions.*

Remark 1. *When there exist systems of the form described above, whose parameters are all greater than 0, then the system is uniformly asymptotically stable at the origin* $(0,0)$*.*

Proof. First, construct the Lyapunov function:

$$V(k_1, k_2) = \int_0^{z_1} a * Softsign(\theta)d\theta + \frac{a}{2} * K_1^2 + \frac{1}{2} * K_2^2 \tag{5}$$

Theorem 1 gives $a > 0$ and $a * Softsign(k_1) > 0$ when $k_1 > 0$, and $a * Softsign(k_1) < 0$ when $k_1 < 0$. Then, from the median integral theorem, we have

$$\int_0^{z_1} a * Softsign(\theta)d\theta = a * Softsign(\sigma) * k_1 > 0 \tag{6}$$

where $0 < \sigma < k_1$; thus, it is obtained that $\int_0^{z_1} a * Softsign(\theta)d\theta > 0$.

When $k_1 \neq 0$ and $k_2 \neq 0$, then $\frac{a}{2} * k_1^2 > 0$ and $\frac{1}{2} * k_2^2 > 0$; therefore, we obtain $V(k_1, k_2) > 0$. Take the derivative of the Lyapunov function:

$$\begin{aligned} \dot{V}(k_1, k_2) &= (a * Softsign(k_1) + a * k_1) * \dot{k}_1 + k_2 * \dot{k}_2 \\ &= a * k_2 * Softsign(k_1) + a * k_1 * k_2 + k_2 * [-a* \\ &(k_1 + Softsign(k_1)) - b * (k_2 + Softsign(k-2))] \\ &= -b * k_2^2 - b * k_2 * Softsign(k_2) \end{aligned} \tag{7}$$

Since Theorem 1 gives $b > 0$, it can be obtained that $b * k_2 * Softsign(k_2) > 0$ and $\dot{V}(k_1, k_2) \leq 0$. It follows that $\dot{V}(k_1, k_2) = 0$ if and only if $k_2 = 0$ near (0.0). Therefore, according to Lyapunov's second theorem, the system of Theorem 1 is asymptotically stable and so Theorem 1 is valid. □

Lemma 1. *Due to the nature of the Softsign function, the system can be divided into two action phases when it comes to the specific role:*

1. *When $|k| > 2.5$, the system is far away from the equilibrium position, $|k| > |Softsign(k)|$. This has a major role in driving system (4), described by:*

$$\begin{cases} \dot{k}_1(t) = k_2(t) \\ \dot{k}_2(t) = -a * k_1(t) - b * k_2(t) \end{cases} \tag{8}$$

2. *When $|k| < 2.5$, the system is far away from the equilibrium position, $|k| < |Softsign(k)|$. This has a major role in driving system (4), described by:*

$$\begin{cases} \dot{k}_1(t) = k_2(t) \\ \dot{k}_2(t) = -a * Softsign(k_1(t)) - b * Softsign(k_2(t)) \end{cases} \tag{9}$$

Remark 2. *The tracking differentiator constructed from the Softsign excitation function can be divided into two action phases in the specific role, which are also each asymptotically stable and allow the system state to always converge quickly and steadily. The change in state is shown in Figure 2, where the origin (0,0) is the equilibrium point.*

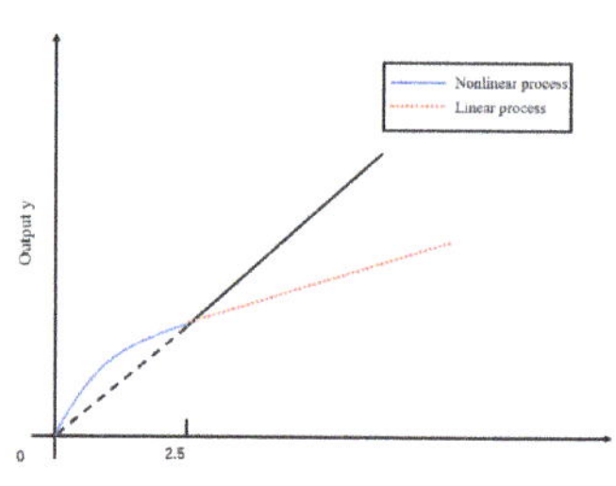

Figure 2. Status changes in the two phases of action of the tracking differentiator.

Proof. From Theorem 1 and Theorem 2 in [28] it is easy to prove that Equations (8) and (9) are also asymptotically stable and enable the fast and stable convergence of the system state.

Therefore, a nonlinear–linear tracking differentiator based on the Softsign excitation function, referred to as SN-LTD, is designed. The control block diagram is shown in Figure 3, and the mathematical model is shown in Equation (10). □

Theorem 2. *For the following system:*

$$\begin{cases} \dot{k}_1(t) = k_2(t) \\ \dot{k}_2(t) = (a * ((x_1(t) - v(t)) + Softsign(x_1(t) - v(t))) \\ +b * (\frac{x_2(t)}{R} + Softsign(\frac{x_2(t)}{R}))) * (-R^2) \end{cases} \tag{10}$$

where $v(t)$ is the input signal, $x_1(t)$ and $x_2(t)$ are variables, and $x_1(t)$ is the input signal after filtering, $x_2(t)$ is the extracted differential signal, and a and b are parameters greater than 0. The solution of this tracking differentiator system satisfies $\lim_{x\to\infty} \int_0^T |x_1(t) - v(t)dt = 0|$ *at $T > 0$.*

According to Theorem 1 and Lemma 1, it is easy to prove that Theorem 2 holds.

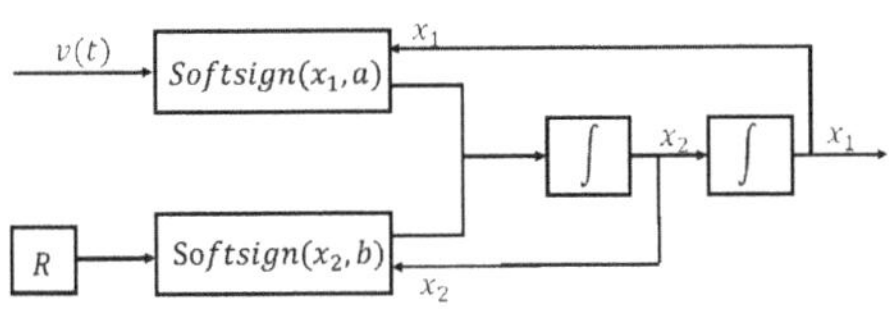

Figure 3. Control block diagram based on the Softsign linear–nonlinear tracking differentiator.

Through the derivation of the formula for the Softsign-based tracking differentiator, it can be found that the output of the Softsign function tends to be 1 when the difference between $x_1(t)$ and $v(t)$ is large, and at this time, the output of the Softsign-based tracking differentiator mainly depends on the difference between $x_1(t)$ and $v(t)$. The method can effectively track the changes in the system. At the same time, the Softsign-based tracking differentiator can better overcome the differential explosion problem because the Softsign function has the ability to effectively filter out noise in edge detection. Also, the range of output signals of the Softsign-based tracking differentiator is wide, effectively avoiding the output saturation problem. This means that the Softsign-based tracking differentiator has better adaptability in dealing with nonlinear systems.

Theorem 3. *The theoretical analysis of the Softsign-based tracking differentiator is compared with an ordinary linear tracking differentiator, and the results show that the Softsign-based tracking differentiator has a faster convergence speed and smaller errors. The theoretical analysis process is as follows:*

The state space equation for the linear tracking differentiator used for the comparison is:

$$\begin{cases} \dot{x}_1(t) = x_2(t) \\ \dot{x}_2(t) = (-a_0 * (x_1(t) - v(t)) - a_1 * \varepsilon * x_2(t)) * \varepsilon^{-2} \end{cases} \tag{11}$$

where $v(t)$ is the input signal, $x_1(t)$ and $x_2(t)$ are the variables, and a_0, a_1, and ε are parameters greater than 0, where $a_1 = 2$, $a_0 = 1$, $\varepsilon = 0.004$, and $a = 7$ and $b = 8$ in Equation (4).

Define the Lyapunov function as $V = x_1^2 + v_2^2$, and for a linear tracking differentiator, calculate the derivative of the Lyapunov function as:

$$\begin{aligned} \dot{V}_1 &= 2 * x_1 * \dot{x}_1 + 2 * x_2 * \dot{x}_2 \\ &= 2 * x_1 * x_2 + 2 * x_2 * (-2 * (x_1 - v) - 0.004 * x_2)/0.004^2 \\ &= 2 * x_1 * x_2 + 2 * x_2 * (-125000 * (x_1 - v) - 250 * x_2) \end{aligned} \tag{12}$$

For the Softsign-based tracking differentiator, the derivative of the Lyapunov function is calculated as:

$$\begin{aligned} \dot{V}_1 =& 2 * x_1 * \dot{x}_1 + 2 * x_2 * \dot{x}_2 \\ =& 2 * x_1 * x_2 + 2 * x_2 * (-2520000 * (x_1 - v) - 3000 * x_2 - 2520000 * softsign(x_1 - v) \\ & - 1800000 * softsign(x_2/600)) \end{aligned} \tag{13}$$

A comparison shows that $\dot{V}_1 > \dot{V}_2$ and the Softsign-based tracking differentiator has faster convergence and more minor errors according to the Lyapunov stability theorem. Similarly, it can be proven that the Softsign-based tracking differentiator has more advantages over other differentiators.

Remark 3. *Compared to the command filters mentioned in the literature [23,24], the use of a Softsign-based tracking differentiator, which uses the input $v(t)$ to obtain $x_1(t)$ and $\dot{x}_1(t)$, improves not only the transient performance of the filtered signal but also the differentiation performance at higher frequencies.*

Remark 4. *Compared with various types of tracking differentiators mentioned in the literature [27–34], the use of the Softsign-based tracking differentiator, which uses the input $v(t)$ to obtain $x_1(t)$ and $\dot{x}_1(t)$, simplifies the structure of the tracking differentiator and reduces the number of parameters while taking into account the rapidity and stability of signal convergence, better solving the "complexity explosion" problem.*

3.2. Design of Backstepping Control Based on the Softsign Tracking Differentiator

Consider the following nth-order single-input and single-output electro-optical tracking system:

$$\begin{cases} \dot{x}_1 = x_2 + f_1(x_1) \\ \dot{x}_2 = x_3 + f_2(x1, x2) \\ \cdots\cdots \\ \dot{x}_{n-1} = x_n + f_{n-1}(x_1, x_2, \ldots, x_{n-1}) \\ \dot{x}_n = u + f_n(x_1, x_2, \ldots, x_n) \end{cases} \tag{14}$$

where $x_1, x_2, \ldots, x_n$ is the state quantity, u is the control rate, the input is x_{1d}, and the output is x_1. In addition, $x_{2d}, x_{3d}, \ldots, x_{nd}$ is the virtual input quantity and also the expected value of the state quantity.

Design the control rate according to the conventional backstepping control design idea. In each x_{i+1} subsystem, $\dot{x}_i = x_{i+1} + f_i(x_1, x_2, \ldots x_i)$ is virtual control, and appropriate virtual feedback $x_{i+1} = x_{(i+1)d}$, where $i = 1, 2, \ldots, n-1$, makes the previous state of the system reach asymptotic stability, but the solution of the system generally does not satisfy $x_{i+1} = x_{(i+1)d}$. For this reason, error variables are introduced in the hope that some asymptotic properties between x_{i+1} and the virtual feedback $x_{(i+1)d}$ can be achieved through the action of control, and thus the asymptotic stability of the whole system can be achieved.

The derivative $\dot{x}_{1d}, \dot{x}_{2d}, \ldots, \dot{x}_{nd}$ of the virtual control quantity $x_{1d}, x_{2d}, \ldots, x_{nd}$ designed in the following is substituted by the differential signal $\dot{\hat{x}}_{1d}, \dot{\hat{x}}_{2d}, \ldots, \dot{\hat{x}}_{nd}$ extracted by a linear–nonlinear tracking differentiator based on the Softsign excitation function which is provided to the backstepping controller.

Assume that the n tracking errors are: $z_1, z_2, \ldots, z_n$

$$\begin{aligned} z_1 &= x_1 - x_{1d} \\ z_2 &= x_2 - x_{2d} \\ &\cdots\cdots \\ z_n &= x_n - x_{nd} \end{aligned} \tag{15}$$

Dividing the nth-order system into n first-order subsystems, n Lyapunov functions are defined in turn:

$$V_1 = \frac{1}{2} * z_1^2 \tag{16}$$

$$V_2 = \frac{1}{2} * z_1^2 + \frac{1}{2} * z_2^2 \tag{17}$$

......

$$V_n = \frac{1}{2} * z_{n-1}^2 + \frac{1}{2} * z_n^2 \tag{18}$$

Then, design the backstepping control process from top to bottom in turn. The design process needs to ensure that $\dot{V}_n \leq 0$, which ensures system stability.

Differentiate the n Lyapunov functions in turn, as follows:

$$\dot{V}_1 = z_1 * \dot{z}_1 = z_1 * (\dot{x}_1 - \dot{\hat{x}}_{1d}) \tag{19}$$

$$\dot{V}_2 = -k_1 * z_1^2 + z_2 * \dot{z}_2 = -k_1 * z_1^2 + z_2 * (\dot{x}_2 - \dot{\hat{x}}_{2d}) \tag{20}$$

......

$$\dot{V}_n = -k_{n-1} * z_{n-1}^2 + z_n * (z_{n-1} + \dot{z}_n) \tag{21}$$

Subsystem 1: In order for the system to be stable, $\dot{V}_1 \leq 0$, i.e., negative definite. At this point, the above requirement can be satisfied as long as $\dot{x}_1 - \dot{x}_{1d}$ tends to $-k_1 * z_1, (k_1 > 0)$, so that the following is obtained:

$$\dot{V}_1 = -k_1 * z_1^2 \tag{22}$$

At this point, set x_{2d} as the tracking object of x_2, and because the system itself $\dot{x}_1 = x_2 + f_1(x_1)$, we can get:

$$x_{2d} = \dot{\hat{x}}_{1d} - k_1 * z_1 - f_1(x_1) \tag{23}$$

Subsystem 2: In order for the system to be stable, $\dot{V}_2 \leq 0$, i.e., negative definite. At this point, the above requirement can be satisfied as long as $\dot{x}_2 - \dot{x}_{2d}$ converges to $-k_2 * z_2, (k_2 > 0)$, so that the following is obtained:

$$\dot{V}_2 = -k_1 * z_1^2 - k_2 * z_2^2 \tag{24}$$

At this point, set x_{3d} as the tracking object of x_3, and because the system itself $\dot{x}_2 = x_3 + f_2(x_1, x_2)$, we can get:

$$x_{3d} = \dot{\hat{x}}_{2d} - k_2 * z_2 - f_2(x_1, x_2) \tag{25}$$

......

Subsystem n: Here, to ensure system stability, it is necessary that $\dot{V}_n < 0$, i.e., negative definite.

As $z_{n-1} + \dot{z}_n = -k_n * z_n$, where $k_n > 0$, we can obtain:

$$\dot{V}_n = -k_{n-1} * z_{n-1}^2 - k_n * z_n^2 \tag{26}$$

Solving the equation $z_{n-1} + \dot{z}_n = -k_n * z_n$ yields:

$$z_{n-1} + (\dot{x}_n - \dot{\hat{x}}_{nd}) = -k_n * z_n \tag{27}$$

$$z_{n-1} + u + f_n(x_1, x_2, \ldots, x_n) - \dot{\hat{x}}_{nd} = -k_n * z_n \tag{28}$$

The final control rate is then obtained:

$$u = \dot{\hat{x}}_{nd} - k_n * z_n - z_{n-1} - f_n(x_1, x_2, \ldots, x_n) \tag{29}$$

Because the errors are exponentially asymptotically stable, and thus so is the control rate designed above, the original nonlinear system is guaranteed to be exponentially asymptotically stable.

To sum up, the design block diagram of the proposed control strategy is summarized in Figure 4.

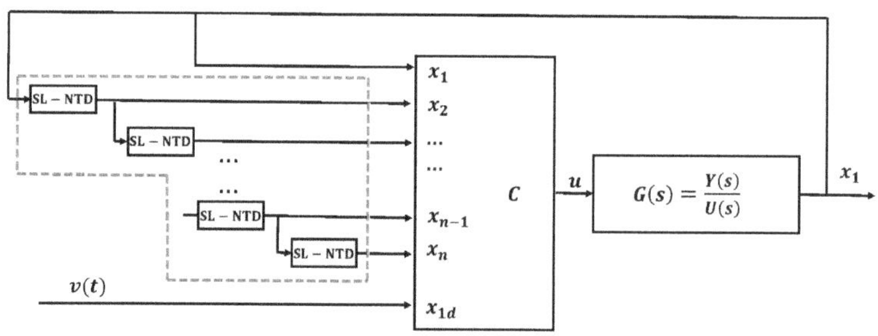

Figure 4. Backstepping control block diagram based on the Softsign linear–nonlinear tracking differentiator.

Remark 5. *The modified tracking differentiator SL-NTD in Figure 4 is used to generate alternative differential signals. In fact, to facilitate the implementation process and reduce the number of control parameters in the developed control strategy, all SL-NTDs can implemented with the same structure.*

Remark 6. *The convenience of the Softsign tracking differentiator-based backstepping control method proposed in this paper is that the traditional backstepping design process can still be used, and when the differentiation step is encountered, the differentiation extracted by the Softsign tracking differentiator is used instead, without breaking the traditional design steps. The tracking differentiator and backstepping control are independent of each other and can be used separately, which is more convenient.*

3.3. *Proof of Stability*

Lemma 2. *The second method of Lyapunov (direct method).*

Stability theorem: For a continuous nonlinear system, if one can construct a scalar function $V(x)$ with continuous first-order partial derivatives with respect to x, $V(0) = 0$ and an attraction region Ω is present around the origin of the state space such that all nonzero states $x \in \Omega$ satisfy the following conditions.

1. $V(x) > 0$.
2. $\dot{V}(x) \leq 0$.
3. For any nonzero $x \in \Omega$, $\dot{V}(\varnothing(t; x_0, 0)) \not\equiv 0$.

Then, the original system equilibrium state $x = 0$ is asymptotically stable in the Ω region.

Proof. The nth-order system is divided into n first-order systems using n Lyapunov functions to maintain stability, as shown in Equations (16)–(18), where there are n-3 more equations between Equations (17) and (18). Looking at Equations (16)–(18), we can see that $V_1, V_2, \ldots, V_n$ are all constant and greater than or equal to zero.

Deriving the above n equations separately, after introducing the virtual control quantity, we obtain Equations (22), (24) and (26), where there are n-3 equations between

Equations (24) and (26). Observe that in Equation (22), since k_1 is defined to be a constant greater than zero, it is guaranteed that $\dot{V}_1$ is constantly less than or equal to zero. Observe that in Equation (24), since k_1 and k_2 are defined as constants greater than zero, it is guaranteed that $\dot{V}_2$ is constantly less than or equal to zero. Similarly, the n-3 equations between Equations (24) and (26) also prove that $\dot{V}_3, \dot{V}_4, \ldots, \dot{V}_{n-1}$ are constantly less than or equal to zero. Substituting the control rate obtained from Equation (29) into the solved Equation (18), we can obtain Equation (26), since k_{n-1} and k_n are defined as constants greater than zero, so we can guarantee that $\dot{V}_n$ is constantly less than or equal to zero.

Summing up the above conclusions, the following equations can be obtained:

$$V_1 = \frac{1}{2} * z_1^2 \geq 0 \tag{30}$$

$$V_2 = \frac{1}{2} * z_1^2 + \frac{1}{2} * z_2^2 \geq 0 \tag{31}$$

$$\cdots\cdots$$

$$V_n = \frac{1}{2} * z_{n-1}^2 + \frac{1}{2} * z_n^2 \geq 0 \tag{32}$$

$$\dot{V}_1 = -k_1 * z_1^2 \leq 0 \tag{33}$$

$$\dot{V}_2 = -k_1 * z_1^2 - k_2 * z_2^2 \leq 0 \tag{34}$$

$$\cdots\cdots$$

$$\dot{V}_n = -k_{n-1} * z_{n-1}^2 - k_n * z_n^2 \leq 0 \tag{35}$$

According to Lyapunov's stability theory, the controller is asymptotically stable at the origin, which ensures the stability of the whole system. Lemma 2 is proven. □

3.4. Parameter Selection Guidelines

The parameters involved in the design of the backstepping control of the optoelectronic tracking system based on the Softsign linear–nonlinear tracking differentiator include the parameters a and b of the improved tracking differentiator and the parameters k_1, k_2 and k_3 of the backstepping control. These parameters need to be considered in the design of the backstepping control to improve the tracking performance of SL-NTDBSC so that the differential signal generated by the Softsign linear–nonlinear tracking differentiator will not cause the "complexity explosion" problem during the whole process and meet the stability requirements. The differential signals generated by the Softsign linear–nonlinear tracking differentiator should be considered in the backstepping control design so that the whole backstepping process will not have the problem of a "complexity explosion" and satisfy the stability requirements. To make the differential signal generated by Softsign linear–nonlinear tracking differentiator closer to the real differential signal and ensure that is does not have the "complexity explosion" problem after multiple differentiations, the tracking differentiator should be designed to combine the fast signal convergence and stability. The parameter tuning guidelines are summarized as follows.

1. Firstly, when designing SL-NTDBSC, it is necessary to ensure that the adjusted parameters can ensure the stability of the system. According to Theorem 2 and Lemma 2, it can be obtained that the system is stable on the premise of $k_1 > 0$, $k_2 > 0$, $k_3 > 0$, $a > 0$ and $b > 0$.
2. Then, to meet the requirements related to the response speed and overshoot, the following must be considered. In the Softsign linear–nonlinear tracking differentiator, adjusting a and b can affect the convergence speed of the system, adjusting a too much will result in a large overshoot, increasing b can speed up the response time of the system, and a too small value of b will reduce the convergence speed of the system, and vice versa will lead to system instability. Increasing k_1 can speed up the system's response in backstepping control, but a too large k_1 may lead to system instability.

Similarly, adjusting k_2 and k_3 can affect the system's overshoot, and suitable values of k_2 and k_3 can reduce the amount of overshoot.

3. Finally, the parameters involved above are adjusted appropriately to optimize their quality, and the control parameters are configured as $a = 7, b = 4, k_1 = 100, k_2 = 30$ and $k_3 = 15$ (verified in the experiments in Section 4). Note that the parameter selection in this method is based on an empirical approach.

4. Simulation and Experimental Verification

4.1. Description of the Backstepping Design of the Specific Controlled Object

In this section, the performance of the backstepping controller based on the improved Softsign tracking differentiator is validated using a precision tracking platform, as shown in Figure 5. The stabilized tracking platform consists of a laser, a charge-coupled device (CCD), a motor, a static mirror, a control module, and other accessories. In this experimental setup, the laser light source is a 635 nanometer fiber-coupled laser diode produced by Thorlabs. The motor used is a voice coil motor with the model number BEI KIMCO LA12-17-000A, and the CCD model is TMC-6740CL, manufactured by NI Corporation based in Austin, TX, USA. The reflective mirror is a custom-designed component. Additionally, the control board employed is the PC104LX3073 from Shenzhen Shenlanyu Technology Co., Ltd. (Shenzhen, China), equipped with the real-time operating system VxWorks. Throughout the experiment, all sampling sensors were operated at a sampling rate of 5000 Hz.

Firstly, driven by the voice coil motor, the target light signal is reflected in the CCD by the tilted mirror, and the target off-target quantity is obtained through filtering and decoding the signal processing module, which is externally fed back to the controller. Secondly, the position, speed and acceleration information of the optical platform controlling the light beam is measured by the sensors mounted on the precision tracking platform in real time, which is internally fed back to the controller. The controller receives the feedback quantity and outputs a control signal to the driver. Finally, after receiving the feedback, the controller outputs a control signal to the driver, which controls the motor to drive the optical platform so that the tracking target is in the center of the CCD target surface to achieve the purpose of precise tracking [36,37].

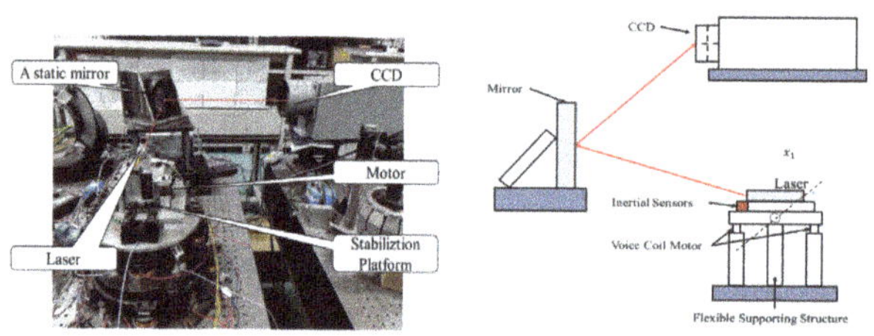

(a) Physical Representation of the Precision Tracking Platform (b) Schematic Diagram of the Precision Tracking Platform

Figure 5. Precision tracking platform.

The system shown in Figure 5 is nonlinear, and its nonlinear characteristics are given as shown in Figure 6. Backstepping control can handle the nonlinear system well. Firstly, the nonlinear part is uniformly classified as a nonlinear term, then the nonlinear part is estimated by designing a new Lyapunov function. Finally, it is compensated for so that the system is unaffected by the nonlinear part. The nonlinear part is estimated to be γz_3, where z_3 is the third sub-system error during the backstepping design process and γ is a gain that is greater than zero. To simplify the expression, in this work, the nonlinear system is linearized directly without considering the nonlinear part.

By inputting the frequency sweep signal into the precision tracking platform, the position-controlled object is measured as shown in Figure 7, where the blue solid line is the actual measured curve and the red dashed line is the fitted curve. After executing the object recognition technique, the expression of the position-controlled object in the frequency domain is derived as:

$$G(s) = \frac{Y(s)}{U(s)} = \frac{13390}{0.1508 * s^3 + 40.73 * s^2 + 973.2 * s + 53580} \quad (36)$$

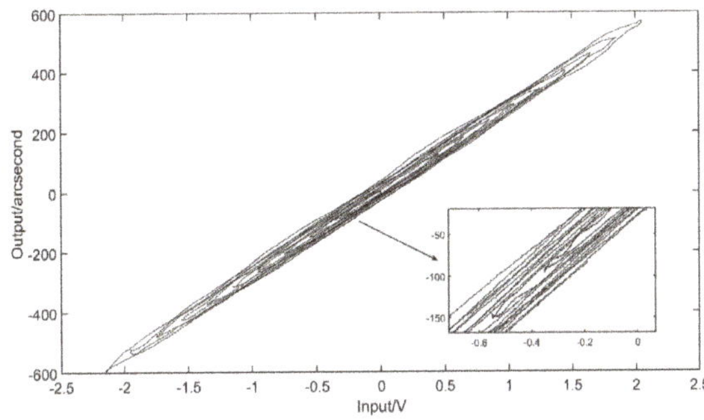

Figure 6. Experimental research on system nonlinearity.

The method proposed in this paper will be compared with the command filter, existing tracking differentiators, etc., using the following specific controlled objects.

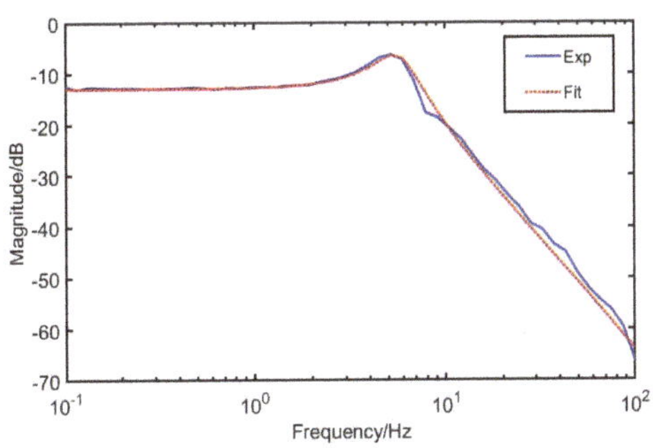

Figure 7. Controlled objects of the precision tracking platform.

By dividing the third-order system into three first-order systems and designing the Lyapunov functions in turn, the control rate is finally obtained according to the backstepping design rule as:

$$u = \frac{53580 * x_1 + 973.2 * x_2 + 40.73 * x_3}{13390} + \frac{0.1508 * (\dot{x}_3 d - k_3 * z_3 - z_2)}{13390} \quad (37)$$

where, in the Softsign improved tracking differentiator-based backstepping controller, $\dot{x}_3 d$ denotes the approximate differential signal of x_3 extracted by the Softsign improved tracking differentiator.

4.2. Simulation Analysis

Firstly, the differentiation effect of the Softsign linear–nonlinear tracking differentiator designed in this paper is verified by MATLAB/Simulink simulations.

The differentiation effect of the Softsign-based linear–nonlinear tracking differentiator in this paper is first simulated and compared with the differentiation effect of command filtering. The three plots in Figure 8a–c are given for a comparison of the differentiation trajectories at the input signal frequencies of $1/(2*pi)$ $10/(2*pi)$, and $20/(2*pi)$, respectively, and it can be seen that the differentiation based on the Softsign linear–nonlinear tracking differentiator is better at the same frequency. Table 1 shows the root mean square values of the errors generated by the command filter and the Softsign linear–nonlinear tracking differentiator when the differentiation is extracted at different input frequencies, and it can be seen that as the frequency increases, the differentiation of the command filter becomes less and less effective compared to the differentiation of the Softsign linear–nonlinear tracking differentiator.

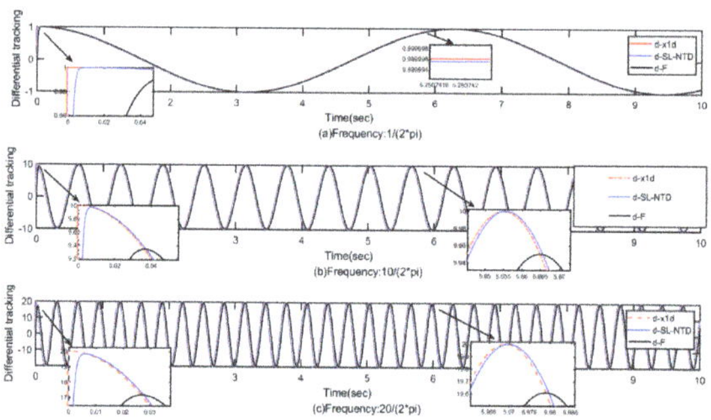

Figure 8. Differential trajectory of the Softsign linear–nonlinear tracking differentiator (d_SL-NTD) and the command filter (d_CF) at different input signal frequencies.

Table 1. RMS values for CF and SL-NTD errors extracted from differentiation at different input frequencies.

Different Methods		SL-NTD	CF
RMS Value	1/(2*pi)	0.00082	0.00705
	10/(2*pi)	0.08192	0.70152
	20/(2*pi)	0.32698	2.76532

The input signal was differentiated twice by a Softsign linear–nonlinear tracking differentiator, a command filter, and a conventional differentiator, respectively. As shown in Figure 9, it can be seen that the conventional differentiator exhibits a large overshoot of 10^{29}, while the Softsign linear–nonlinear tracking differentiator exhibits an overshoot of 10^{1}. It can be seen that the Softsign linear–nonlinear tracking differentiator can greatly mitigate the "complexity explosion" problem.

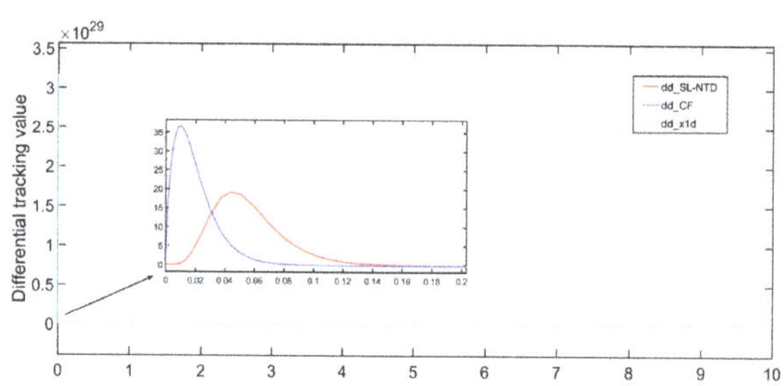

Figure 9. Tracking results of the SL-NTD, CF, and traditional differentiator after two differentiations.

When comparing the trajectory tracking of conventional backstepping control and command-filter-based backstepping control with the trajectory tracking of the Softsign excitation function-based tracking differentiator, as shown in Figure 10a, it can be seen that all three can achieve trajectory tracking, but the command-filter-based backstepping control and the conventional backstepping control fall behind at the beginning stage. As can be seen from Figure 10b, their trajectory tracking errors are more prominent, with the traditional backstepping control exhibiting the most significant errors, followed by command-filter-based backstepping control, which has a larger overshoot at the beginning and takes longer to stabilize and still has periodic error fluctuations after stabilization. This shows that the trajectory tracking effect of the Softsign excitation function-based tracking differentiator for backstepping control proposed in this paper is significantly better than that of command-filter-based backstepping control and traditional backstepping control.

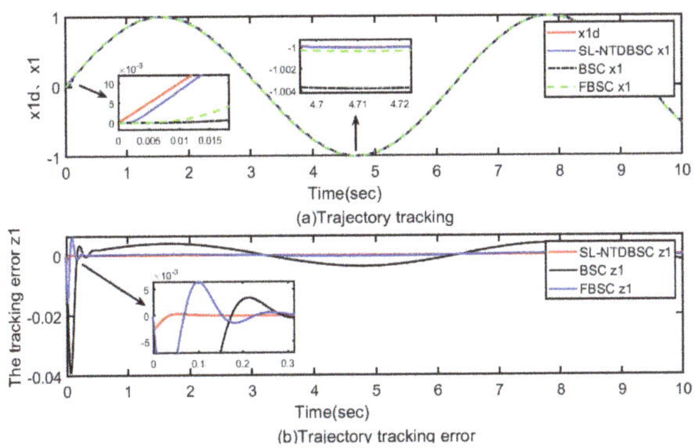

Figure 10. Comparison of trajectory tracking under FBSC and SL-NTDBSC.

To more strongly illustrate that the backstepping control of linear–nonlinear tracking differentiator based on the Softsign excitation function in this paper has a better trajectory tracking performance, it was compared with backstepping control using other tracking differentiators. The formulas for the tracking differentiators used for comparison are as follows:

Tracking differentiator 1: NTD [33]:

$$\begin{cases} \dot{x}_1(t) = x_2(t) \\ \dot{x}_2(t) = -R^2 * a * [x_1(t) - v(t) + tanh(x_1(t) - v(t))] \\ -R^2 * b * [x_2(t)/R + tanh(x_2(t)/R)] \end{cases} \tag{38}$$

Tracking differentiator 2: LTD [27]:

$$\begin{cases} \dot{x}_1(t) = x_2(t) \\ \varepsilon^2 * \dot{x}_2(t) = -a_0 * (x_1(t) - v(t)) - a_1 * \varepsilon * x_2(t) \end{cases} \tag{39}$$

Tracking differentiator 3: HSTD [28]:

$$\begin{cases} \dot{x}_1(t) = x_2(t) \\ \varepsilon^2 * \dot{x}_2(t) = -a_0 * (x_1(t) - v(t)) \\ -a_1 * |x_1(t) - v(t)|^m * sgn(x_1(t) - v(t)) - b_0 * \varepsilon \\ x_2(t) - b_1 * |\varepsilon * x_2(t)|^n * sgn(\varepsilon * x_2(t)) \end{cases} \tag{40}$$

The values of the parameters associated with the linear–nonlinear tracking differentiator based on the Softsign excitation function introduced in this paper and each of the other comparative tracking differentiators are as shown in Table 2.

Comparing the information in Table 2, it can be seen that the number of parameters of the linear–nonlinear tracking differentiator based on the Softsign excitation function is smaller compared to other comparative tracking differentiators, which makes the overall parameter adjustment process of the backstepping control designed in this paper easier.

Figure 11a shows a comparison of the tracking results of the differential signal for the backstepping control of the linear–nonlinear tracking differentiator based on the Softsign excitation function (SL-NTDBSC) with those of the backstepping control based on the linear tracking differentiator (LTDBSC), the backstepping control based on the high-speed tracking differentiator (HSTDBSC) and the backstepping control based on the new nonlinear–linear tracking differentiator (NTDBSC). Referring to Figure 11a,b, it is obvious that SL-NTDBSC achieves the smallest trajectory tracking error with no excessive jitter and periodic errors, minimizing the error in this range.

Table 2. Parameter values of different tracking differentiators.

Different Tracking Differentiators	Parameter Values
SL-NTD	$a = 7, b = 5$
NTD	$R = 150, a_0 = 7, a_1 = 5$
LTD	$\varepsilon = 0.004, a_0 = 2, a_1 = 1$
HSTD	$\varepsilon = 0.004, a_0 = 5, a_1 = 0.5,$ $b_0 = 2, b_1 = 0.5, m = n = 0.5$

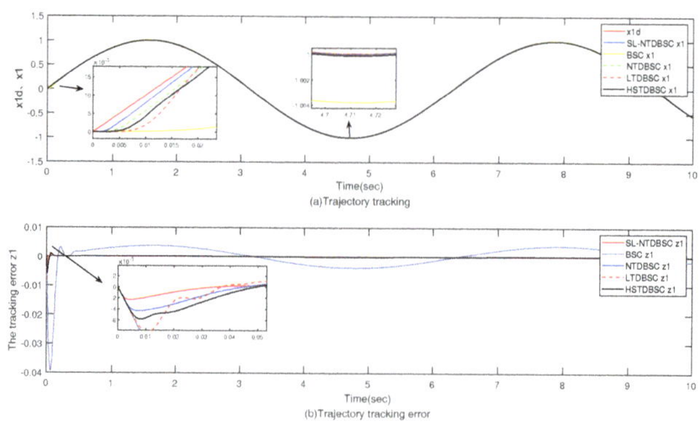

Figure 11. Comparison of trajectory tracking under backstepping control based on different tracking differentiators.

In order to quantitatively analyze the enhancement effect, two indicators were selected, as shown in Table 3 below.

$$I_{ITAE} = \int_0^t t|z_1(t)|dt \quad (41)$$

$$I_{ISE} = \int_0^t [z_1(t)]^2 dt \quad (42)$$

Table 3. Performance indexes of experimental data in Figure 11.

Indicators / Controllers	ITAE Values	ISE Values
SL-NTDBSC	9.7414	0.00052
NTDBSC	17.3161	0.0016
SL-NTDBSC lifting effect	43.74%	67.5%
LTDBSC	20.7765	0.0025
SL-NTDBSC lifting effect	53.11%	79.2%
HSTDBSC	28.1929	0.0032
SL-NTDBSC lifting effect	65.45%	83.75%

The ITAE (integrated time absolute error) in Equation (41) is the absolute value of the error multiplied by the integration of the time term over time, which reflects both the magnitude of the error and the speed of error convergence, taking into account the control accuracy and the speed of convergence. The ISE in Equation (42) is the integral squared error, which indicates a large error in the suppression transition process. $z_1(i)$ are the position control errors at the i-th step of the control cycle. The control accuracy of SL-NTDBSC is the best and its convergence speed is the fastest, judging from the values of the I_{ITAE} metric of SL-NTDBSC relative to other backstepping methods. In addition, the error in the transition process of SL-NTDBSC is almost zero, judging from the enhancement in the I_{ISE} metric of SL-NTDBSC relative to other backstepping methods.

4.3. Experimental Results

To verify the results obtained from the simulations, experimental validation was carried out by using an accurate tracking platform and is detailed in this section. It is evident from the simulation analysis that the trajectory tracking performance of LTDBSC and HSTDBSC in backstepping control based on the improved tracking differentiator which was used for comparison is significantly weaker than that of NTDBSC. Therefore, the experiments in this paper compare the trajectory tracking performance of SL-NTDBSC, BSC, NTDBSC, and FBSC at an input signal amplitude of 100 and frequencies of 0.1, 0.5, 1.0, and 2.0.

As shown in Figure 12a, both conventional backstepping control and improved backstepping control can realize the tracking of the target trajectory when the frequency of the input signal is 0.1 Hz. However, the trajectory tracking performance of SL-NTDBSC and NTDBSC is better than that of the other methods, and the trajectory tracking errors of SL-NTDBSC and NTDBSC are minimized, which is also clearly seen in Figure 12b. When the frequency of the input signal is 0.5 Hz, as shown in Figure 13a, there is a decreasing trend in the trajectory accuracy of conventional backstepping control, and from Figure 13b, it can be seen that the amplitude of the trajectory tracking error of SL-NTDBSC is minimized. When the input signal frequency is 1.0 Hz, as shown in Figure 14a, the conventional backstepping controller cannot effectively control the trajectory accuracy, and overshooting occurs when the input waveform reaches the top or the bottom. It is clear from Figure 14b that the trajectory tracking error of conventional backstepping control is more than 20. The trajectory tracking errors of FBSC and NTDBSC are still larger than that of SL-NTDBSC. When the frequency of the input signal is 2.0 Hz, it can be seen from Figure 15 that conventional backstepping control is no longer able to achieve trajectory tracking. In Figure 15b, it can be seen that the active interval of the trajectory tracking error of FBSC and NTDBSC is almost two times that of SL-NTDBSC.

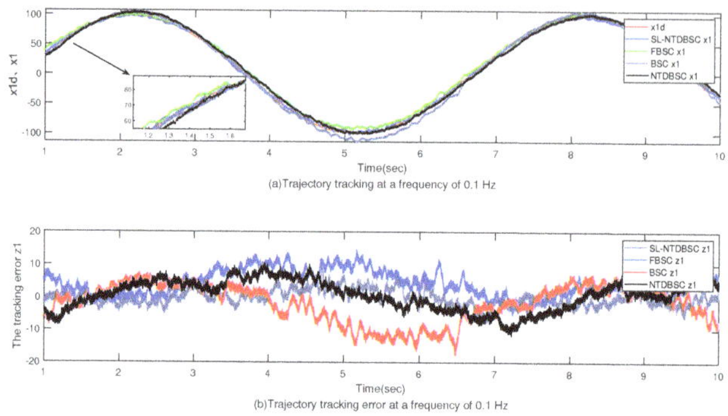

Figure 12. Trajectory tracking of different types of backstepping control at a frequency of 0.1 Hz based on an experimental platform.

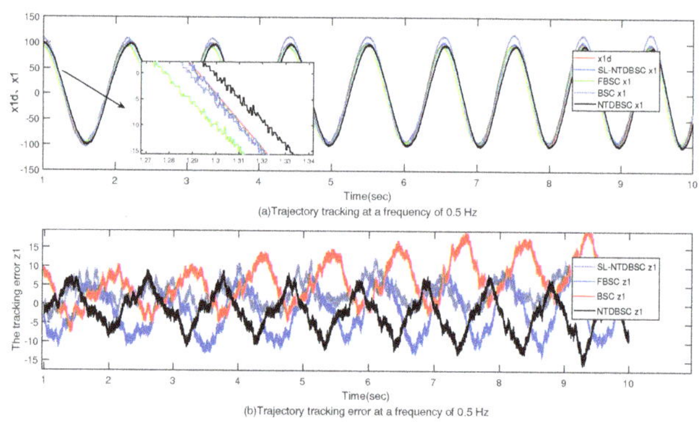

Figure 13. Trajectory tracking of different types of backstepping control at a frequency of 0.5 Hz based on an experimental platform.

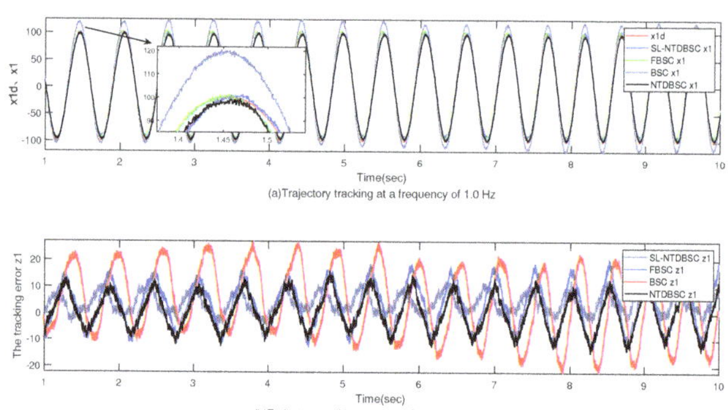

Figure 14. Trajectory tracking of different types of backstepping control at a frequency of 1.0 Hz based on an experimental platform.

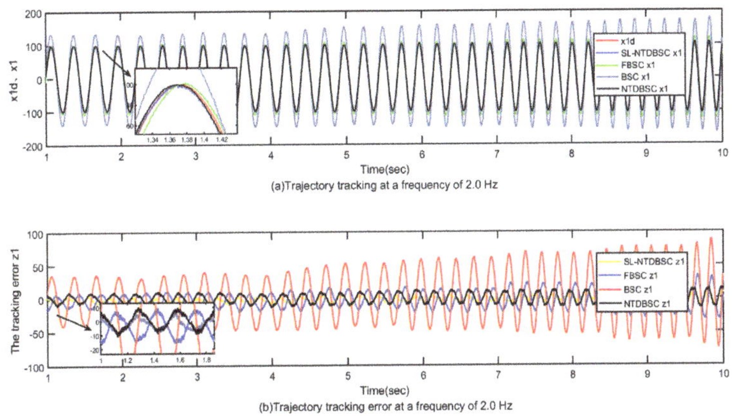

Figure 15. Trajectory tracking of different types of backstepping control at a frequency of 2.0 Hz based on an experimental platform.

In addition to the two quantitative analysis indexes used in the simulation, two new indexes were added to the experiment. All of these indexes can be used to measure the performance of a trajectory tracking control system from slightly different perspectives.

$$I_{MSE} = \frac{1}{N}\sum_{i=1}^{N}[z_1(i)]^2 \tag{43}$$

$$I_{APE} = \max_{i=1,\dots,N} |z_1(i)| \tag{44}$$

where I_{MSE} denotes the mean square error index. It measures the average error of the system concerning the reference trajectory. The smaller the I_{MSE}, the smaller the average error of the system and the better the tracking performance. I_{APE} is the absolute peak error index. The role of trajectory tracking is mainly to measure the difference between the pulse width of the control system output and the pulse width of the reference trajectory, which can reflect the accuracy of the control system. If the value of the I_{APE} index is small, it means that the control accuracy of the system is high, and accordingly, the trajectory tracking performance is better. If the value of the I_{APE} index is larger, it means that the control accuracy of the system is lower, and accordingly, the trajectory tracking performance may be poorer. $z_1(i)$ are the position control errors at the i-th step control cycle, and N denotes the number of tracking error data for calculation.

Table 4 below shows the performance indexes for the data shown in Figures 12–15. Firstly, the I_{ITAE} index was analyzed, from 0.1 Hz to 2.0 Hz. The values for SL-NTDBSC were smaller than those of the other backstepping methods used for comparison, which indicates that the system under SL-NTDBSC has the lowest overall degree of deviation from the reference trajectory and has the best tracking performance. At 2.0 Hz, the performance of SL-NTDBSC is improved by 70.92%, 90.19%, and 48.29%, respectively, compared to that of FBSC, BSC and NTDBSC. Secondly, when analyzing the I_{ISE} index, from 0.1 Hz to 2.0 Hz, the value for SL-NTDBSC is consistently smaller than that of the other backstepping methods used for comparison, which indicates that the error is minimized over the whole time range under SL-NTDBSC and the system has the best tracking performance. Then, the I_{MSE} index was analyzed, from 0.1 Hz to 2.0 Hz. The values for SL-NTDBSC are consistently smaller than those of other backstepping control methods, which indicates that the system has the smallest average error for the reference trajectory and the best trajectory tracking performance under SL-NTDBSC. Finally, after analyzing the I_{APE} index from 0.1 Hz to 2.0 Hz, it is shown that the values for Sl-NTDBSC are still the lowest. The maximum tracking error of SL-NTDBSC is 12.8234 at 2.0 Hz, while the maximum tracking errors of the other backstepping methods are 34.7270, 89.0730, and 15.7490, respectively. Compared with BSC, the maximum tracking error of SL-NTDBSC is reduced by 85.60%, which indicates that the system has the highest control accuracy and the best trajectory tracking performance under SL-NTDBSC.

Comprehensively analyzing Figures 12–15 and Table 4, when the frequency of the input signal gradually increases, the trajectory tracking error of SL-NTDBSC changes less compared to other backstepping methods, where the change in the trajectory tracking error of BSC is especially large. The error that occurs at 2.0 Hz is larger, which shows that trajectory tracking was not realized.

When the input signal is a step signal, as shown in Figure 16, SL-NTDBSC has a faster response and a higher tracking accuracy when input signals are rapidly changed. However, in practice, due to the presence of noise and other influencing factors, these control methods will have jitter during tracking.

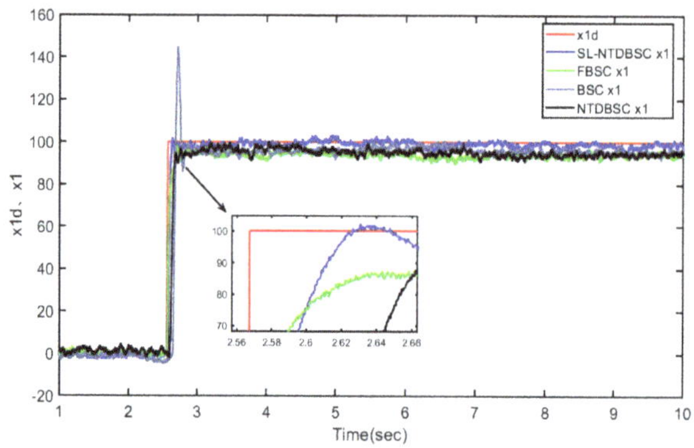

Figure 16. Trajectory tracking of various types of backstepping control based on the experimental platform when the input is a step signal.

Table 4. Performance indexes of experimental data in Figures 12–15.

		I_{APE}		
	SL-NTDBSC	FBSC	BSC	NTDBSC
0.1 Hz	10.9908	14.7040	17.5949	11.8499
0.5 Hz	12.3875	13.9040	21.1608	15.9570
1.0 Hz	14.0081	20.7113	26.9787	16.0001
2.0 Hz	12.8234	34.7270	89.0730	15.7490
		I_{ITAE}		
	SL-NTDBSC	FBSC	BSC	NTDBSC
0.1 Hz	7.6038	18.2346	19.6941	15.5492
0.5 Hz	15.1563	22.5381	30.6350	18.7924
1.0 Hz	19.0437	32.7671	49.3542	24.7648
2.0 Hz	14.8539	51.0840	151.4505	28.7252
		I_{MSE}		
	SL-NTDBSC	FBSC	BSC	NTDBSC
0.1 Hz	5.5064	32.5650	35.4257	20.3340
0.5 Hz	19.3630	36.8641	66.5298	27.5336
1.0 Hz	30.7469	79.3472	184.1473	45.6778
2.0 Hz	18.0900	162.8390	1443.10	52.0943
		I_{ISE}		
	SL-NTDBSC	FBSC	BSC	NTDBSC
0.1 Hz	4.1425	24.4990	26.6511	15.2975
0.5 Hz	14.5670	27.7332	50.0511	20.7138
1.0 Hz	23.1312	59.6937	138.5359	34.3638
2.0 Hz	13.6093	122.5054	1085.70	39.1911

5. Discussion

The proposed backstepping control method in this paper based on a Softsign linear–nonlinear tracking differentiator addresses the "complexity explosion" issue present in traditional backstepping control. It reduces the number of parameters, simplifies parameter adjustment, and simultaneously balances signal convergence speed and stability, further enhancing the trajectory tracking performance of backstepping control. As electro-optical tracking systems have increasingly diverse applications, with a variety of disturbance types and frequencies emerging, there is a need for further improvements to backstepping control to enhance the tracking and disturbance rejection capabilities. From the perspective of backstepping control design theory, fundamental nonlinear disturbances and uncertainties are treated as uncertain terms for suppression. Faced with diverse disturbance problems, one approach is to consider each disturbance as a new uncertain term and introduce a new disturbance estimation error. This can be achieved by using Lyapunov theory to estimate these uncertain terms, aiming to improve the disturbance rejection and tracking accuracy.

Furthermore, superior parameter adjustment methods will further enhance the tracking performance of optical tracking systems. Therefore, future work will involve designing a parameter adjustment method based on improved fuzzy control, combining it with improved backstepping control to simultaneously enhance the tracking precision and disturbance rejection capabilities in electro-optical tracking systems. This holds significant application value for electro-optical tracking systems.

6. Conclusions

To address the problems caused by electro-optical tracking systems with uncertainties, this paper employs backstepping control. However, traditional backstepping control is susceptible to the "complexity explosion" issue. Therefore, this paper introduces, for the first time, a backstepping control design based on a Softsign linear–nonlinear tracking differentiator. A novel tracking differentiator is designed using the Softsign function, enhancing the differentiation effect and providing some filtering capabilities. Simulation and experimental results confirm the effectiveness and superiority of this control method. The approach proposed in this paper overcomes the "complexity explosion" issue associated with traditional backstepping control by using a Softsign linear–nonlinear tracking differentiator to approximate the traditional differentiation process. After two consecutive differentiations, the overshoot peak is only $1/10^{28}$ of that of traditional differentiation, significantly reducing the likelihood of a complexity explosion. Moreover, this method reduces the number of parameters, simplifies parameter adjustment, and simultaneously balances signal convergence speed and stability, thereby improving the trajectory tracking performance of backstepping control. According to the ITAE index, over the frequency range of 0.1 Hz to 2.0 Hz, the tracking performance of this approach is improved by 65.88% compared to traditional backstepping control, 50.96% compared to command-filter-based backstepping control, and 35.46% compared to NTD-based backstepping control. Additionally, stability design using Lyapunov theory ensures the stability of this method, and it also guarantees the boundedness of system signals.

However, as the application scenarios of electro-optical tracking systems evolve, there will be an increased demand for control methods to have improved tracking precision characteristics and disturbance rejection capabilities. To meet the practical needs of electro-optical tracking systems, the next phase of research will focus on optimizing the method presented in this paper to further address the issue of the reduced disturbance rejection performance in multiple scenarios. Furthermore, forthcoming work will also consider the parameter adjustment problem in backstepping control, aiming to make parameter tuning more accurate and straightforward. In summary, the method proposed in this paper holds significant value for electro-optical tracking systems, and future research efforts will strive to enhance its performance and applicability in various real-world scenarios.

Author Contributions: Theoretical analysis: J.L. and S.Z.; Designing experiments and analyzing data: J.L. and H.W.; Conducting simulations: J.L. and S.Z.; Writing the paper: J.L.; Revising the paper: J.D. and Y.M. All authors have read and agreed to the published version of the manuscript.

Funding: This work was supported by the Special Research Assistant Program, Chinese Academy of Sciences, China (Grant No. E329691C21) and Natural Science Foundation of Sichuan Province for Youths, China (Grant No. 24NSFSC3777).

Institutional Review Board Statement: Not applicable.

Informed Consent Statement: Not applicable.

Data Availability Statement: Data are contained within the article.

Conflicts of Interest: The authors declare no conflicts of interest.

References

1. Zhang, B.; Nie, K.; Chen, X.; Mao, Y. Development of sliding mode controller based on internal model controller for higher precision electro-optical tracking system. *Actuators* **2022**, *11*, 16. [CrossRef]
2. Boroson, D.M.; Robinson, B.S.; Burianek, D.A.; Murphy, D.V.; Biswas, A. Overview and status of the lunar laser communications demonstration. In Proceedings of the Free-Space Laser Communication Technologies XXIV, SPIE, San Francisco, CA, USA, 24–26 January 2012; Volume 8246, pp. 69–78.
3. Bi, Z.; Tian, Z.; Luo, T.; Fu, S. Study on Shipborne Video Electro-Optical Tracking System Based on FPGA. In Proceedings of the 5th International Conference on Electrical Engineering and Automatic Control, Changsha, China, 4–6 November 2016; Springer: Berlin/Heidelberg, Germany, 2016, pp. 521–529.
4. Huang, L.; Song, J.; Zhang, C.; Cai, G. Design and performance analysis of landmark-based INS/Vision Navigation System for UAV. *Optik* **2018**, *172*, 484–493. [CrossRef]
5. Bilal, D.K.; Unel, M.; Tunc, L.T.; Gonul, B. Development of a vision based pose estimation system for robotic machining and improving its accuracy using LSTM neural networks and sparse regression. *Robot. Comput.-Integr. Manuf.* **2022**, *74*, 102262. [CrossRef]
6. Yang, L.; Etsuko, K. Review on vision-based tracking in surgical navigation. *IET Cyber-Syst. Robot.* **2020**, *2*, 107–121. [CrossRef]
7. Ma, D.; Chen, J.; Liu, A.; Chen, J.; Niculescu, S.I. Explicit bounds for guaranteed stabilization by PID control of second-order unstable delay systems. *Automatica* **2019**, *100*, 407–411. [CrossRef]
8. Zhang, M. Finite-time model-free trajectory tracking control for overhead cranes subject to model uncertainties, parameter variations and external disturbances. *Trans. Inst. Meas. Control* **2019**, *41*, 3516–3525. [CrossRef]
9. Tong, W.; Zhao, T.; Duan, Q.; Zhang, H.; Mao, Y. Non-singleton interval type-2 fuzzy PID control for high precision electro-optical tracking system. *ISA Trans.* **2022**, *120*, 258–270. [CrossRef]
10. Ding, Z.; Zhao, F.; Lang, Y.; Jiang, Z.; Zhu, J. Anti-disturbance neural-sliding mode control for inertially stabilized platform with actuator saturation. *IEEE Access* **2019**, *7*, 92220–92231. [CrossRef]
11. Jang, S.G.; Yoo, S.J. Predefined-time-synchronized backstepping control of strict-feedback nonlinear systems. *Int. J. Robust Nonlinear Control* **2023**, *33*, 7563–7582. [CrossRef]
12. Kanchanaharuthai, A.; Mujjalinvimut, E. Nonlinear disturbance observer-based backstepping control for a dual excitation and steam-valving system of synchronous generators with external disturbances. *Int. J. Innov. Comput. Inf. Control* **2018**, *14*, 111–126.
13. Ayadi, M.; Derbel, N. Nonlinear adaptive backstepping control for variable-speed wind energy conversion system-based permanent magnet synchronous generator. *Int. J. Adv. Manuf. Technol.* **2017**, *92*, 39–46. [CrossRef]
14. Yue, F.; Li, X.; Chen, C.; Tan, W. Adaptive integral backstepping sliding mode control for opto-electronic tracking system based on modified LuGre friction model. *Int. J. Syst. Sci.* **2017**, *48*, 3374–3381. [CrossRef]
15. Wang, N.; He, H. Adaptive homography-based visual servo for micro unmanned surface vehicles. *Int. J. Adv. Manuf. Technol.* **2019**, *105*, 4875–4882. [CrossRef]
16. Liu, Z.; Lai, G.; Zhang, Y.; Chen, X.; Chen, C.L.P. Adaptive neural control for a class of nonlinear time-varying delay systems with unknown hysteresis. *IEEE Trans. Neural Netw. Learn. Syst.* **2014**, *25*, 2129–2140.
17. Bribiesca-Argomedo, F.; Krstic, M. Backstepping-forwarding control and observation for hyperbolic PDEs with Fredholm integrals. *IEEE Trans. Autom. Control* **2015**, *60*, 2145–2160. [CrossRef]
18. Liu, Y.H.; Huang, L.; Xiao, D. Adaptive dynamic surface control for uncertain nonaffine nonlinear systems. *Int. J. Robust Nonlinear Control* **2017**, *27*, 535–546. [CrossRef]
19. Ma, Z.; Ma, H. Improved adaptive fuzzy output-feedback dynamic surface control of nonlinear systems with unknown dead-zone output. *IEEE Trans. Fuzzy Syst.* **2020**, *29*, 2122–2131. [CrossRef]
20. Chen, P.; Zhang, T. Adaptive dynamic surface control of stochastic nonstrict-feedback constrained nonlinear systems with input and state unmodeled dynamics. *Int. J. Adapt. Control Signal Process.* **2020**, *34*, 1405–1429. [CrossRef]
21. Liang, X.; Qu, X.; Wang, N.; Zhang, R.; Li, Y. Three-dimensional trajectory tracking of an underactuated AUV based on fuzzy dynamic surface control. *IET Intell. Transp. Syst.* **2020**, *14*, 364–370. [CrossRef]

22. Jiang, M.; Yang, Y.; Gao, C.; Wu, L. Command filtered backstepping-based containment control for a class of multi-agent systems with uncertainty. *Int. J. Robust Nonlinear Control* **2023**, *33*, 10177–10863. [CrossRef]
23. Yu, J.; Shi, P.; Zhao, L. Finite-time command filtered backstepping control for a class of nonlinear systems. *Automatica* **2018**, *92*, 173–180. [CrossRef]
24. Han, Y.; Yu, J.; Zhao, L.; Yu, H.; Lin, C. Finite-time adaptive fuzzy control for induction motors with input saturation based on command filtering. *IET Control Theory Appl.* **2018**, *12*, 2148–2155. [CrossRef]
25. Ma, J.; Zheng, Z.; Li, P. Adaptive dynamic surface control of a class of nonlinear systems with unknown direction control gains and input saturation. *IEEE Trans. Cybern.* **2014**, *45*, 728–741. [CrossRef]
26. Fu, C.; Wang, Q.G.; Yu, J.; Lin, C. Neural network-based finite-time command filtering control for switched nonlinear systems with backlash-like hysteresis. *IEEE Trans. Neural Netw. Learn. Syst.* **2020**, *32*, 3268–3273. [CrossRef]
27. Zhao, Z.L.; Guo, B.Z. On convergence of nonlinear active disturbance rejection control for SISO nonlinear systems. *J. Dyn. Control Syst.* **2016**, *22*, 385–412. [CrossRef]
28. Shao, X.; Wang, H. Back-stepping robust trajectory linearization control for hypersonic reentry vehicle via novel tracking differentiator. *J. Frankl. Inst.* **2016**, *353*, 1957–1984. [CrossRef]
29. Bu, X.W.; Wu, X.Y.; Chen, Y.X.; Bai, R.Y. Design of a class of new nonlinear disturbance observers based on tracking differentiators for uncertain dynamic systems. *Int. J. Control Autom. Syst.* **2015**, *13*, 595–602. [CrossRef]
30. Feng, H.; Li, S. A tracking differentiator based on Taylor expansion. *Appl. Math. Lett.* **2013**, *26*, 735–740. [CrossRef]
31. Dong, X.; Zhang, P. Design and phase plane analysis of an arctangent-based tracking differentiator. *Control Theory Appl.* **2010**, *27*, 533–537.
32. Liu, Z.; Jiang, Y. Design of a modified tracking differentiator. *World J. Eng. Technol.* **2017**, *5*, 668–674. [CrossRef]
33. Fan Binyao, Z.Y. Design of New Fast and Highly Stable Nonlinear Linear Tracking Differential Device. *Ind. Control Comput.* **2021**, *34*, 79–82.
34. Fan, Y.; Jing, W.; Bernelli-Zazzera, F. Nonlinear Tracking Differentiator Based Prescribed Performance Control for Space Manipulator. *Int. J. Control Autom. Syst.* **2023**, *21*, 876–889. [CrossRef]
35. Zou, Q.; Huang, T.L.Z. Adaptive trajectory tracking of robotic manipulators using a new tracking differentiator. *Control Eng. Pract.* **2015**, *42*, 31–41.
36. Xia, W.; He, Q.; Duan, Q.; Zhou, X.; Deng, J.; Mao, Y. Equivalent acceleration feedforward based on sensor optimization and robust prediction. *Opto-Electron. Eng.* **2021**, *48*, 210153.
37. Liu, C.; Mao, Y.; Qiu, X. Disturbance-Observer-Based LQR Tracking Control for Electro-Optical System. *Photonics* **2023**, *10*, 900. [CrossRef]

photonics

Article

Research on Ground Object Echo Simulation of Avian Lidar

Zhigang Su [1,*], Le Sang [1], Jingtang Hao [1], Bing Han [1], Yue Wang [2] and Peng Ge [3]

[1] Sino-European Institute of Aviation Engineering, Civil Aviation University of China, Tianjin 300300, China; 2021122034@cauc.edu.cn (L.S.); jthao@cauc.edu.cn (J.H.); b-han@cauc.edu.cn (B.H.)
[2] Information Countermeasure Technology Laboratory, Beijing Research Institute of Telemetry, Beijing 100076, China; wangyue@brit.com.cn
[3] The 38th Research Institute of China Electronics Technology Group Corporation, Hefei 230093, China; gepeng09@gmail.com
* Correspondence: ssrsu@vip.sina.com

Abstract: The clutter suppression effect of ground objects significantly impacts the detection and tracking performance of avian lidar on low-altitude bird flock targets. It is imperative to simulate the point cloud data of ground objects in lidar to explore effective methods for suppressing clutter caused by ground objects in avian lidar. The traditional ray-tracing method is enhanced in this paper to efficiently obtain the point cloud simulation results of ground objects. By incorporating a beam constraint and a light-energy constraint, the screening efficiency of effective rays is improved, making them more suitable for simulating large scenes with narrow lidar beams. In this paper, a collision detection scheme is proposed based on beam constraints, aiming to significantly enhance the efficiency of ray-tracing collision detection. The simulation and experimental results demonstrate that, in comparison with other conventional simulation methods, the proposed method yields the point cloud results of ground objects that exhibit greater conformity to the actual lidar-collected point cloud results in terms of shape characteristics and intensity features. Additionally, the simulation speed is significantly enhanced.

Keywords: lidar; point cloud; simulation; collision detection; ground clutter

Citation: Su, Z.; Sang, L.; Hao, J.; Han, B.; Wang Y.; Ge, P. Research on Ground Object Echo Simulation of Avian Lidar. *Photonics* **2024**, *11*, 153. https://doi.org/10.3390/photonics11020153

Received: 22 December 2023
Revised: 25 January 2024
Accepted: 1 February 2024
Published: 5 February 2024

1. Introduction

The lidar system enables the rapid and precise acquisition of accurate positional information regarding obstacles or targets within its environment, making it a valuable tool in environmental perception, intelligent navigation, and other related fields [1,2]. However, when monitoring bird activity in the low-altitude airspace near airports, the presence of ground object echoes significantly interferes with the detection and tracking of bird flocks by the lidar system. Methods based on deep learning, such as GACNet [3], SSTNet [4], CVANet [5], etc., are commonly employed for point cloud target segmentation, which involves the separation of ground object targets from lidar-received point clouds and the subsequent suppression of the ground object echo. However, deep learning methods necessitate an ample number of labeled samples to train their networks, a requirement that cannot be met by the limited quantity of actual collected data. Therefore, it becomes imperative to simulate lidar's ground object echoes and generate sufficient annotated samples in order to effectively train relevant networks and explore effective approaches for suppressing ground object echoes.

The main methods for simulating lidar target echoes include ray tracing [6–8], physical optics simulation [9–11], radar scattering field simulation [7,12,13], and other approaches. The ray-tracing method is capable of accurately simulating the propagation and scattering process of light, thereby generating high-precision echo signals. However, due to its computational complexity and slow execution speed, ray-tracing methods are generally more suitable for scenarios involving small targets and simple scenes. The physical optical simulation method is based on a precise physical optical model to accurately simulate the optical characteristics of the target and generate high-precision echo signals. However, this method is highly dependent on the accuracy of the model and optical parameters, making it challenging to simulate complex scenes. The radar scattering field simulation method utilizes numerical computing technology to accurately simulate the interaction between the target and its surrounding environment, thereby generating efficient and precise echo signals. This approach necessitates the detailed modeling of the electromagnetic characteristics of the target, which entails substantial data and computational resources.

The limited storage capacity and reading and writing speed of the system restrict avian lidar from simultaneously improving the monitoring range and spatial resolution. Therefore, in order to ensure the coverage of the monitored airspace, avian lidar typically sacrifices spatial resolution performance and adopts a lower spatial resolution. Therefore, it is unnecessary to excessively focus on the detailed information of ground objects when simulating the echo of a ground object in avian lidar. In essence, a simplified model can be employed to describe the ground object in the echo simulation, thus rendering the ray-tracing method suitable for effectively simulating the ground object echo detected by avian lidar.

In ray-tracing methods, the rays emitted by and received by lidar are considered effective rays. Clearly, when tracking the rays emitted by lidar, the light path becomes more complicated due to reflections generated upon collision with object surfaces; however, the proportion of effective rays remains relatively low. Conversely, non-effective rays consume significant resources and impede the efficiency of ray-tracing methods. Common approaches to addressing the issue of extensive computational complexity in ray-tracing methods involve enhancing hardware computing capabilities or improving effective ray-screening techniques. Typically, GPUs serve as prominent hardware devices for boosting computing performance [14,15]. The parallel computing prowess of GPUs can significantly enhance computational efficiency. Furthermore, devices incorporating the Compute Unified Device Architecture (CUDA) can further optimize the GPU's computing power utilization. Enhanced methods for effective ray screening primarily involve reverse ray tracing or bidirectional ray tracing [6]. Reverse ray tracing is based on backward path tracing from the receiving end of the lidar to the transmitting end, ensuring that all rays are effective. However, this method does not accurately account for light refraction. Bidirectional ray tracing involves simultaneously calculating the light paths in both directions from both ends of the lidar, thereby improving the proportion of effective rays and enhancing performance.

In order to enhance the efficiency of ground object echo simulation for avian lidar and reduce the hardware calculation requirements of the simulation system, an improved ray-tracing method is proposed in this paper. The method incorporates the bidirectional ray-tracing concept, enabling the efficient screening of effective rays through a beam constraint and light-energy constraint, thereby avoiding the tracking of non-effective rays. Additionally, the method employs a batch processing approach for the batch collision detection of ray clusters, thus improving the collision detection efficiency in ground object echo simulation. The key contributions of this paper are as follows:

1. The beam constraint and light-energy constraint are determined based on the narrow beam of lidar and the lowest responsive light-energy level. In ground object echo simulation, these constraints can significantly enhance the efficiency of effective light screening and reduce the hardware calculation performance required by the simulation method.
2. The proposed collision detection scheme enables the simultaneous detection of all rays within the narrow lidar beam, thereby significantly reducing the time required for collision detection in the ray-tracing process.

The remaining sections of this paper are structured as follows: Section 2 provides an in-depth description of the models involved in laser pulse propagation. In Section 3, we introduce the batch collision detection scheme. Subsequently, we summarize the concrete implementation steps of the simulation method proposed in this paper in Section 4.

The simulation results are presented in Section 5, while a comprehensive discussion is provided in Section 6. Finally, the conclusion of this paper is presented in Section 7.

2. Associated Models of Ray-Tracing Method

A typical lidar structure comprises a base, a rotor, and a laser mounted on the rotor, as depicted in Figure 1a. The laser consists of a vertically and pitch-adjustable multi-channel transmitter/receiver. To avoid mutual interference between different channels during scanning, the laser transmitter/receiver is activated in a specific staggered sequence, as illustrated in Figure 1b. Consequently, the channel interference is not considered in this paper when simulating ground object echoes.

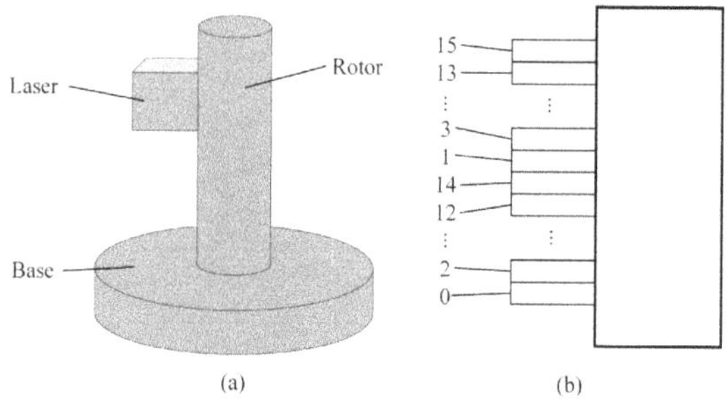

Figure 1. A schematic diagram of lidar: (**a**) Typical structure. (**b**) Activation sequence of the channels.

The entire lidar detection process encompasses the emission of laser pulses from the transmitter, their reflection off the object's surface, and their reception by the receiver. Hence, this process can be divided into five stages: emission, forward propagation, reflection, backward propagation, and reception, as depicted in Figure 2. The lidar ground object echo simulation aims to replicate the trajectory alteration and energy attenuation of laser pulses emitted by the lidar system across five distinct stages. By conducting a comprehensive scene scan, the simulation achieves a faithful representation of ground object echoes throughout the entire scene. During this process, the Monte Carlo beamwidth model is employed to simulate the transmitted beam of the lidar, while an atmospheric attenuation model analyzes light-energy loss during laser pulse propagation in the air. Additionally, a surface light energy reflectance model describes interactions between the ray and object surfaces, and a receiver noise model further characterizes the noise impact on ground object echoes.

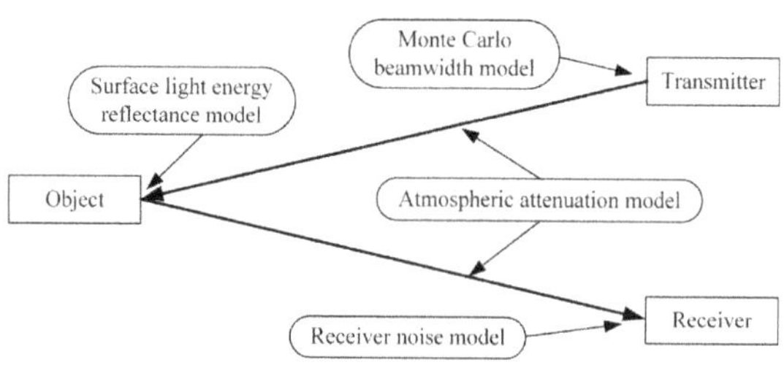

Figure 2. The process of laser pulse propagation.

The simulation of ground object echoes involves constructing a three-dimensional (3-D) virtual scene that represents the lidar working environment through computer modeling. The world origin, located at the center of this virtual scene, serves as the coordinate origin for establishing an East–North–Up coordinate system within the Cartesian coordinate framework. The object mesh grid is imported into the scene for further scanning. Each object in the scene is defined by its position and rotation information. The position information is represented using a 3-D coordinate vector, while the rotation information can be expressed either through an orientation vector or a rotation matrix. By altering the lidar's direction, we are able to achieve the scanning of the 3-D scene and obtain point cloud results that simulate echoes from all objects within this scene.

2.1. *Monte Carlo Beamwidth Model*

The laser pulse emitted by lidar propagates over a certain distance with a specific beamwidth. In the ray-tracing method, the propagation of the laser pulse is described using rays; however, this representation fails to fully capture the beam characteristics of laser pulse radiation. To address this limitation, a Monte Carlo beamwidth model is employed to describe the transmitting end of the lidar system. This model simulates a beam with a predetermined beamwidth [6,7,11] by utilizing multiple randomly distributed rays within that range, as illustrated in Figure 3.

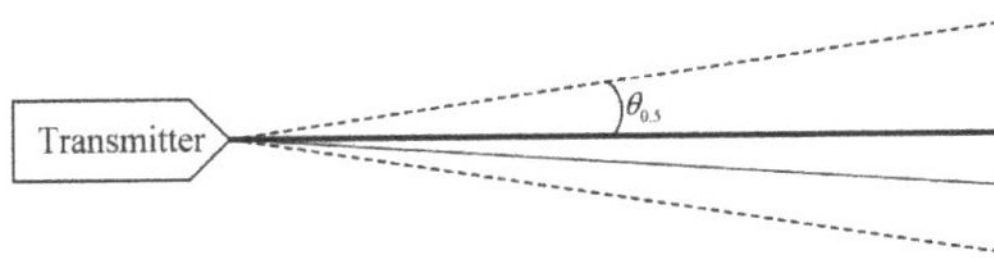

Figure 3. Monte Carlo beamwidth model.

The laser pulse is assumed to be a rectangular pulse with a pulse width of τ_e and an energy of Q_S. The total energy of N random rays within the beamwidth is denoted as

$$Q_s = \sum_{n=1}^{N} Q(\theta_n) \tag{1}$$

where the offset angle, θ_n, represents the angular difference between the n-th random ray and the direction vector $\mathbf{d}$ of the beam center. This can be specifically expressed as

$$\theta_n = \widehat{\mathbf{d}, \mathbf{d}_n} \tag{2}$$

with the direction vector $\mathbf{d}_n$ representing the direction of the n-th random ray. If the half-beamwidth is denoted by $\theta_{0.5}$, $\theta_n \in [-\theta_{0.5}, \theta_{0.5}]$. In Equation (1), the function $Q(\theta_n)$ denotes the energy of the ray at the offset angle θ_n, which describes how the energy of the emitted ray is distributed with respect to this offset angle. Assuming that the radiant energy density is uniform across the beamwidth, we have

$$Q(\theta_n) = \frac{1}{N} Q_s \tag{3}$$

2.2. *Surface Light Energy Reflectance Model*

After the ray is reflected by the surface of the object, there exists a relationship between the reflected light energy Q_{out} and the incident light energy Q_{in}:

$$Q_{out} = Q_{in} f_{sur}(\theta_i, \theta_r) \tag{4}$$

where $f_{sur}(\theta_i, \theta_r)$ is the surface light energy reflectance function, θ_i is the incidence angle, and θ_r is the reflected angle. The value range of θ_i is $[0, \pi/2]$, while the value range of θ_r is $[-\pi/2, \pi/2]$. When the reflected ray and the incident ray are located on the same side of the normal vector of the object surface, θ_r is negative, and vice versa. The surface light energy reflectance function ($f_{sur}(\theta_i, \theta_r)$) depends not only on the incidence angle and reflected angle but also on the roughness of the material on the object's reflecting surface. The surface light energy reflectance function ($f_{sur}(\theta_i, \theta_r)$) comprises a diffuse reflectance component and a specular reflectance component:

$$f_{sur}(\theta_i, \theta_r) = f_d(\theta_i) + f_s(\theta_i, \theta_r) \tag{5}$$

where $f_d(\theta_i)$ is the diffuse reflectance component, while $f_s(\theta_i, \theta_r)$ is the specular reflectance component. The schematic diagrams in Figure 4a,b illustrate the phenomena of diffuse reflection and specular reflection of a ray on an object's surface, respectively. Within Figure 4, the vectors $\mathbf{d}_i$, $\mathbf{d}_r$, and $\mathbf{d}_t$ represent the direction vectors for the incident ray, reflected ray, and specular reflection exit directions, respectively.

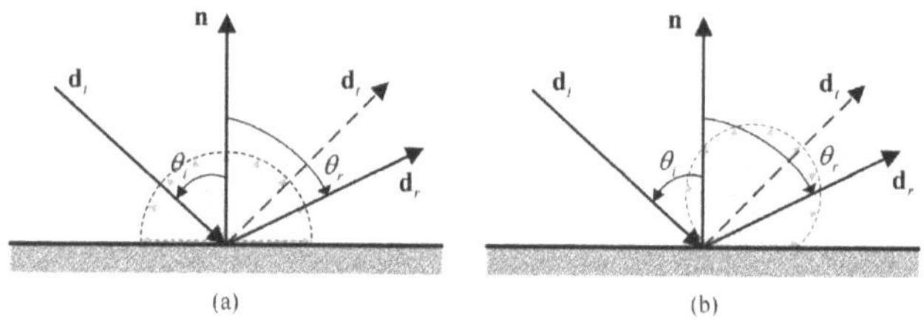

Figure 4. Ray reflection on surface: (**a**) Diffuse reflection. (**b**) Specular reflection.

According to the Lambert Cosine Law, the diffuse reflectance component can be described as [16]

$$f_d(\theta_i) = k_d \cos\theta_i \tag{6}$$

where k_d is the diffuse coefficient. Equation (6) demonstrates that the diffuse reflectance component is contingent upon the incident angle θ_i. As the incident angle increases, the diffuse reflectance component diminishes.

In the case where the surface of an object behaves like an ideal mirror, the ray will exit in a direction known as $\mathbf{d}_t$. However, due to imperfections on the object's surface, the exiting ray will reflect diffusion, resulting in a region centered around $\mathbf{d}_t$, as illustrated in Figure 4b. According to the Blinn–Phong model, the specular reflectance component is defined as [16]

$$f_s(\theta_i, \theta_r) = k_s \max\left\{0, \cos^{k_p}(\theta_r - \theta_i)\right\} \tag{7}$$

where k_s is the specular coefficient, while k_p is the glossiness coefficient. The specular reflectance component, as indicated by Equation (7), achieves its maximum value k_s in the direction of $\mathbf{d}_t$ and gradually decreases with increasing offset angle relative to $\mathbf{d}_t$. The glossiness coefficient k_p determines the effective distribution range of reflected light during specular reflection; a larger value corresponds to a smaller effective distribution range.

Substituting Equations (6) and (7) into Equation (5) yields

$$f_{sur}(\theta_i, \theta_r) = k_d \cos\theta_i + k_s \max\left\{0, \cos^{k_p}(\theta_r - \theta_i)\right\} \tag{8}$$

The parameters k_d, k_s, and k_p in Equation (8) for the surface material of the object can be determined by fitting the reflectance data from various material surfaces in the MERL BRDF material database [17]. Given these surface material parameters, the surface light energy reflectance function $f_{sur}(\theta_i, \theta_r)$ is dependent on both the incidence angle θ_i and the reflected angle θ_r. When a ray is reflected off the object's surface, the reflected ray exists within a range of $[-\pi/2, \pi/2]$ centered around the normal direction of the surface.

In the simulation of ground object echoes, our focus is solely on the ray that is reflected by the target and received by the lidar or adjacent targets of interest. Therefore, the direction ($\mathbf{d}_r$) of a reflected ray is determined based on the positional relationship between the lidar or target of interest and the reflecting surface. Subsequently, the corresponding reflected angle (θ_r) is obtained. By utilizing Equation (8), the value of $f_{sur}(\theta_i, \theta_r)$ is calculated, which represents the reflectance function for light energy. Finally, substituting $f_{sur}(\theta_i, \theta_r)$ into Equation (4) allows us to determine Q_{out}, representing the reflected light energy.

2.3. Atmospheric Attenuation Model

The attenuation effect of light propagating in the atmospheric medium cannot be disregarded. Previous studies [18,19] have all examined the form of attenuation for light propagation in this medium. The atmosphere is modeled as an isotropic uniform medium, and the attenuation coefficient (μ) is assumed to remain constant, resulting in an exponential model for atmospheric attenuation.

$$Q_a(l) = Q_0 \exp(-\mu l) \tag{9}$$

where Q_0 is the initial energy of the light, while $Q_a(l)$ represents the energy of the laser pulse after traveling a distance of l.

2.4. Receiver Noise Model

The Monte Carlo beamwidth model employed by lidar at the transmitting end utilizes N random rays to simulate the transmission beam. Consequently, it is imperative that N random rays are also reflected back to the receiver at the receiving end for unified processing, ensuring that the light energy received by the receiver is

$$Q_{rec}[m] = \sum_{n=1}^{N} Q_r(\theta_n, \tau_n) + Q_n[m] \tag{10}$$

where m corresponds to a specific range unit within lidar, measured in terms of sampling time intervals (τ_s); $Q_{rec}[m]$ signifies the amount of light energy detected by the m-th range unit; $Q_r(\theta_n, \tau_n)$ quantifies how much light energy is captured when an emitted laser pulse is reflected by an object's surface via the n-th random ray; θ_n and τ_n represent, respectively, both an offset angle for the n-th random ray and a temporal delay for receiving its corresponding laser pulse; and $Q_n[m]$ stands for noise energy picked up in the m-th range unit, where it is defined as

$$Q_n[m] = b[m]\tau_s \cdot h\nu \tag{11}$$

where $b[m]$ represents the quantity of noise photons detected in the m-th range unit, h denotes the Planck constant, and ν corresponds to the frequency of the laser. Generally, the distribution of $b[m]$ follows a Poisson distribution with the parameter $\bar{b} \cdot \tau_s$, where $\bar{b}$ represents the noise photon rate of the environment in which the lidar receiver is situated.

The dynamic range of the lidar receiver is typically limited, meaning that the receiver has acceptable upper and lower thresholds for receiving optical signals. These thresholds are recorded as Q_{high} (upper energy limit) and Q_{low} (lower energy limit), respectively. As a result, the received light energy can only fluctuate within the range from Q_{low} to Q_{high}. Any light energy ($Q_{rec}[m]$) exceeding the upper limit will be clamped, while $Q_{rec}[m]$ below the lower limit will not be perceived by the receiver. Consequently, after entering the receiver, the light energy becomes

$$Q_{en}[m] = \begin{cases} Q_{high}, & Q_{rec}[m] \geq Q_{high} \\ Q_{rec}[m], & Q_{low} < Q_{rec}[m] < Q_{high} \\ 0, & Q_{rec}[m] \leq Q_{low} \end{cases} \tag{12}$$

2.5. Received Light Energy Presentation Model

When conducting ground object echo simulation, it is necessary to visually represent the intensity information of the laser signal received by the receiver in color. To achieve a more enhanced rendering effect, the Hue-Saturation-Brightness (HSB) color mode, which closely emulates the human visual experience, is employed to accurately represent the intensity information of point clouds. The saturation and brightness of the color should be set to 1 while mapping the dynamic range of the received light energy $[Q_{low}, Q_{high}]$ to the hue range from blue to red, as illustrated in Figure 5.

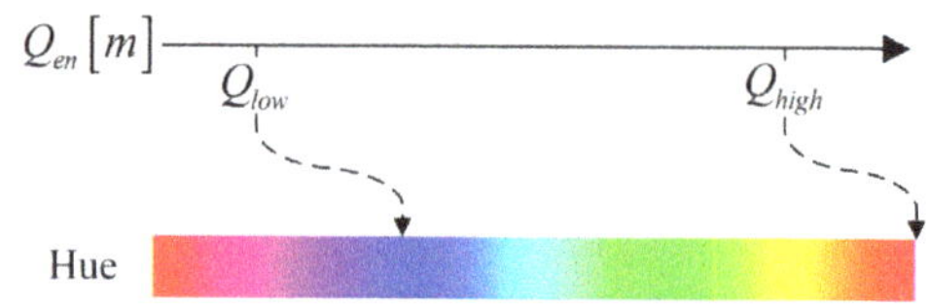

Figure 5. The mapping relationship between the received light energy and hue in the HSB mode.

As can be seen from Figure 5, the intensity of the received light energy is represented by the color tone. Because the hue values for cool blue and warm red are 0° and 240°, respectively, the hue value corresponding to the incoming laser energy at the receiver is

$$h[m] = \frac{Q_{high} - Q_{en}[m]}{Q_{high} - Q_{low}} 240^\circ \tag{13}$$

3. Collision Detection Scheme

The collision detection between the ray and the object surface in the scene poses a significant challenge when employing the ray-tracing method to simulate ground object echoes. The efficiency of the ray-tracing method is directly influenced by the quality of the collision detection scheme.

The term "collision detection" refers to the process of determining whether the emitted ray intersects with any objects in the scene based on its starting point and direction. If a collision occurs, it indicates that the ray illuminates an object along its propagation path within the scene; otherwise, the ray will not illuminate any objects and will continue infinitely. In terms of collision detection results, the closest point to the ray's starting point represents where it shines onto an object's surface.

After the incidence of a ray on an object's surface, it undergoes reflection. As per the surface light energy reflectance model outlined in Section 2.2, the reflected rays are distributed within the range $[-\pi/2, \pi/2]$ centered around the normal vector $\mathbf{n}$ of the object's surface. A portion of these reflected rays directly return to the lidar receiver, while some no longer intersect and vanish into infinity. Additionally, certain portions collide with objects in the scene once again and are reflected. The aforementioned processes continue until all corresponding light energy is depleted. Since each emitted ray can undergo multiple reflections, conducting multiple collision detections for each ray becomes necessary. Optimizing the collision detection scheme directly enhances efficiency in ray-tracing methods.

3.1. *Construction of Collision Detection Tree*

The objects within a 3-D scene can be perceived as an arrangement of triangular surfaces. Collision detection between the ray and the object surface in the scene essentially involves detecting collisions between the ray and these triangular surfaces. A straightforward approach to collision detection would involve traversing all triangular surfaces within the scene to identify any collisions with the ray, but this method is computationally demanding and impractical. To enhance collision detection efficiency, it becomes necessary to construct bounding volumes of varying sizes based on the triangular planes present in the scene, utilizing these bounding volumes to establish a collision detection tree that enables the rapid identification of collisions between the ray and triangular planes.

The bounding volume is a simple geometric structure containing multiple triangular surfaces, which enables the ray to initially assess the intersection with the bounding volume and subsequently determine the precise points of potential intersections within its respective scope. The bounding volume serves as an efficient means for rapid area screening. Commonly employed types of bounding volumes include the Axis-Aligned Bounding Box (AABB) [20], Oriented Bounding Box (OBB) [21], and Sphere Bounding Volume (SBV) [22].

The collision detection tree is a tree structure based on bounding volumes, designed to optimize the calculation of collision detection. Commonly used tree structures include the Bounding Volume Hierarchy (BVH) tree [23], Binary Space Partitioning (BSP) tree [24], and k-dimensional (k-D) tree [25]. The collision detection tree consists of a root node, intermediate nodes, and leaf nodes. Each node in the collision detection tree is denoted as

$$node_i = \{B_i, F_i, p_{i,1}, p_{i,2}\} \tag{14}$$

where B_i is the bounding volume corresponding to this node, while F_i represents the triangular plane inside the leaf node. The numbers $p_{i,1}$ and $p_{i,2}$ represent the two child nodes of this node and are excluded in the case of a leaf node. The set $S_{node} = \{node_i\}$ comprises all nodes that form the collision detection tree.

The construction process of the collision detection tree involves determining each node of the tree sequentially, as depicted in Algorithm 1. Steps 1 and 2 in Algorithm 1 serve as initialization processes for the relevant parameters. Firstly, it is essential to construct a bounding volume B_i for each element F_i within the set S_{flat} and incorporate it into the bounding volume set S_1. Secondly, the variables involved in constructing the collision detection tree are initialized. Steps 3 to 18 outline the detailed construction process of the collision detection tree. When the set S_{layer} is not empty, nodes need to be constructed from its elements. For each element L_p in the set S_{layer}, if it contains more than one bounding volume, a corresponding root node or intermediate node should be created; if it contains only one bounding volume, a leaf node needs to be constructed. In the case of a root node or intermediate node, all bounding volumes within L_p should be merged into a larger bounding volume B_{j0}, which will serve as the bounding volume for this particular node. Additionally, all bounding volumes within L_p must be divided into two subsets based on the construction rules of the respective tree structure and temporarily stored in the set S_{temp}. The BVH tree construction in step 9 allows for the direct division of the set into its first and second parts. On the other hand, when constructing a k-D tree, it is necessary to iteratively select dimensions as axes for partitioning and split the set into two approximately equal parts along these dimensions. Node $node_{j0}$ integrates B_{j0} with child node numbers j_1 and $j_1 + 1$, which are then included in the collection S_{node}. For leaf nodes, both the last bounding volume of the node and its corresponding triangular plane are integrated into node $node_{j0}$ and added to the set S_{node}. The set S_{layer} is updated with the set S_{temp} until all elements have been traversed, resulting in an empty set S_{layer}. Ultimately, the set S_{node} encompasses all nodes of the collision detection tree, thus completing its construction.

The collision detection between a ray and the object surface in the scene is achieved by utilizing the intersection of the ray and the bounding volume. The collision detection tree enables the efficient retrieval of the leaf node that intersects with the ray, ultimately providing information about the triangular surface where the collision occurs.

3.2. Intersection of Beam and Bounding Volume

The Monte Carlo beamwidth model is employed to describe the emitted beam of lidar, resulting in the confinement of the corresponding N random rays within a narrow beamwidth. During the collision detection between rays and surface elements, there exists a high probability that these N random rays will collide with the bounding volume in either the same leaf node or adjacent leaf nodes. Consequently, this inevitably leads to an increased number of identical intermediate nodes among these N random rays. Therefore, utilizing beams for retrieving leaf nodes can effectively reduce the computation required for each random ray and enhance the retrieval efficiency.

When retrieving leaf nodes using a beam, it is essential to judge the intersection between the conical beam and the bounding volume. The judgment only needs to be whether the beam and the bounding volume intersect, without necessitating the identification of the specific point or line of intersection.

Algorithm 1 Construction of the collision detection tree.

Input: The collection of triangular surfaces $S_{flat} = \{F_i\}$
Output: The set of nodes in the collision detection tree $S_{node} = \{node_i\}$

The minimum bounding volume B_i is constructed for each element F_i in set S_{flat}. The type of bounding volume can be selected based on the requirements, such as the AABB, OBB, or SBV. All the generated bounding volumes constitute the set $L_1 = \{B_i\}$.
The parameters of the collision detection tree are initialized. The current node j_0 is set to 1 and the child node j_1 is set accordingly. Complex sets $S_{layer} = \{L_p\}$ and $S_{temp} = \{L_q\}$ are constructed, which consist of bounding volume sets. It is ensured that S_{layer} contains only one element, namely, L_1, while S_{temp} and S_{node} are set as empty sets. The number of elements in the set S_{layer} is assigned to the variable P.
while $P > 0$ **do**
  Let $q = 1$ and $j_1 = j_0 + P$
  for $p = 1$ **to** P **do**
    The variable Q is assigned the number of bounding volumes in the p-th element L_p of the set S_{layer}.
    if $Q > 1$ **then**
      The smallest bounding volume B_{j0} that can surround all bounding volumes in L_p is built.
      According to the construction rules of tree structures, such as a BVH tree or k-D tree, the elements in L_p are partitioned into two subsets, namely, L_q and L_{q+1}, which are then added to the set S_{temp} as the q-th and $q+1$-th elements, respectively.
      The j_0-th node $node_{j0} = \{B_{j0}, \ \ , j_i, j_1 + 1\}$ is constructed. As this node is not a leaf node, it does not contain any triangular plane information. The node $node_{j0}$ is included in the collection S_{node} as the j_0-th element.
      Let $q = q + 2$, $j_1 = j_1 + 2$
    else
      The leaf node $node_{j0} = \{B_{j0}, F_{j0}, \ \ , \ \ \}$ is constructed by assigning the unique bounding volume in L_p to B_{j0} and associating the corresponding triangular surface with F_{j0}. The leaf node will no longer possess child node numbers. The node $node_{j0}$ is included in the collection S_{node} as the j_0-th element.
    end if
    Let $j_0 = j_0 + 1$
  end for
  Let $S_{layer} = S_{temp}$ and $S_{temp} = \emptyset$
end while
return S_{node}

The intersection assessment between the beam and the bounding volume involves assessing the intersection between two geometric structures in 3-D space. This process can be decomposed into evaluating the intersection between the beam angle and the rectangle on each of the three orthogonal faces in a two-dimensional (2-D) plane, as illustrated in Figure 6. If there are overlaps between the projected beams in all three planes and the projected rectangle of the bounding volume, then it can be concluded that there is an intersection between the beam and the bounding volume.

The intersection assessment between the beam and the rectangle in a 2-D plane can be further decomposed into assessing the intersection between the two rays that form the beam and the rectangle, as illustrated in Figure 7. In a 2-D plane, a beam intersects with a rectangle only if at least one ray intersects with it or if the two rays are positioned on opposite sides of the rectangle. By considering the end point of each ray as its origin, we define the vector pointing from this origin to each vertex of the rectangle as a rectangular vertex vector. The cross-product operation is performed between the direction vector of the ray and each vertex vector of the rectangle. If the resulting cross-product vectors have opposite directions, it indicates an intersection between the particular ray and the rectangle.

Consequently, determining whether there is an intersection between a beam and bounding volume can be decomposed into evaluating whether there is an intersection among multiple rays and rectangles.

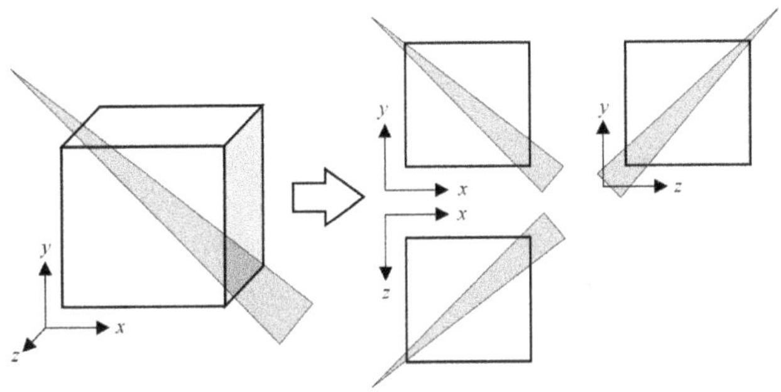

Figure 6. The intersection assessment between the beam and the bounding volume.

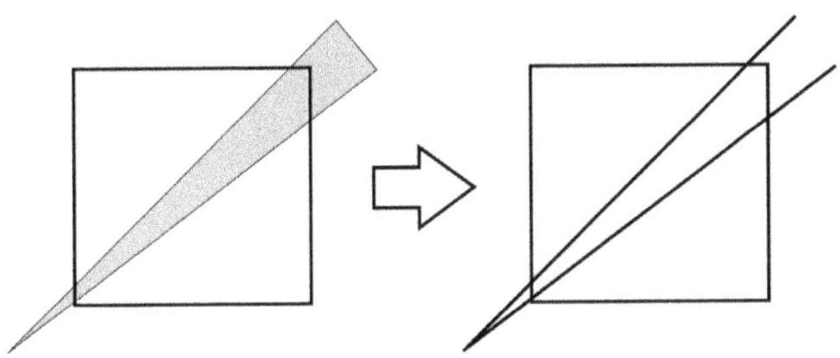

Figure 7. Intersection assessment between beam and rectangle.

The judgment of the beam-bounding volume intersection for deeper intermediate nodes within this bounding volume needs to be made if the initial judgment is true, until the leaf node that intersects the beam is retrieved.

3.3. *Intersection Between Ray and Triangular Surface*

By applying a beam constraint, the collision detection tree locates all leaf nodes that may intersect with random rays in the beam and constructs a set of potential collision surface elements ($S_{potential}$) from corresponding triangular surfaces. Each random ray is utilized to determine the intersection point of relevant triangular planes within the set.

The geometric configuration of the intersection assessment between a ray and a triangular surface is illustrated in Figure 8. Cartesian coordinates are established with O as the origin of the ray, where **d** represents the direction vector of the ray. The position vectors of the three vertices A, B, and C on the triangular surface are denoted by $\mathbf{p}_A$, $\mathbf{p}_B$, and $\mathbf{p}_C$, respectively. Additionally, $\mathbf{p}_I$ represents the position vector of the intersection point I between the ray and the triangular surface.

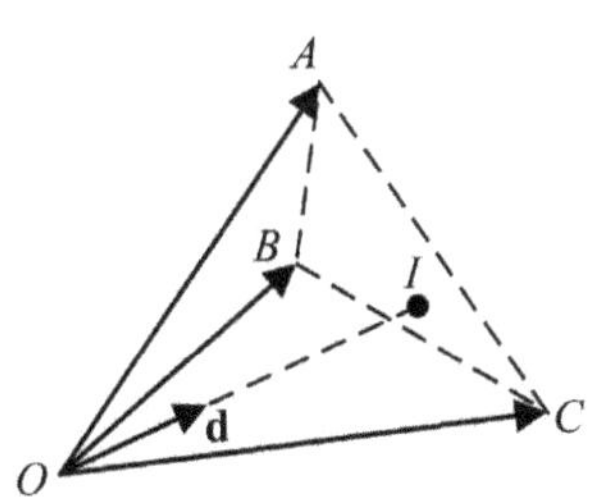

Figure 8. The intersection point between the ray and the triangular surface.

According to the geometric relationship shown in Figure 8, we can have

$$\mathbf{p}_I = w\mathbf{d} \tag{15}$$

where w is a positive real number. Meanwhile, the direction vector $\mathbf{d}$ can also be decomposed into

$$\mathbf{d} = \alpha\mathbf{p}_A + \beta\mathbf{p}_B + \gamma\mathbf{p}_C \tag{16}$$

with the coefficients α, β, and γ being real numbers representing the decomposition of the direction vector $\mathbf{d}$ into $\mathbf{p}_A$, $\mathbf{p}_B$, and $\mathbf{p}_C$, respectively. In geometry, point I being on the surface of triangle ABC is a necessary and sufficient condition, denoted by

$$w(\alpha + \beta + \gamma) = 1 \tag{17}$$

The sufficient and necessary condition for the direction vector $\mathbf{d}$ inside the tetrahedron OABC is that α, β, and γ are all non-negative and at least one of them is non-zero.

The values of α, β, and γ can be computed based on Equation (16). If the direction vector $\mathbf{d}$ lies within the tetrahedron $OABC$, then it is necessary for the ray to intersect with the triangular surface ABC. By substituting Equation (17) into Equation (15), we can determine the position vector of the intersection point I:

$$\mathbf{p}_I = \frac{1}{\alpha + \beta + \gamma}\mathbf{d} \tag{18}$$

3.4. *The Problem of Secondary Reflection*

When a ray is reflected on the surface of an object, according to the surface light energy reflectance model, the reflected rays are distributed within the range of $[-\pi/2, \pi/2]$, centered around the normal vector $\mathbf{n}$ on the object's surface. Apart from some of these reflected rays being directly reflected and received by a receiver, other portions may undergo secondary reflection upon colliding with other objects in the environment.

The reflection point of the secondary reflection must fall within the lidar receiving beam in order for the ray to enter the lidar receiver. Hence, when tracking secondary reflections, it is sufficient to consider only those arising from triangular surfaces among the set of potential collision surface elements ($S_{potential}$).

Each reflected ray undergoes significant attenuation compared to the incident ray. The increase in the number of reflections further amplifies the attenuation of light energy. A portion of the rays, after undergoing secondary reflection and reaching the receiver, fail to reach the level of sensitivity that allows for perception. Consequently, based on the sensitivity of the lidar receiver, backward precomputation is employed to calculate the minimum energy at each reflection point. If the energy of the reflected ray falls below this minimum threshold, it will no longer be tracked. This approach effectively avoids extensive, non-effective ray tracing.

After secondary reflection, there arises the issue of subsequent ray reflection once again, necessitating the repetition of the secondary reflection process. However, due to the attenuation of light energy during multiple reflections, the lidar receiver typically fails to perceive the laser pulse after undergoing multiple reflections. Therefore, in simulation processes, only the problem of secondary reflection is usually taken into consideration.

4. Proposed Simulation Method

On the basis of the aforementioned models of laser propagation and the collision detection scheme, the implementation process of the improved ray-tracing method for simulating ground object echoes can be explained, as depicted in Algorithm 2.

In the simulation method proposed in this paper, the collision detection tree is constructed using Algorithm 1 based on 3-D virtual scene information and radar parameters. Subsequently, during the simulation process of ground object echoes, the corresponding points' 3-D position and intensity distribution information are acquired according to the beam. For a given beam, N random rays are generated using the Monte Carlo beamwidth model. Afterward, the collision information between these random rays and the triangular surface of the ground object is obtained through batch collision detection. The reflection direction of each ray is determined based on the triangular surface information of the ground object within the beam irradiation range and the position of the lidar receiver. If the reflected ray encounters another triangular surface, the information within the beam irradiation range and the position of the lidar receiver are reused to determine its reflection direction again, until it reaches the lidar or collides with a triangular surface outside of the beam irradiation range. For the ray received by lidar after one or more reflections, the light energy of the laser pulse transmitted along this ray is calculated at each crucial node. The received ray is then screened by the light-energy constraint. By summing up the energy corresponding to all the rays received by lidar within the beam and incorporating additive noise, the received light energy is transformed into $Q_{rec}[m]$. Finally, based on the received light energy presentation model, it is compared with the upper and lower limits to generate $Q_{en}[m]$, which is then mapped to points of different colors. The simulation of the interior object echo of a single beam is now complete. Upon sweeping through the entire 3-D scene, we can obtain the point cloud distribution, where distinct colors indicate varying intensities of ground object echoes at corresponding locations.

Algorithm 2 Proposed simulation method

Input: A 3-D virtual scene, parameter values for various materials and lidar
Output: The simulated point cloud of the 3-D virtual scene

```
Based on the information from the 3D virtual scene and lidar parameters, the collision
detection tree is constructed using Algorithm 1.
for traversing all beam directions in the ground object echo simulation do
    Generate N random rays according to Monte Carlo beamwidth model.
    By utilizing the collision detection tree, a batch collision detection is performed for
    N random rays in the beam.
    for traversing N random rays do
        while the ray collides with the triangular surface of the ground object within
        the beam irradiation range do
            The effective ray is filtered by applying the beam constraint. The direction of
            the reflected ray is determined based on the triangular surface information of
            the ground object within the illuminating range of the beam and the position
            of the lidar receiver.
        end while
        Specifically for the ray whose light path terminates in the lidar receiver, the
        models in Section 2 are utilized to compute the light energy of the laser pulse
        emitted by lidar at each crucial node along the propagation path.
        if the light of the laser pulse at each crucial note satisfies the light-energy con-
        straint, then
            The lidar is capable of receiving the laser pulse propagated along this light
            path while preserving its light-energy information Q_r(θ_n, τ_n).
        else
            The laser pulse propagating along the light path is insufficient to activate the
            lidar receiver, thus necessitating the disregard of this particular ray.
        end if
    end for
    All retained energy is added up, and additive noise is introduced to form the
    received light energy Q_rec[m] in lidar. Based on the upper and lower limits of
    receivable light energy, the light energy Q_en[m] is formed upon entering the receiver.
    According to the received light energy presentation model, the received energy is
    mapped to points of varying colors.
end for
```

5. Simulation Results

The reliability and high efficiency of the ground object echo simulation method proposed in this paper are demonstrated through a series of simulation experiments conducted in this section.

5.1. Selection of Bounding Volume Type and Collision Detection Tree Structure

The BVH tree [23] or k-D tree [25] is constructed using the AABB [20] or SBV [22] bounding volumes to investigate the efficiency of collision detection for different combinations. The total time required for constructing the collision detection tree and performing collision detection under various combinations varies with the number of triangular surfaces, as depicted in Figure 9. It can be observed from Figure 9 that as the number of triangular surfaces increases, both the construction time of the collision detection tree and the total time spent on collision detection increase proportionally. Notably, when adopting the SBV, there is a significantly higher growth rate in total time compared to the AABB, which remains consistently high throughout. Under identical bounding volume conditions, the k-D tree exhibits higher construction and detection efficiency. Therefore, in this paper, the AABB is utilized to construct the k-D tree in ground object echo simulation.

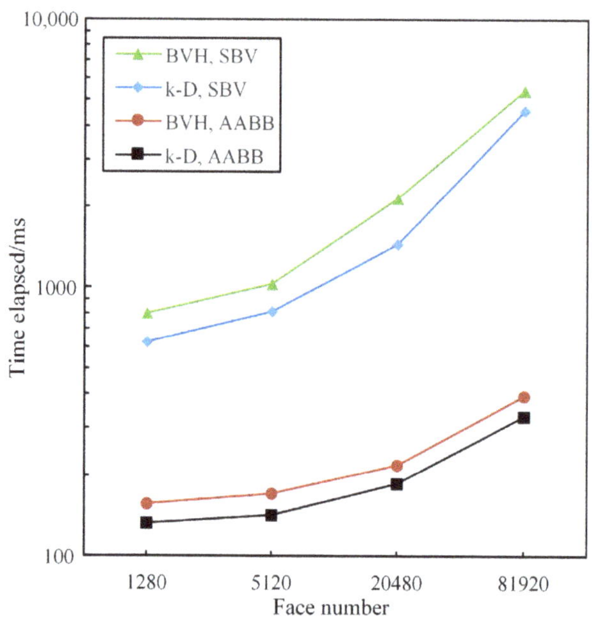

Figure 9. The time elapsed for constructing the collision detection tree and performing collision detection under various combinations.

The efficiency of the point cloud simulation method proposed in this paper is further analyzed here. In the simulation experiment, a spherical surface composed of a specific number of triangular surfaces was used as the simulation target. Figure 10 illustrates the point cloud simulation time for the proposed method, the method based on the bidirectional BRDF model [10], and the method based on the Phong model [11] with varying numbers of triangular surfaces. From Figure 10, it can be observed that when there are fewer triangular surfaces, the proposed method takes more time compared to the other two methods. However, as the number of triangular surfaces increases, the rate at which time consumption grows for simulating point clouds using our proposed method is significantly lower than those of both alternative methods. Consequently, as we increase the number of triangular surfaces, our proposed method exhibits an increasingly advantageous reduction in time consumption compared to both alternative methods. When dealing with point cloud simulations in complex environments, our proposed method demonstrates superior advantages over these two alternatives.

The present study introduces a beam constraint, a light-energy constraint, and a batch collision detection scheme to expedite the screening of effective rays in the ray-tracing process and reduce the computational complexity of collision detection throughout the simulation. To further analyze the impact of these enhancements on simulation efficiency, it is beneficial to draw upon ablation experiments, commonly employed in neural networks, for discussion. In this experiment, a spherical surface composed of a specific number of triangular surfaces remains as the simulation target. Various combinations of the beam constraint, light-energy constraint, and batch collision detection schemes are incorporated based on the original ray-tracing method to assess time consumption during the point cloud simulation under different numbers of triangular surfaces. The specific results are illustrated in Figure 11. The combination of the original ray-tracing method with the beam constraint, the light-energy constraint, or the batch collision detection scheme resulted in a noticeable decrease in simulation time, as depicted in Figure 11. When the number of triangular surfaces is low, the combination of the original ray-tracing method and batch collision detection scheme exhibits the shortest execution time. However, when dealing with a high number of triangular surfaces, employing the combination of the original ray-tracing method and the beam constraint proves to be more efficient. Furthermore, incorporating any two aspects with the original ray-tracing method further reduces the simulation time. Notably, among these combinations, utilizing both the beam constraint and the batch collision detection scheme yields superior performance.

After combining the original ray-tracing method with the aforementioned three aspects to form the enhanced approach proposed in this paper, it is evident that the corresponding simulation time exhibits optimal performance. In conclusion, by addressing three key aspects of improvement within the original ray-tracing method, the proposed approach effectively enhances computational efficiency in simulations.

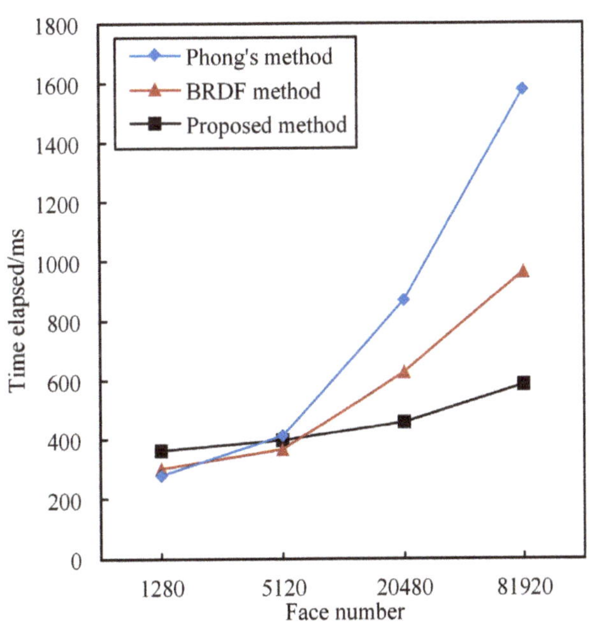

Figure 10. Time elapsed for different methods.

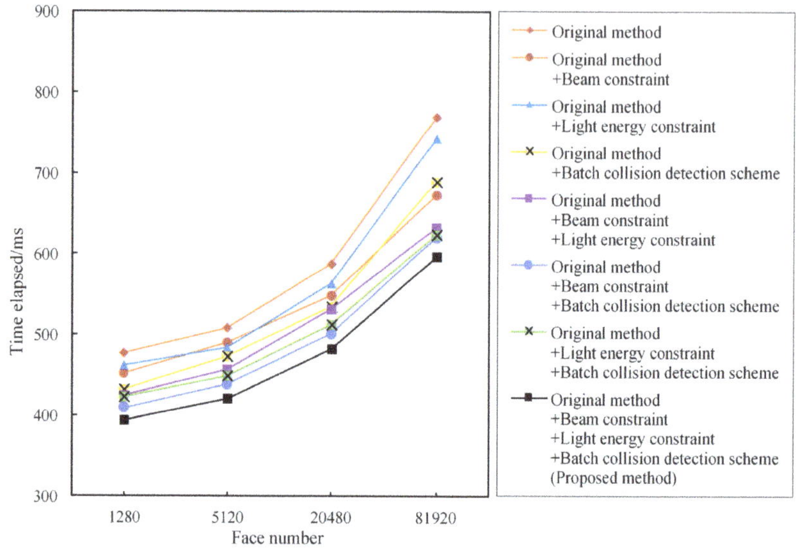

Figure 11. Results of ablation experiments.

5.2. Simulation Results of Simple Small Scene

The proposed method was utilized for simulating the point cloud of a simple small scene and validating its effectiveness by comparing it with the actual lidar-collected data. The experimental area selected is the corridor section within the laboratory, encompassing various elements, such as windows, doors, stairwells, walls, and long corridors, as depicted in Figure 12a. Based on the real scene information, a 3-D virtual scene was constructed, as shown in Figure 12b. The diffuse coefficient (k_d), specular coefficient (k_s), and glossiness coefficient (k_p) of different materials in the 3-D virtual scene were determined by fitting data from the MERL BRDF material library [17], with the specific values presented in Table 1.

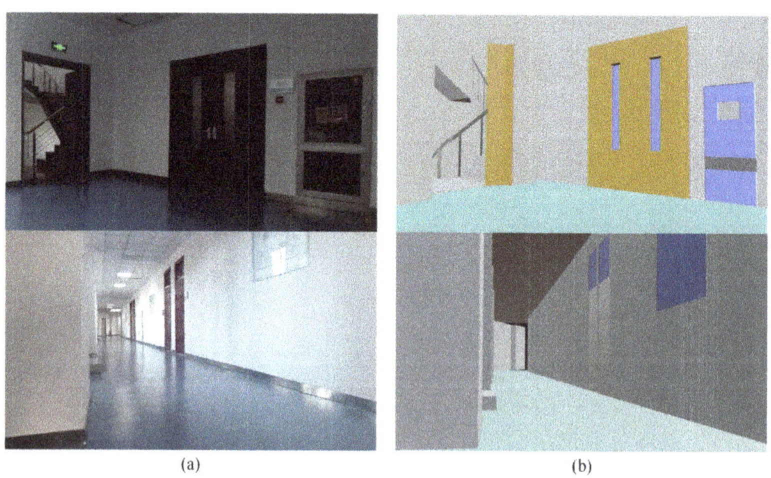

Figure 12. Experimental scene: (**a**) Real scene. (**b**) Virtual scene.

Table 1. The parameter values for various materials.

Material	k_d	k_s	p
Wall	0.146	0.054	112
Glass	0.020	0.852	8046
Metal	0.075	0.634	6803
Wood	0.107	0.100	82

The lidar utilized in this experiment is the Laser Intelligent C16 multi-line mechanical lidar. The horizontal resolution and vertical resolution of the lidar are 0.09° and 2.0°, respectively, with the vertical resolution being approximately 22 times greater than the horizontal resolution. To ensure a balanced point cloud data resolution in both the vertical and horizontal directions, the actual radar data collected during the experiment were downsampled by a factor of 22 along the horizontal direction. In order to compare the actual collected point cloud with the simulated point cloud, an object echo simulation was conducted using radar with set horizontal and vertical resolutions of 2.0° each. Other relevant parameter settings employed in this simulation can be found in Table 2. The simulation device utilizes an Intel i7-10510U CPU operating at a frequency of 1.8 GHz and is equipped with 8 GB of memory. All program codes were implemented in the Java language without utilizing GPU acceleration. Whether the point cloud was obtained through actual collection or simulated generation, the color tone represents the light-energy intensity of each corresponding point, while the diameter of the point indicates its distance distribution. The simulation method proposed in this paper takes 83 ms when simulating the simple small scene, whereas the original ray-tracing method requires approximately 130 ms. The proposed approach demonstrates a time reduction of about 36.2% compared to the original method.

Table 2. Configurations of model parameters.

Notation	Explanation	Value
$\theta_{0.5}$	Half-beamwidth	5.6×10^{-3} rad
N	Total number of random rays within beam	100
μ	Atmospheric attenuation coefficient	0.088
$\bar{b}$	Noise photon rate	20 kHz
h	Planck constant	6.626×10^{-34} J · s
ν	Frequency of the laser	3.313×10^{14} Hz
Q_s	Energy of the emitted laser pulse	2.5×10^{-4} J
Q_{high}	Receiver upper energy limit	6.25×10^{-5} J
Q_{low}	Receiver lower energy limit	2×10^{-6} J

Three typical areas are selected as the judgment indicators to evaluate the consistencies of the shape characteristics and intensity distribution between the real and simulated results.

The first area to consider is depicted in Figure 13a, featuring a wooden door with a window, a glass cabinet housing the fire hydrant, and adjacent walls. As indicated by the parameters in Table 1, wood and wall materials have higher diffuse coefficients (k_d), while glass materials exhibit a higher specular coefficient (k_s) and glossiness coefficient (k_p). Consequently, the reflection of wood and wall materials is predominantly diffuse, whereas glass materials primarily undergo specular reflection with weak diffuse reflection. Therefore, for a lidar system whose transmitter and receiver are located at the same position, only when ray irradiates the surface at a very small incidence angle can it receive specular reflection energy; otherwise, the received energy is mainly from diffuse reflections. Figure 13b,c depict the light intensity distribution of actual radar-acquired point clouds and simulated point clouds, respectively. In comparison with Figure 13a, it is evident that the wall echo exhibits maximum strength, followed by the wood echo, while the glass echo is the weakest; furthermore, material boundaries are clearly discernible. Notably, the middle region of the wooden door reflects stronger echoes than its sides due to its surface being nearly perpendicular to the incident light, resulting in specular reflection.

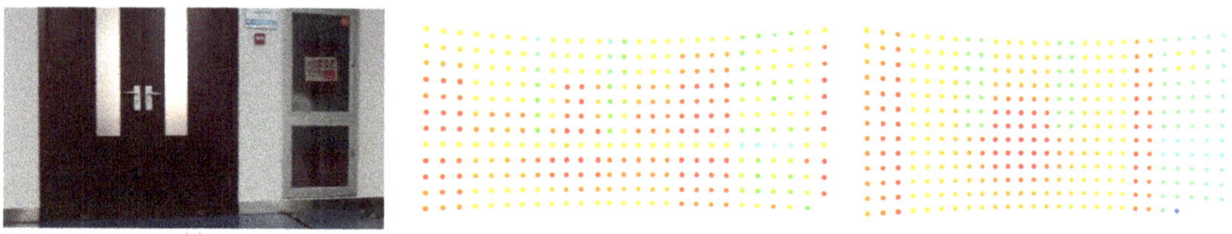

(a) (b) (c)

Figure 13. Door area: (**a**) Area photo. (**b**) Actual collected point clouds. (**c**) Simulated point clouds.

To accurately assess the distribution consistency between the actual collection point cloud and the simulated point cloud, the Hausdorff distance [26] and Root Mean Square Error (RMSE) [27] were employed as quantitative metrics. The Hausdorff distance, a commonly used measure of dissimilarity between two point clouds, considers the distances between each point in both clouds and their nearest neighbors, selecting the maximum value among them. The specific expression for calculating the Hausdorff distance is denoted by

$$D_{Hausdorff} = \max\left\{\max_{\mathbf{p}_a}\left(\min_{\mathbf{p}_s}\|\mathbf{p}_a - \mathbf{p}_s\|\right), \max_{\mathbf{p}_s}\left(\min_{\mathbf{p}_a}\|\mathbf{p}_s - \mathbf{p}_a\|\right)\right\} \tag{19}$$

where $\mathbf{p}_a$ and $\mathbf{p}_s$ represent the coordinates of any point in the actual collected point cloud and simulated point cloud, respectively. Due to a discrepancy in the total number of points between the two clouds, the Root Mean Square Error (RMSE) is utilized to measure the distance between the nearest neighbors in each cloud, i.e.,

$$\mathrm{RMSE} = \sqrt{\frac{1}{n}\sum_{a=1}^{n}\min_{\mathbf{p}_s}\|\mathbf{p}_a - \mathbf{p}_s\|^2} \tag{20}$$

where n represents the total number of points in the simulated point cloud. The smaller the values of the Hausdorff distance and RMSE, the higher the similarity between the two point clouds. Based on Figure 13b,c, the Hausdorff distance is measured to be 0.164, while the RMSE is calculated as 0.04.

The second selected area is the stairwell, as depicted in Figure 14a. As evident from Figure 14a, this area exhibits a relatively intricate geometric structure and comprises two different materials: metal and wall. The stair handrail is constructed of metal, while the wall and steps can be treated based on the wall material. As demonstrated by the point cloud results in Figure 14b,c, the complex structure of this region poses a challenge for fully capturing its structural information when point clouds are sparse. However, based on the distribution of point clouds, Figure 14b,c exhibit good similarity to each other. The Hausdorff distance and RMSE values are 0.908 and 0.33, respectively, indicating their strong resemblance, but it is weaker than that shown in Figure 13.

The final region analyzed is the long corridor region, as depicted in Figure 15a. As evident from Figure 15a, an acrylic bulletin board is installed in this area. The reflective characteristics of the material resemble those of glass, thus substituting the parameters of glass material for simulation purposes. Comparing Figure 15b,c, it can be observed that the intensity of point clouds reflecting the bulletin-board area is significantly weaker than that of point clouds on the surrounding wall. With increasing distance, the intensity of wall point clouds gradually diminishes. In even more distant regions, where the echo intensity falls below the receiver's threshold and goes unnoticed, the corresponding areas are absent from the point cloud results. The Hausdorff distance and RMSE in this region are 0.373 and 0.10, respectively. Evidently, the point cloud similarity is superior in the long corridor region compared to the stairwell region but inferior to that in the region depicted in Figure 13a.

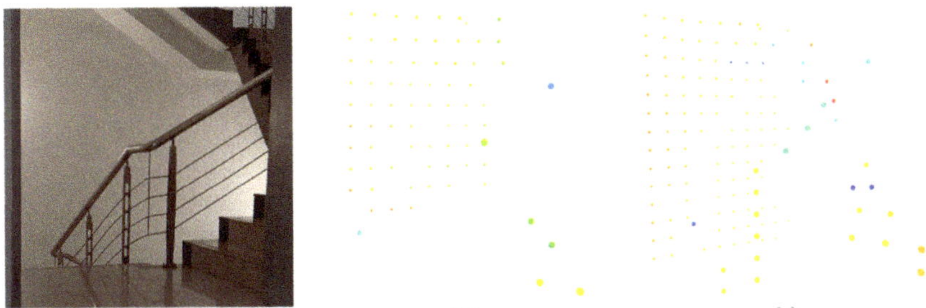

(a) (b) (c)

Figure 14. Stairwell area: (**a**) Area photo. (**b**) Actual collected point clouds. (**c**) Simulated point clouds.

The previous analysis of the point cloud consistency and intensity changes in key areas within the experimental scene confirms that the proposed point cloud simulation method in this paper effectively yields highly accurate results consistent with reality.

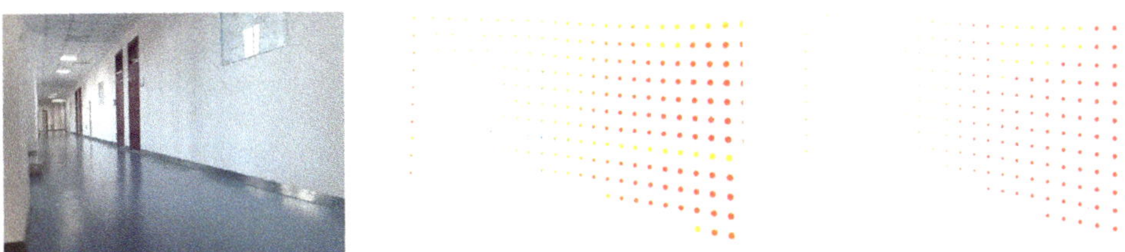

(a) (b) (c)

Figure 15. Corridor area: (**a**) Area photo. (**b**) Actual collected point clouds. (**c**) Simulated point clouds.

5.3. Simulation Results of Complex Large Scene

In this section, based on the technical parameters of avian lidar, the point cloud of the ground object echoes is simulated in a complex large scene. The selected scene is Shanghai Hongqiao International Airport. Figure 16a shows its satellite thumbnail, and the top view of the mesh grid of the main airport buildings is illustrated in Figure 16b. The lidar system is positioned near the center of the runway, indicated by a red "+" symbol in both Figure 16a,b. For this experiment, we adopted identical model parameters to those used to simulate the simple small scene, except for setting the atmospheric attenuation coefficient μ to 0.035 and transmitted pulse power Q_s to 3.6×10^{-4} J.

During the simulation of ground object echoes, the lidar's vertical coverage range is 0 to 6 degrees, which is divided into either 16 or 32 channels. This means that the lidar has a vertical resolution of either 0.375 degrees or 0.1875 degrees, and its horizontal resolution is equal to the vertical resolution. The same hardware as described in Section 5.2 was utilized for this ground object echo simulation experiment. When using 16 channels, our proposed method completes the simulation for the entire airport in approximately 1122 ms, whereas the original ray-tracing method takes around 1439 ms. On the other hand, when employing all available 32 channels on the lidar system, our proposed method requires about 1633 ms compared to 1831 ms with the original ray-tracing approach.

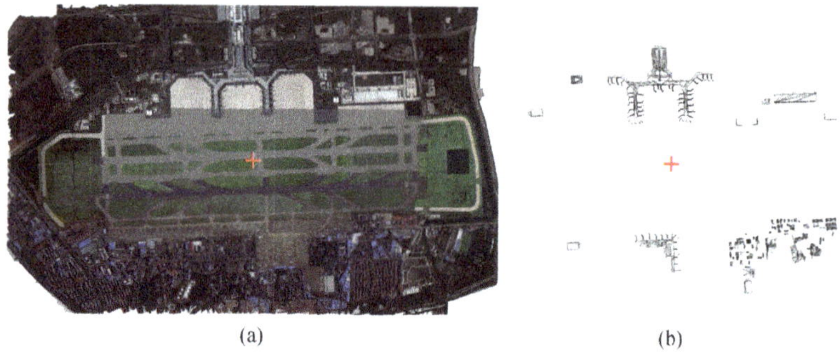

Figure 16. Shanghai Hongqiao international airport overview: (a) Satellite thumbnail. (b) Top view of mesh grid of main buildings.

Taking the T1 terminal of the airport as an example, the top views of simulated point clouds obtained by lidar using 16 channels and 32 channels are plotted in Figure 17a,b, respectively. In Figure 17, the black network represents the corresponding architectural structure based on the spatial positioning of the T1 terminal in the 3-D virtual scene. Simulated point clouds are represented by colored dots, indicating their coordinate positions within the same 3-D virtual space. Each point's coordinates within the cloud are determined by factors such as the lidar location, beam direction, and laser pulse travel distance. It can be observed from Figure 17a,b that both cases effectively capture the facade shape of the building toward the lidar. However, in Figure 17a,b, these point clouds appear blue due to their relatively low light energy compared to other colors on a color–energy mapping scale. This is primarily attributed to two factors: firstly, there is a significant distance between the lidar and the T1 terminal; secondly, considering this distance, the points on the T1 terminal are relatively close together compared to their distance from lidar. By comparing Figure 17a,b, it can be further observed that the lidar employs a greater number of channels to acquire denser point clouds, thereby enabling a more precise depiction of target details.

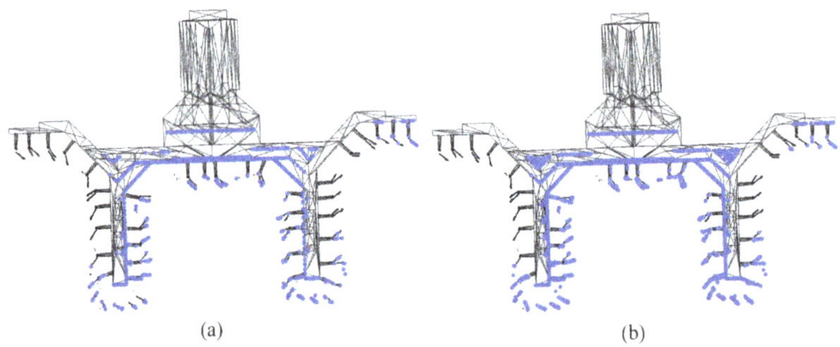

Figure 17. Top view for results of different vertical channels: (a) Sixteen channels. (b) Thirty-two channels.

By utilizing simulation techniques, a multitude of specific target point clouds can be conveniently generated under varying conditions, such as different noise levels, viewing angles, and resolutions. These generated point clouds can serve as annotated samples for training neural networks, enabling the trained network to accurately identify and eliminate ground objects from lidar-received point clouds. This ensures that the detection and tracking of bird targets are not impeded by the presence of ground object data.

6. Discussion

Avian lidar is an effective means for monitoring bird activity in low-altitude airspace. However, the presence of ground object echoes seriously affects the detection and tracking performance of the avian lidar system. To effectively suppress these ground object echoes, it is necessary to simulate them accurately in the avian lidar system. In this paper, an improved ray-tracing method is proposed to simulate ground object echoes in airport scenes. Our proposed method utilizes bidirectional ray tracing and incorporates beam constraint and light-energy constraint techniques to efficiently filter out irrelevant rays during the simulation process. Additionally, our method employs batch collision detection to enhance the collision efficiency between multiple random rays and ground objects within a beam. By optimizing specific strategies within the simulation process, our proposed method significantly improves the simulation efficiency without compromising the resolution of the simulated point cloud.

The advantage of the ray-tracing method [6–8] lies in its ability to accurately simulate the propagation and scattering process of light, as well as generate high-precision ground object echo signals. However, when dealing with complex scenes, the calculation load of ray tracing significantly increases due to multiple reflections of light, resulting in low simulation efficiency. The introduction of a beam constraint and a light-energy constraint in this method improves the selection of effective light and greatly reduces the tracking of unnecessary rays. Compared to the hardware acceleration method mentioned in the literature [14,15], our proposed method enables a complex scene simulation without relying on expensive GPU resources. A comparative analysis with the BRDF method [10] and Phong's method [11] reveals that our proposed method exhibits superior simulation efficiency for ground object echoes within environments featuring large surface element scales. In small-scale scene experiments, the simulated point cloud generated in this paper demonstrates a strong realistic effect, closely resembling actual lidar-collected point clouds both structurally and in terms of intensity distribution. Furthermore, when simulating an airport scene, the generated point cloud effectively describes the distribution characteristics of ground object surfaces toward radar; particularly noteworthy is its improved descriptive capability after increasing the number of vertical lidar channels.

The research presented in this paper successfully achieved the efficient simulation of target echoes in complex environments. However, due to the limited availability of experimental conditions, there is currently no actual lidar system for acquiring ground object echo data in large scenes. Therefore, it is not possible to directly compare the simulated point cloud data with real point cloud data collected from such scenes. Consequently, there remains a need for more robust methods to validate the rationality of the simulated ground object echo point cloud.

7. Conclusions

In this paper, an enhanced and efficient ray-tracing simulation method is presented for simulating ground object echoes in airport avian lidar. By optimizing the effectiveness of light screening and collision detection between light and targets, the proposed method can be applied to simulate target echoes in large and complex scenes. Simulation experiment results demonstrate that the optimized ray-tracing and collision detection methods presented in this paper exhibit superior overall performance compared to classical algorithms. Under the similar control of other parameters, the running time and simulation speed of the proposed target simulation method are notably improved when compared with other methods, particularly for scenarios involving a high number of surfaces. Furthermore, based on actual lidar-collected point cloud results, the simulated results from our model accurately reflect both the physical shape and echo intensity characteristics of objects within scanned scenes.

Author Contributions: Conceptualization, Z.S.; methodology, Z.S.; software, L.S. and J.H.; validation, Z.S. and J.H.; formal analysis, Z.S. and L.S.; investigation, L.S. and J.H.; resources, Z.S., Y.W. and P.G.; data curation, L.S., B.H., Y.W. and P.G.; writing—original draft preparation, Z.S. and L.S.; writing—review and editing, Z.S. and L.S.; supervision, Z.S.; project administration, Z.S.; funding acquisition, Z.S. All authors have read and agreed to the published version of the manuscript.

Funding: This research was funded by the Tianjin Municipal Education Commission under Grant No. 2022KJ059, the Major Science and Technology Projects in Anhui Province under Grant No. 202103a13010006, and the Fundamental Research Funds for the Central Universities under Grant No. 3122017111.

Institutional Review Board Statement: Not applicable.

Informed Consent Statement: Not applicable.

Data Availability Statement: Data are contained within the article.

Conflicts of Interest: Author Peng Ge was employed by the company the 38th Research Institute of China Electronics Technology Group Corporation. The remaining authors declare that the research was conducted in the absence of any commercial or financial relationships that could be construed as a potential conflict of interest.

Abbreviations

The following abbreviations are used in this manuscript:

GPU	Graphic Processing Unit
CUDA	Compute Unified Device Architecture
3-D	Three-dimensional
HSB	Hue–Saturation–Brightness
AABB	Axis-Aligned Bounding Box
OBB	Oriented Bounding Box
SBV	Sphere Bounding Volume
BVH	Bounding Volume Hierarchy
BSP	Binary Space Partitioning
k-D	k-dimensional
2-D	Two-dimensional

References

1. Zhao, J.; Li, Y.; Zhu, B.; Deng, W.; Sun, B. Method and Applications of Lidar Modeling for Virtual Testing of Intelligent Vehicles. *IEEE Trans. Intell. Transp. Syst.* **2021**, *22*, 2990–3000. [CrossRef]
2. Rougeron, G.; Garrec, J.L.; Andriot, C. Optimal positioning of terrestrial LiDAR scanner stations in complex 3D environments with a multiobjective optimization method based on GPU simulations. *ISPRS J. Photogramm. Remote Sens.* **2022**, *193*, 60–76. [CrossRef]
3. Zhang, W.; Li, Z.; Li, G.; Zhuang, P.; Hou, G.; Zhang, Q.; Li, C. GACNet: Generate Adversarial-Driven Cross-Aware Network for Hyperspectral Wheat Variety Identification. *IEEE Trans. Geosci. Remote Sens.* **2023**, *62*, 5503314. [CrossRef]
4. Zhang, W.; Li, Z.; Sun, H.H.; Zhang, Q.; Zhuang, P.; Li, C. SSTNet: Spatial, spectral, and texture aware attention network using hyperspectral image for corn variety identification. *IEEE Geosci. Remote Sens. Lett.* **2022**, *19*, 5. [CrossRef]
5. Zhang, W.; Zhao, W.; Li, J.; Zhuang, P.; Sun, H.; Xu, Y.; Li, C. CVANet: Cascaded visual attention network for single image super-resolution. *Neural Netw.* **2024**, *170*, 622–634. [CrossRef] [PubMed]
6. Yang, X.; Wang, Y.; Yin, T.; Wang, C.; Lauret, N.; Regaieg, O.; Xi, X.; Gastellu-Etchegorry, J.P. Comprehensive LiDAR simulation with efficient physically-based DART-Lux model (I): Theory, novelty, and consistency validation. *Remote Sens. Environ.* **2022**, *272*, 112952. [CrossRef]
7. Gastellu-Etchegorry, J.P.; Yin, T.G.; Lauret, N.; Grau, E.; Rubio, J.; Cook, B.D.; Morton, D.C.; Sun, G.Q. Simulation of satellite, airborne and terrestrial LiDAR with DART (I): Waveform simulation with quasi-Monte Carlo ray tracing. *Remote Sens. Environ.* **2016**, *184*, 418–435. [CrossRef]
8. Esmorís, A.M.; Yermo, M.; Weiser, H.; Winiwarter, L.; Höfle, B.; Rivera, F.F. Virtual LiDAR Simulation as a High Performance Computing Challenge: Toward HPC HELIOS++. *IEEE Access* **2022**, *10*, 105052–105073. [CrossRef]
9. Linnhoff, C.; Rosenberger, P.; Winner, H. Refining Object-Based Lidar Sensor Modeling—Challenging Ray Tracing as the Magic Bullet. *IEEE Sens. J.* **2021**, *21*, 24238–24245. [CrossRef]
10. Chai, G.; Zhang, J.; Huang, X.; Guo, B.; Tian, L. Study of the dynamic scenes model for laser imaging radar simulation. *Xi'an Dianzi Keji Daxue Xuebao/J. Xidian Univ.* **2014**, *41*, 107–113. [CrossRef]

11. Yang, J.S.; Li, T.J. Simulation of Space-Based Space Target Scene Imaging. *Laser Optoelectron. Prog.* **2022**, *59*, 253–260.
12. Neumann, T.; Kallage, F. Simulation of a Direct Time-of-Flight LiDAR-System. *IEEE Sens. J.* **2023**, *23*, 14245–14252. [CrossRef]
13. Winiwarter, L.; Esmorís Pena, A.M.; Weiser, H.; Anders, K.; Martínez Sánchez, J.; Searle, M.; Höfle, B. Virtual laser scanning with HELIOS++: A novel take on ray tracing-based simulation of topographic full-waveform 3D laser scanning. *Remote Sens. Environ.* **2022**, *269*, 112772. [CrossRef]
14. López, A.; Ogayar, C.J.; Jurado, J.M.; Feito, F.R. A GPU-Accelerated Framework for Simulating LiDAR Scanning. *IEEE Trans. Geosci. Remote Sens.* **2022**, *60*, 3000518. [CrossRef]
15. Richa, J.P.; Deschaud, J.-E.; Goulette, F.; Dalmasso, N. AdaSplats: Adaptive Splatting of Point Clouds for Accurate 3D Modeling and Real-Time High-Fidelity LiDAR Simulation. *Remote Sens.* **2022**, *14*, 6262. [CrossRef]
16. Tan, K.; Cheng, X. Specular reflection effects elimination in terrestrial laser scanning intensity data using Phong model. *Remote Sens.* **2017**, *9*, 8. [CrossRef]
17. Matusik, W.; Pfister, H.; Brand, M.; McMillan, L. A Data-Driven Reflectance Model. *ACM Trans. Graph. (TOG)* **2003**, *22*, 759–769. [CrossRef]
18. Ceolato, R.; Berg, M.J. Aerosol light extinction and backscattering: A review with a lidar perspective. *J. Quant. Spectrosc. Radiat. Transf.* **2021**, *262*, 107492. [CrossRef]
19. Liang, B.; Niu, J.; He, S.; Liu, H.; Qin, C. Tunnel lighting calculation model based on bidirectional reflectance distribution function: Considering the dynamic changes in light transmittance in road tunnels. *Tunn. Undergr. Space Technol.* **2023**, *140*, 105313. [CrossRef]
20. Xing, Y.S.; Liu, X.P.; Xu, S.P. Efficient collision detection based on AABB trees and sort algorithm. In Proceedings of the 2010 8th IEEE International Conference on Control and Automation (ICCA 2010), Xiamen, China, 9–11 June 2010; pp. 328–332. [CrossRef]
21. Zhao, Y.; Huang, J.; Li, W. Fitting oriented bounding box algorithm for accurate positioning of transformer wiring terminals. In Proceedings of the 2023 35th Chinese Control and Decision Conference (CCDC), Yichang, China, 20–22 May 2023; pp. 4058–4061. [CrossRef]
22. Huang, X.; Zhang, S.; Zhou, L.; Huang, L. A hybrid algorithm for the minimum bounding sphere problem. *Oper. Res. Lett.* **2022**, 50, 150–154. [CrossRef]
23. Gu, Y.; He, Y.; Fatahalian, K.; Blelloch, G. Efficient BVH construction via approximate agglomerative clustering. In Proceedings of the—High-Performance Graphics 2013, HPG 2013, Anaheim, CA, USA, 19–21 July 2013; pp. 81–88. [CrossRef]
24. Ize, T.; Wald, I.; Parker, S.G. Ray tracing with the BSP tree. In Proceedings of the RT'08-IEEE/EG Symposium on Interactive Ray Tracing 2008, Proceedings, Los Angeles, CA, USA, 9–10 August 2008; pp. 159–166. [CrossRef]
25. Shan, Y.X.; Li, S.; Li, F.X.; Cui, Y.X.; Li, S.; Zhou, M.; Li, X. A Density Peaks Clustering Algorithm With Sparse Search and K-d Tree. *IEEE Access* **2022**, *10*, 74883–74901. [CrossRef]
26. Liu, K.; Ma, H.; Zhang, L.; Cai, Z.; Ma, H. Strip Adjustment of Airborne LiDAR Data in Urban Scenes Using Planar Features by the Minimum Hausdorff Distance. *Sensors* **2019**, *19*, 5131. [CrossRef]
27. Chen, J.H.; Zhao, D.; Zheng, Z.J.; Xu, C.; Pang, Y.; Zeng, Y. A clustering-based automatic registration of UAV and terrestrial LiDAR forest point clouds. *Comput. Electron. Agric.* **2024**, *217*, 108648. [CrossRef]

photonics

Article

Super-Resolution Imaging Enhancement through a 2D Scanning Galvanometer: Algorithm Formulation and Application in Aerial Optoelectronic Systems

Tianxiang Ma [1,2], Chao Liang [2], Yuting Han [2,3], Fang Yuan [2,3], Lingtong Meng [2], Yongsen Xu [2], Honghai Shen [2] and Yunqing Liu [1,*]

1 Changchun University of Science and Technology, Changchun 130000, China
2 Key Laboratory of Airborne Optical Imaging and Measurement, Changchun Institute of Optics, Fine Mechanics and Physics, Chinese Academy of Sciences, Changchun 130033, China
3 University of Chinese Academy of Sciences, Beijing 100039, China
* Correspondence: mzliuyunqing@163.com; Tel.: +86-1384-316-3761

Abstract: As the fields of aviation and aerospace optics continue to evolve, there is an increasing demand for enhanced detection capabilities in equipment. Nonetheless, in applications where both optical and mechanical constraints are stringent, the continuous expansion of optical aperture and focal length is impractical. Given the existing technological landscape, employing super-resolution algorithms to enhance the imaging capability of optical systems is both practical and highly relevant. This study capitalizes on using a 2D scanning galvanometer in optical systems to acquire micro-displacement information. Initially, an imaging model for optical systems equipped with a 2D scanning galvanometer was established, and the displacement vectors for both forward and sweep image motions were defined. On this foundation, we incorporated micro-displacement information that can induce high-frequency aliasing. Subsequently, the motion paths of the galvanometer were planned and modeled. To align image sequences with micro-displacement correlations, the Lucas–Kanade (L-K) optical flow method was employed with multi-layer pyramid iteration. Then, super-resolution reconstruction was performed using kernel regression techniques. Ultimately, we tested the algorithm on an aeronautical optoelectronic pod to evaluate its impact on optical resolution and imaging quality. Compared with the original images, the 16-frame image demonstrated a 39% improvement in optical resolution under laboratory conditions. Moreover, the algorithm exhibited satisfactory performance under both nighttime and daytime conditions, as well as during aerial tests.

Keywords: aerial optoelectronic systems; super-resolution; 2D scanning galvanometer; kernel regression

Citation: Ma, T.; Liang, C.; Han, Y.; Yuan, F.; Meng, L.; Xu, Y.; Shen, H.; Liu, Y. Super-Resolution Imaging Enhancement through a 2D Scanning Galvanometer: Algorithm Formulation and Application in Aerial Optoelectronic Systems. *Photonics* **2023**, *10*, 1203. https://doi.org/10.3390/photonics10111203

Received: 13 September 2023
Revised: 20 October 2023
Accepted: 24 October 2023
Published: 27 October 2023

1. Introduction

As the aviation and aerospace photoelectric payloads' imaging distance extends and the resolution increases, challenges arise. Under constraints involving volume, weight, and other factors, the aperture and focal length of the payload's optical system cannot be increased infinitely. In electronics, high-resolution images correspond to a higher pixel count. The detector must arrange a denser pixel distribution per unit area, imposing very stringent demands on the design and manufacture of the detector, and the transmission capability of the data link [1]. Research focusing on obtaining higher-resolution images at the current level of optoelectronic technology has intensified [2–6].

In the field of aerospace and aviation, super-resolution imaging typically involves the photoelectric load acquiring low-resolution image information of a single frame or a sequence of multiple frames for a scene. Algorithms are then employed to reconstruct high-resolution images. For image detail accuracy, multi-frame super-resolution algorithms have found widespread use [7–9].

Multi-frame super-resolution imaging encompasses two vital steps. The first involves the procurement of low-resolution images containing high-frequency aliasing information. Primary technical paths include micro-scanning and sub-pixel stitching. The former employs a micro-motion mechanism, moving the optical-system-formed image on the detector at a 1/N pixel pitch, to capture several low-resolution images containing high-frequency aliasing information.

A rapid micro-scanning imaging device was developed by Fortin and Chevrette [10]. Comprising three main sections—the micro-scanning head, the controller, and the power amplifier—the system enables four operation modes. These modes are fixed position, 2 × 2, 3 × 3, and 4 × 4 micro-scan. In the case of a given staring array imager, the resolution is doubled.

Utilization of a piezoelectric nanopositioner for micro-displacements was demonstrated by Wiltse and Miller [11]. This approach provides substantial enhancements in minimum resolvable temperature/minimum resolvable contrast curves, allowing resolution beyond the sensor's native Nyquist frequency. Increases in target range discrimination of approximately 18% are attributed to four-point micro-scanning.

An infrared micro-scanning lens was designed by Zhang et al. [12] for constructing an infrared micro-scanning optical system. The entire field of view (FOV)'s displacement uniformity and ideal image quality on the image plane are ensured by this system. Optimal results were achieved with 2 × 2 magnification.

However, the micro-scanning method, requiring a micro-motion mechanism for aliased image acquisition, poses high demands on volume, power consumption, processing, and displacement accuracy. Therefore, its application range may be subject to some limitations.

Sub-pixel stitching technology, another approach, employs sub-pixel-level detector stitching to enhance the imaging system's resolution. Initially applied on the French SPOT-5 satellite [13], this technique enabled an imaging resolution of up to 2.5 m. The method entails placing two linear array detectors in dislocation (0.5-pixel displacement) on the optical system's image plane. With the optical system's resolution designed as 5 m, an increase to 2.5 m is achieved. The Swiss Leica Company and the German Aerospace Center's joint development of the ADS40 aerial surveying and mapping camera also utilized detector splicing to achieve super-resolution in the imaging system [14].

Tang et al. [15] introduced a mosaic technique that is grounded in the geometric structure of TDI CCD sensors and the object space projection plane. The sensors are projected into the object space indirectly, along and across the target trajectory's direction. This method has been applied in the daily processing systems of ZY-1 02C and ZY-3 satellites, yielding positive results. However, the application of sub-pixel stitching is primarily limited to line detector systems.

The second step in the process involves a high-resolution image reconstruction algorithm that uses a multi-frame image sequence. Typically, multi-frame reconstruction algorithms are categorized into the frequency domain method [16,17], non-uniform interpolation method [18,19], iterative back-projection method [20], convex set projection method [21,22], and probability analysis method [23,24]. Although these methods can achieve enhanced super-resolution results, high-performance algorithms often require significant computing resources. Consequently, they are unsuitable for aerospace and aviation platforms with high real-time demands and limited computing capabilities.

In this research, a 2D scanning galvanometer is planned to be utilized for aliasing information acquisition, along with a non-radial kernel function to merge multi-frame images and achieve super-resolution images. Initially, the mathematical relationship among the swing angle of the two-dimensional mirror, the image point's displacement on the image plane, and the exposure time is defined. Then, constraints are given for the super-resolution imaging of the photoelectric system with two-dimensional swinging. A super-resolution algorithm processing flow, based on prior displacement information, is then established. Following this, low-resolution image information is registered, fused, and reconstructed, and the multi-frame sequence depth's influence on the imaging effect is examined. Finally, the algorithm is tested in the laboratory, daytime, nighttime, and real-application scenarios. Experimental evidence indicates that the algorithm enhances image resolution and performs well in the aerial camera when applied.

2. Acquisition of Micro-Displacement

The degradation in aerial camera imaging quality is mainly attributed to image motion. To address this issue, commonly used methods encompass optical, mechanical, electronic, and image compensation techniques. Yet, most optical systems presently employ a 2D scanning galvanometer for image motion compensation.

In this study, the 2D scanning galvanometer is leveraged to carry out image motion compensation, imposing sub-pixel micro-displacement on the sensor to capture multiple frames of low-resolution image sequences with aliasing information. First, an imaging model of the optical system incorporating a 2D scanning galvanometer is created, and the mathematical relationships among the 2D scanning galvanometer oscillating vector, target moving vector, and micro-displacement vector are defined. Subsequently, the 2D scanning galvanometer's usage in the system is detailed, and its trajectory model is established.

2.1. The Imaging System Model

The active changes in the aerial camera's optical axis are primarily associated with two factors: alterations in aircraft attitude and the active swinging of the camera movement mechanism. The aircraft's ground-target relative moving speed is translated into the image-plane coordinate system with the aircraft track coordinate system, body coordinate system, fixed mirror, optical system, and two-dimensional pendulum mirror. Based on the light transmission relationship, a mathematical model is established between the target moving speed and the moving speed of the point on the image plane. The fundamental optical structure of an aerial camera is illustrated in Figure 1.

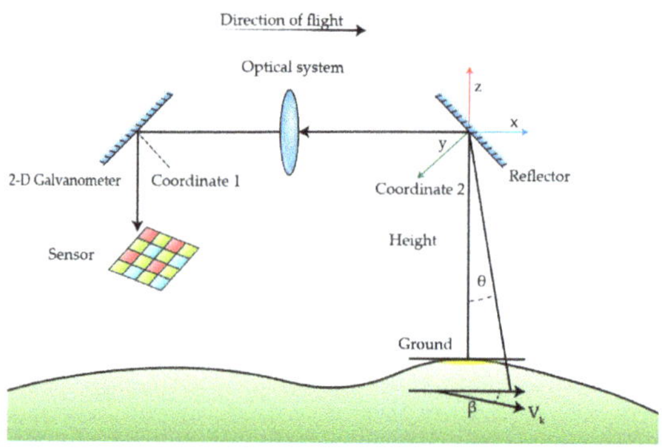

Figure 1. The base optical structure of an aerial camera.

The transformation in the homogeneous coordinate change involves multiple coordinate systems, namely, the ground coordinate system, the fixed-mirror coordinate system, the 2D scanning galvanometer coordinate system, and the image-plane coordinate system. The track angle of attack, sideslip angle, and roll angle are represented by α, β, and γ, respectively. Likewise, the camera's sweep angle, depression angle, and push-broom angle are designated by φ, ϕ, and η, respectively. The transformation of the speed of the image point in the image-plane coordinate system is accomplished through a sequence from the track coordinate system, via the aircraft coordinate system, to the image-plane coordinate system.

In Figure 1, the camera's sweep angle is assumed to be θ, with the focal length being represented by f; the flight height, by H; and the track speed, by v. The coordinate system's center point is defined as the intersection point between the central axis of the lens and the fixed mirror. Under the track coordinate system, the sideslip angle's impact on the image motion vector on the image plane is formulated as

$$v_\beta = T_1 K T_2 \begin{bmatrix} \cos\beta & -\sin\beta & 0 \\ \sin\beta & \cos\beta & 0 \\ 0 & 0 & 1 \end{bmatrix} \begin{pmatrix} v \\ 0 \\ 0 \end{pmatrix}. \tag{1}$$

where T_1 and T_2 are the vector forms of the reflection law and their expressions is

$$T_1 = I - 2C_1 C_1^T, \tag{2}$$

$$T_2 = I - 2C_2 C_2^T. \tag{3}$$

where C_1 and C_2 are identified as the normal vectors of the 2D scanning galvanometer and the mirror, respectively; their expressions are as follows:

$$C_1 = \begin{bmatrix} 1 & 0 & 0 \\ 0 & \cos\phi & -\sin\phi \\ 0 & \sin\phi & \cos\phi \end{bmatrix} \left(-\sin\frac{\pi}{4}\ 0\ -\cos\frac{\pi}{4} \right)^T, \tag{4}$$

$$C_2 = \begin{bmatrix} \cos\varphi & 0 & -\sin\varphi \\ 0 & 1 & 0 \\ \sin\varphi & 0 & \cos\varphi \end{bmatrix} \begin{bmatrix} 1 & 0 & 0 \\ 0 & \cos\eta & -\sin\eta \\ 0 & \sin\eta & \cos\eta \end{bmatrix} \begin{bmatrix} 1 & 0 & 0 \\ 0 & \cos\phi & -\sin\phi \\ 0 & \sin\phi & \cos\phi \end{bmatrix} \left(\cos\frac{\pi}{4}\ 0\ -\sin\frac{\pi}{4} \right)^T. \tag{5}$$

In Equations (2) and (3), the 4 × 4 identity matrix is represented by I, and the refraction transformation matrix of the lens in Equation (1) is denoted by K, as follows:

$$K = \frac{f}{H} \begin{bmatrix} 1 & 0 & 0 \\ 0 & -1 & 0 \\ 0 & 0 & -1 \end{bmatrix}. \tag{6}$$

Similarly, the influence of the flight path angle of attack (α) on the image motion vector is as follows:

$$v_\alpha = T_1 K T_2 \begin{bmatrix} \cos\alpha & 0 & -\sin\alpha \\ 0 & 1 & 0 \\ \sin\alpha & 0 & \cos\alpha \end{bmatrix} \begin{pmatrix} v \\ 0 \\ 0 \end{pmatrix}. \tag{7}$$

Following the sequence of sideslip angle, attack angle, and roll angle, the image motion vector (v_{speed}) caused by the target on the detector target surface can be derived as follows:

$$v_{speed} = T_1 K T_2 \begin{bmatrix} 1 & 0 & 0 \\ 0 & \cos\gamma & \sin\gamma \\ 0 & -\sin\gamma & \cos\gamma \end{bmatrix} \begin{bmatrix} \cos\alpha & 0 & -\sin\alpha \\ 0 & 1 & 0 \\ \sin\alpha & 0 & \cos\alpha \end{bmatrix} \begin{bmatrix} \cos\beta & -\sin\beta & 0 \\ \sin\beta & \cos\beta & 0 \\ 0 & 0 & 1 \end{bmatrix} \begin{pmatrix} v \\ 0 \\ 0 \end{pmatrix} \tag{8}$$

An assumption is made that the camera rotates at constant speed around the x-axis, with the rotation speed being ω. The swing speed on the image plane is then formulated as follows:

$$\mathrm{v_{roll}} = \mathrm{f} \cdot \omega. \tag{9}$$

The vector can be written as

$$v' = [0\ \mathrm{v_{roll}}\ 0]^T. \tag{10}$$

This leads to the synthesis of the vector and the image motion velocity vector of the entire system as follows:

$$v_{system} = v_{speed} + v'. \tag{11}$$

On this foundation, the aliasing required for super-resolution is attainable by adding the micro-displacement vector, and the final vector needed for super-resolution is

$$v_{sr} = v_{speed} + v' + v_{md}. \tag{12}$$

where v_{md} represents the superimposed micro-displacement vector in single-step compensation. Its value might be one-half pixel, one-third pixel, or other step sizes to gather aliasing information. This can be chosen based on the required sequence depth and moving path parameters.

2.2. Two-Dimensional Scanning Galvanometer Movement Mode

In the staring imaging mode, the two-dimensional oscillating mirror is operated multiple times within a restricted range to obtain aliasing information. Using the image sequence depth of 4 as an example, Figure 2 illustrates the change in the optical-axis pointing angle brought by the 2D scanning galvanometer during the imaging process.

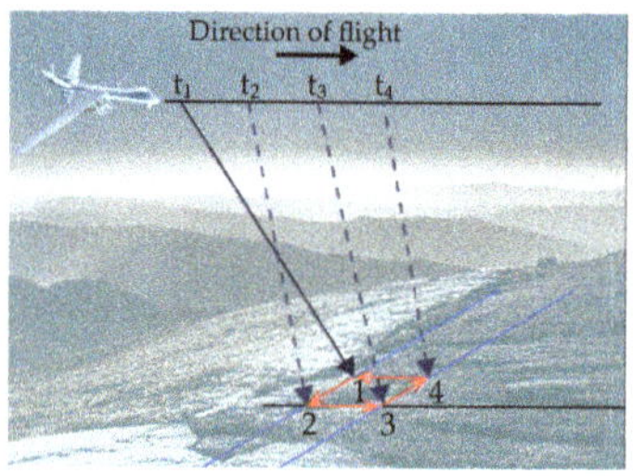

Figure 2. Diagram of boresight pointing at different moments of imaging.

An assumption is further made that the aircraft flies at a constant speed. Figure 3 illustrates sequential moments (t_1, t_2, t_3, and t_4) where the center of the boresight points to specific locations (points 1, 2, 3, and 4). The angle of the optical-axis pointing in the four imaging changes is minimal. In the cross-sectional view of the direction, the displacement of the center of the visual axis between moments t_1 and t_4, and between t_2 and t_3 is Δx. Similarly, the displacement in the side view of the direction between moments t_1 and t_2, and between t_3 and t_4 is Δy.

(**a**) (**b**)

Figure 3. Diagram of micro-displacement and angle. (**a**) The change in the boresight in the cross-sectional view of flight direction; (**b**) the change in the boresight in the side view of flight direction.

These changes in pointing ultimately influence the position of the image point on the image plane. During the imaging process with a sequence depth of 4, the trajectory of the image point on the detector's image plane is depicted in Figure 4.

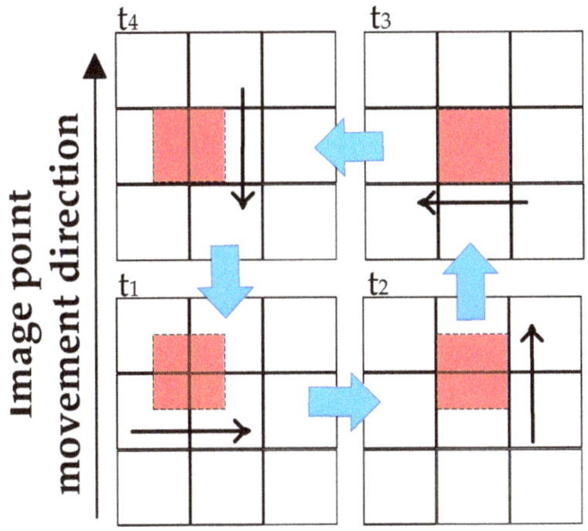

Figure 4. Schematic diagram of the motion of the image point on the image sensor.

From moments t_1 to t_4 of the staring imaging process, to capture the aliasing information of the ground target, the swing mirror not only compensates for the image motion caused by velocity and swing but also adds forward and backward micro-displacements at t_1 and t_3. Additionally, micro-displacements are added in the swing direction at t_2 and t_4. The red area represents the position on the image plane corresponding to the center of the boresight. As illustrated in Figure 4, the camera acquires the first image at the initial position. When forming the second image, the image point corresponding to the center of the visual axis is shifted to the right by 1/2 pixel from its previous position due to the micro-displacement added by the swing mirror. The process then proceeds to acquire the third image, shifting 1/2 pixel to the rear side, and moving the fourth image to the front by 1/2 pixel in a similar fashion.

In summary, the pendulum mirror's swing angle must take into account two effects: compensation for image motion and superimposed micro-displacement to capture aliasing information. The swing path of the pendulum mirror and the micro-displacement step size can vary. The micro-displacement step size might be planned directly according to 1/3 or 1/4 pixel, but the depth of the imaging sequence changes in response. The longer the imaging sequence, the greater the compensation required for image motion by the pendulum mirror. The design of the step size and path must align with the specific system, aiming to gather sufficient aliasing information.

3. Super-Resolution Algorithm Process and Principle

In this chapter, the processing flow of the super-resolution algorithm is presented. Given the optical–mechanical structure characteristics of the aerial visible-light camera, aliased images are achieved with the regular swing of the two-dimensional pendulum mirror. The sequences are subsequently registered, and an anisotropic kernel function is employed for super-resolution reconstruction. This method is similar to the work carried out by Wronski et al. [22], where the data processing includes alignment and anisotropic regression kernel construction. However, in this paper, the micro-displacement relationship between sequence images is defined, greatly enhancing the algorithm's computing speed and robustness.

Initially, the original image sequence that has been acquired is registered according to the reference frame. The micro-displacement vector matrix for each frame in the image sequence relative to the base frame is ascertained. The exact positional relationship of the images within the known sequence enables the exclusion of mis-registered regions and objects, thereby enhancing registration accuracy.

Subsequently, the local gradient information of the image is calculated. The shape of the kernel function is then adjusted based on the local gradient tensor matrix. The anisotropic kernel function alters the shape in accordance with changes at the image's edge, with different weights being allocated to various positions to preserve the details of the reconstructed image.

Finally, these images are weighted and accumulated to produce a super-resolution image. The complete process is visually represented in Figure 5.

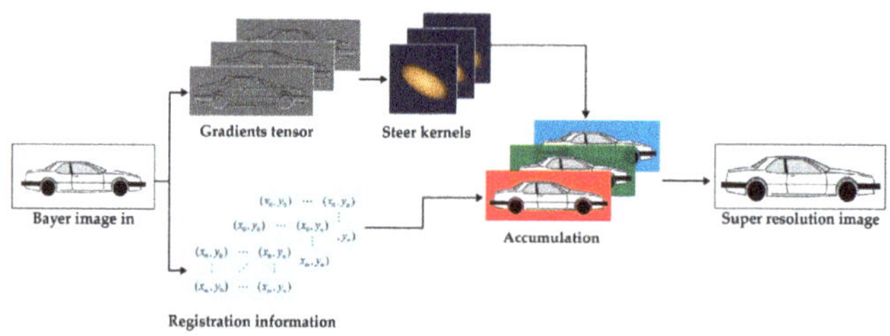

Figure 5. Overview of super-resolution algorithm.

3.1. Image Sequence Alignment Based on Controllable Displacement

The problem of image registration forms the foundation for ensuring the accuracy of image super-resolution. In the image processing flow described in this paper, alignment must first be performed on the image obtained through micro-displacement before it can be input into the image algorithm sequence. Generally, the first image in the sequence is selected as the reference frame, and subsequent images are aligned with this frame. This principle is depicted in Figure 6.

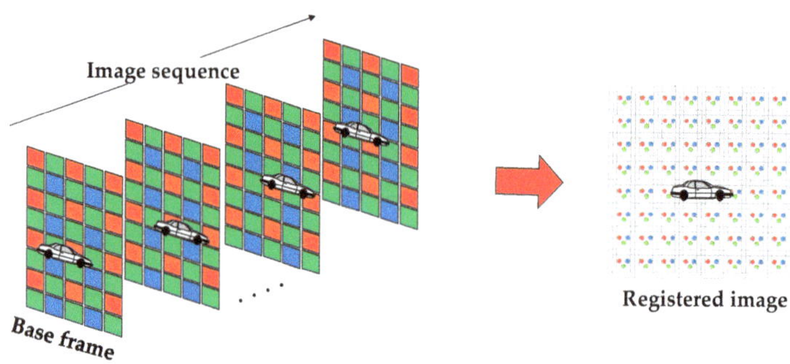

Figure 6. Sub-pixel displacements from 2D scanning galvanometer.

In the alignment of an image sequence with an aliasing relationship to the same reference frame, the camera samples the same position in space N times. Due to the presence of micro-displacement, different sampling positions on the detector's target surface are achieved. Consequently, the color information of the three RGB spectral segments at the same spatial position can be obtained, allowing for the direct generation of color images [25].

The characteristics of aerial imaging include a relatively long imaging distance and minimal sudden or rapid target position changes within the FOV. These characteristics align with the fundamental assumption of the L-K optical flow method [26]. To attain sub-pixel-level registration accuracy, the pyramid method is utilized using multiple iterations. The control over registration accuracy is mainly executed by selecting pyramid layers and the search range of the registration module. Considering the moving distance of the image plane caused by single-step micro-displacement, there is no need to search the registration area extensively. A constrained search range significantly enhances both registration accuracy and computational efficiency. If the calculated displacement vector of the current frame is less than or equal to the current single-step micro-displacement distance, the data are retained. Otherwise, the pixels in that area are discarded, enhancing registration robustness and minimizing the effects of misregistration on image quality, and it is as follows:

$$\begin{bmatrix} S_{x\_Final} \\ S_{y\_Final} \end{bmatrix} = \begin{cases} \begin{bmatrix} S_x \\ S_y \end{bmatrix} & if\ ((S_x \leq \Delta x)||(S_y \leq \Delta y)) \\ discard & else \end{cases}. \tag{13}$$

where Δx and Δy symbolize the displacement triggered by the single-step movement of the micro-motion mechanism on the detector's target surface.

3.2. Super-Resolution Reconstruction Based on Anisotropic Gaussian Kernel

Kernel regression is an approach for fitting a nonlinear model, fundamentally employing the kernel function as a weight function to create a regression model. Unlike the traditional isotropic kernel function, in image applications, there is a desire for the kernel shape to correlate with the image's edge information. The kernel function's shape at various positions is adjusted in accordance with the current point's intensity information, the surrounding point image intensity information, and other factors. This ensures that the edge information in the image after the regression operation is more pronounced, effectively enhancing image authenticity and the signal-to-noise ratio, as well as realizing anisotropy in the edge area [27,28]. The adaptive kernel regression method consists of two basic steps, where the first step is an initial estimate of the image gradients, which is made using some kind of gradient estimator. The second step is to use this estimate to measure the main directions of local gradients in the image. In the second step, this orientation information is used to adaptively "steer" the local kernel, resulting in elongated, elliptical contours spread along the directions of the local edge structure. A steering matrix is defined as follows:

$$H_i^{steer} = h\mu_i C_i^{-\frac{1}{2}}. \tag{14}$$

This matrix is a symmetric covariance matrix of local pixel gray-level differences. By appropriately choosing C_i and h, the algorithm can control the direction of the kernel function and the denoising effect. The local edge structure is related to the gradient covariance. Assuming a Gaussian kernel for the kernel function, the mathematical form of the steering matrix is as follows:

$$K_{H_i^{steer}}(x_i - x) = \frac{\sqrt{\det(C_i)}}{2\pi h^2 \mu_i^2} \exp\left\{ -\frac{(x_i - x)^T C_i (x_i - x)}{2h^2 \mu_i^2} \right\}. \tag{15}$$

where C_i is the local covariance matrix of a kernel function, which can be expressed by Equation (16) and is obtained with the singular-value decomposition (SVD) of the covariance matrix based on the local gradient.

$$C_i = \gamma_i U_{\theta_i} \Lambda_i U_{\theta_i}^T, \tag{16}$$

where U_{θ_i} denotes the function of the rotation matrix, Λ_i represents the function of the stretching matrix, and γ_i is the radius amplitude of the ellipse; their definitions are shown in Equations (17) and (18). By using these three factors, a kernel function constructed from local gradients can control the weight of the kernel function at the image's edge. It provides greater weight to the tangential direction of the edge, thereby enhancing the clarity of the target contour information.

$$U_{\theta_i}^T = \begin{bmatrix} \cos\theta_i & \sin\theta_i \\ -\sin\theta_i & \cos\theta_i \end{bmatrix}, \tag{17}$$

$$\Lambda_i = \begin{bmatrix} \sigma_i & 0 \\ 0 & \sigma_i^{-1} \end{bmatrix}. \quad (18)$$

4. Experimental Verification

4.1. Experimental Device Construction

The algorithm delineated in this paper is implemented in an aerial imaging system, including the processing flow and the super-resolution algorithm. Initially, the algorithm is laboratory-tested, where the two-dimensional swing mirror only performs swing compensation and micro-displacement superposition. Subsequently, in-flight tests are conducted. Speed compensation, swing compensation, and micro-displacement superposition are added to the motion of the 2D scanning galvanometer, validating the algorithm's effectiveness in real-world applications.

4.2. Comparison with Demosaicing Techniques

In this paper, the implemented algorithm exploits the high-frequency aliasing introduced by multi-frame sequence image information to restore the color of the Bayer format image, instead of using interpolation. Several classical interpolation methods are selected for comparison, including VNG [29], FlexISP [30], and ADMM [31] algorithms. Real scenes serve as input images, and the comparative results are illustrated in Figure 7.

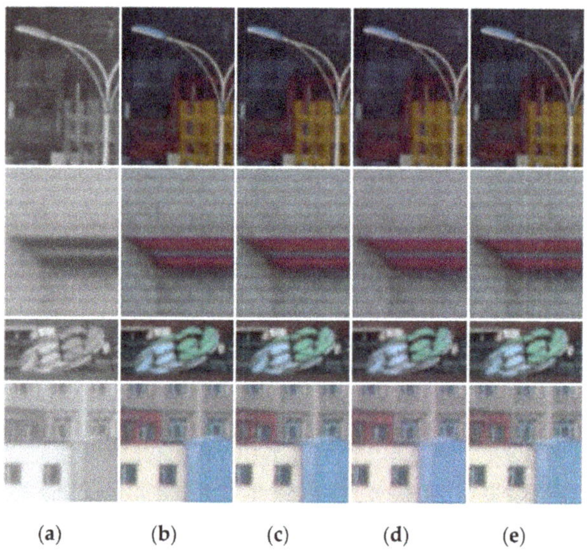

Figure 7. Comparison with demosaicing techniques. (**a**) Bayer; (**b**) VNG; (**c**) FlexISP; (**d**) ADMM; (**e**) ours.

4.3. Resolution Test

The algorithm is evaluated using resolution test targets. The test device is shown in Figure 8.

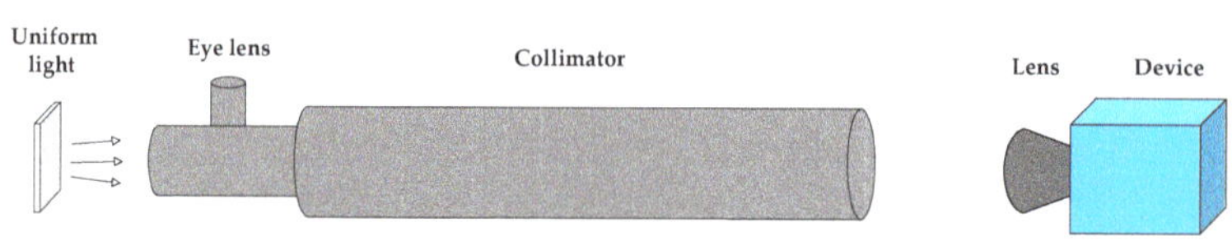

Figure 8. Optical resolution test device.

Using a collimator with a focal length of 1.6 m, the width of the resolution target ranges from 0.04 mm to 0.02 mm from the widest to the narrowest, and the width difference between two adjacent groups of stripes is 0.002 mm. Its dimensions are shown in Figure 9.

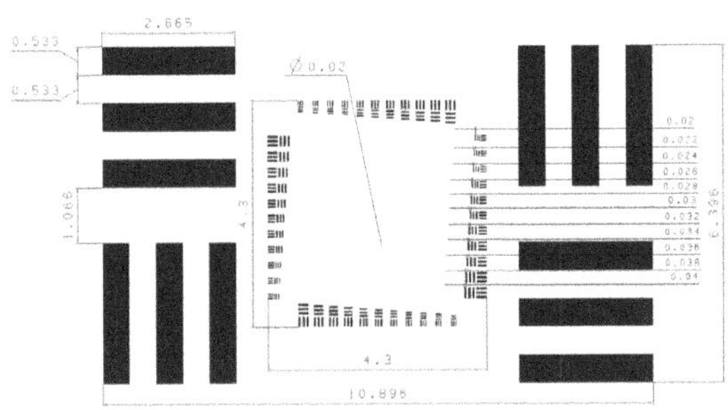

Figure 9. Resolution test target dimension drawing.

The micro-displacement required for super-resolution is derived from the 2D scanning galvanometer in the optical system, with constant exposure time and gain. Under laboratory conditions, the resolution target is photographed, with the original image being depicted in Figure 10.

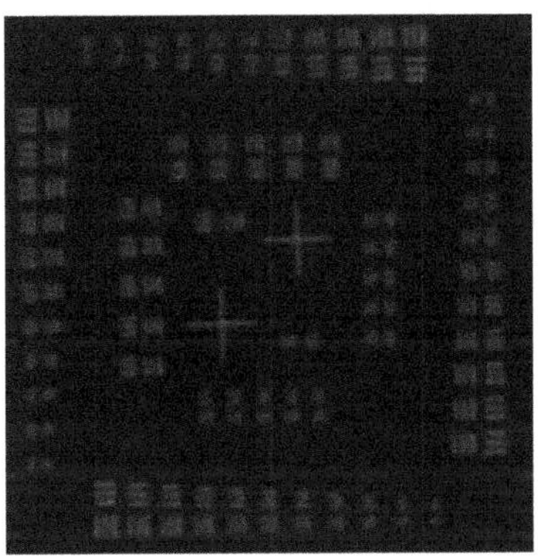

Figure 10. Raw image of the resolution test target.

It can be analyzed from Figure 10 that the resolution of the original image is 13.8 lp/mm and the angular resolution is 0.000045 rad. Different numbers of image sequences are employed to assess the super-resolution algorithm, and the results are presented in Figure 11.

Through visual observation, we find the smallest target that can be resolved in each sequence depth result. It can be seen from Table 1 that when the image sequence depth is changed, the values of optical resolution and angular resolution change accordingly.

Table 1. Resolution statistical table.

Number of Images	Optical Resolution	Angular Resolution	Resolution Improvement Compared with RAW
4	14.7 lp/mm	0.0000425 rad	6%
8	15.6 lp/mm	0.0000400 rad	13%
12	16.7 lp/mm	0.0000375 rad	21%
16	19.2 lp/mm	0.0000325 rad	39%

The tests demonstrate that the multi-frame super-resolution algorithm significantly reduces noise in the image, noticeably enhancing the resolution of details.

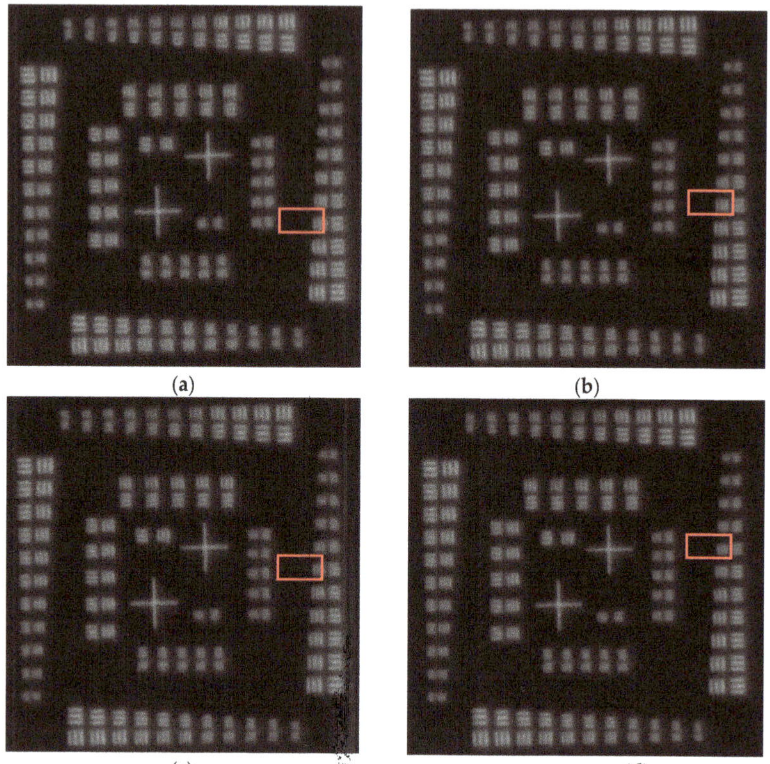

(a) (b) (c) (d)

Figure 11. Experimental results of the algorithm on multi-frame resolution target images. (**a**) The result of 4 frames; (**b**) the result of 8 frames; (**c**) the result of 12 frames; (**d**) the result of 16 frames.

4.4. Spatial Resolution Improvement Test

4.4.1. Daytime Outdoor Scene Imaging Test

To further evaluate the algorithm, both daytime and nighttime scenes are tested. In the daytime scene, a multi-frame sequence of 8 or 12 images is used, with integration time and gain maintained. The imaging scene is portrayed in Figure 12.

Figure 12. Imaging scenes under daytime conditions.

The comparative effects between the bicubic interpolation image and the super-resolution image are displayed in Figure 13, which is the detail in the red box in Figure 12. The super-resolution image exhibits superior edge details and texture restoration capabilities.

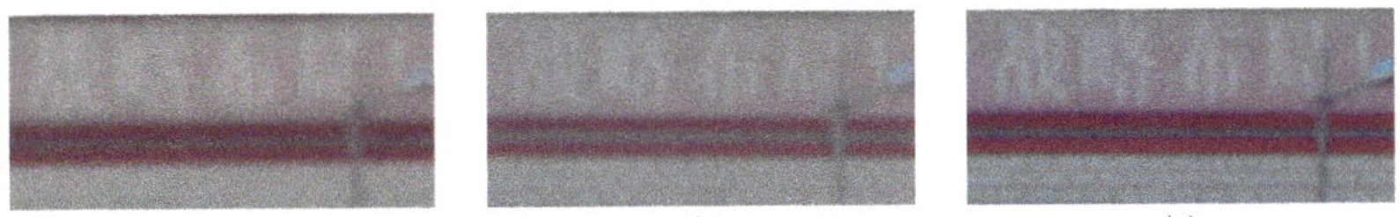

(**a**) (**b**) (**c**)

Figure 13. Imaging scenes under daytime conditions. (**a**) Bicubic interpolation image; (**b**) 8-frame super-resolution image; (**c**) 12-frame super-resolution image.

4.4.2. Nighttime Outdoor Scene Imaging Test

In the night scene, the image sequence retains identical imaging conditions, ensuring consistent exposure time and gain during the process. A total of 12 frames are utilized, and the imaging scene is captured in Figure 14.

Figure 14. Imaging scenes under night conditions.

A comparison of the interpolated original image with the super-resolution image under nighttime imaging conditions reveals superior noise control in the super-resolution image. The restoration of target details in the super-resolution image is significantly more effective than in the original. Details that were previously submerged in noise within the original image were well restored, as shown in Figure 15.

(**a**) (**b**) (**c**)

Figure 15. Comparison of night imaging details. (**a**) Bicubic interpolation image; (**b**) 8-frame super-resolution image; (**c**) 12-frame super-resolution image.

4.4.3. Influence of Accumulation Cycle on Performance

To examine the impact of image sequence length on the super-resolution effect, tests were conducted on both sunlight and night scenes. A selection of 4 to 16 images was used for each test. Figure 16 illustrates the PSNR and SNR of the super-resolution images, highlighting that the number of images in the sequence positively affects these parameters during nighttime. In a static scene, the inclusion of more image sequences yields improved super-resolution effects.

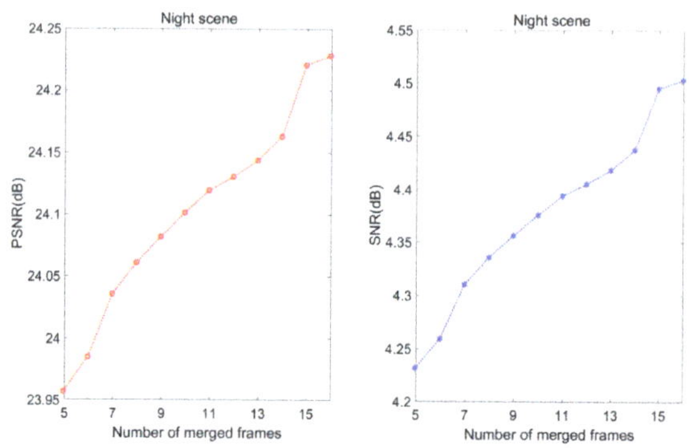

Figure 16. The relationship between the number of frames, and PSNR and SNR in the night scene.

Contrarily, in daytime scenarios, increasing the number of image frames does not continuously enhance the PSNR and SNR of the super-resolution images in Figure 17. An upward trend becomes stagnant or even declines after the number of image frames reaches eight. This stagnation is attributed to the increasing complexity of image registration and the rising probability of misregistration in image details with the addition of image sequences. As a result, the image evaluation factor may remain static or decrease. Therefore, a reasonable control of sequence depth from 8 to 12 frames is suggested based on the test results.

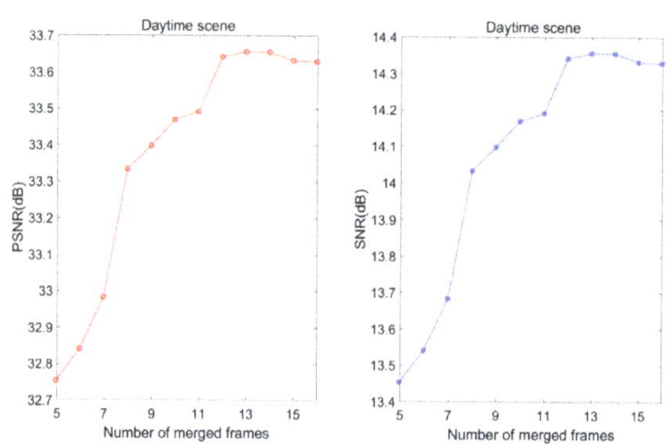

Figure 17. The relationship between the number of frames, and PSNR and SNR in the daytime scene.

4.4.4. Results of Flight Experiments

The algorithm was deployed in a photoelectric load and tested during a flight. The pendulum mirror's movements were set to eight, with a step size of half a pixel, following a circular rectangular path; the speed-to-height ratio of the system was 44.5; the pixel size was 5.5 microns; and the focal length of the lens was 180 mm. Figure 18 depicts the experimental results.

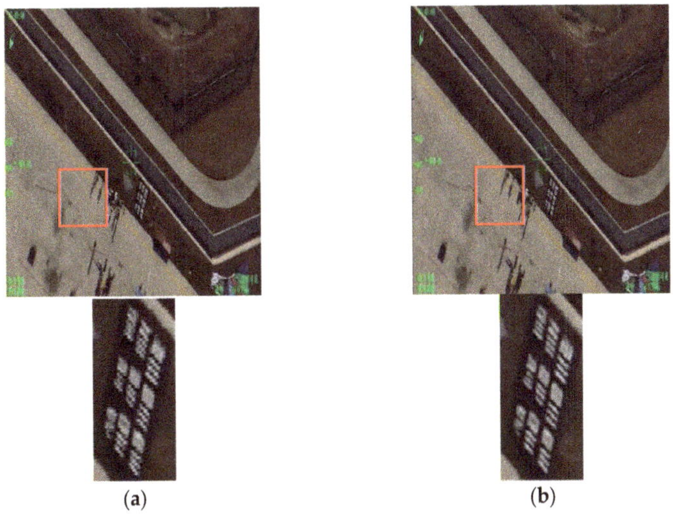

Figure 18. Flight test image. (**a**) Bicubic interpolation image and details; (**b**) 8-frame super-resolution image and details.

The test shows that the algorithm can be applied in the optoelectronic pod, and a careful inspection of local details shows that the algorithm has good efficacy in enhancing image resolution.

5. Conclusions

A super-resolution algorithm has been formulated for aeronautical optoelectronic systems equipped with a 2D scanning galvanometer. Initially, within a standard aerial optical system with a two-dimensional swing mirror, a mathematical model was established between the target displacement and image point on the image plane. The swing path was designed, and the swing vector was analyzed.

Subsequently, based on aerial ground imaging characteristics, a super-resolution algorithm was designed, employing a multi-layer pyramid optical flow method to register sequential images with known micro-displacement, thereby substantially improving registration accuracy and robustness.

An anisotropic kernel function was generated using the gradient tensor matrix, and this kernel function was employed to perform super-resolution reconstruction of the image. This approach significantly preserves image edge details and augments system resolution.

An aeronautical optoelectronic load incorporating the algorithm underwent tests under both daytime and nighttime conditions. The results demonstrate the algorithm's ability to markedly suppress noise and enhance image resolution in night scenes. In a resolution test, 16 frames in our algorithm increased the resolution of the optical system by 39%.

Furthermore, an increase in the sequence's number of images can elevate the algorithm's performance, but beyond a certain limit, gains in SNR and PSNR slow down due to factors such as misregistration and compensation of nonlinear errors caused by increased image motion. Data indicate that a sequence depth of 8–12 images should generally be selected for optimal results.

Author Contributions: Conceptualization T.M.; methodology, T.M.; software, T.M., C.L. and Y.H.; validation, Y.X.; formal analysis, T.M. and Y.L.; investigation T.M. and F.Y.; resources, Y.X. and L.M.; writing—original draft preparation, T.M. and Y.H.; writing—review and editing, T.M. and Y.H.; visualization, T.M. and L.M.; supervision, Y.L. and H.S.; project administration, Y.L., Y.X. and H.S.; All authors have read and agreed to the published version of the manuscript.

Funding: This research received no external funding.

Institutional Review Board Statement: Not applicable.

Informed Consent Statement: Not applicable.

Data Availability Statement: This study did not create new data.

Conflicts of Interest: The authors declare no conflict of interest.

References

1. Gouveia, L.C.P.; Choubey, B.J.S.r. Advances on CMOS image sensors. *Proc. IEEE* **2016**, *36*, 231–239. [CrossRef]
2. Jacobsen, K.J. High resolution satellite imaging systems-an overview. *Photogramm. Fernerkund. Geoinf.* **2005**, *2005*, 487.
3. Moser, G.; Serpico, S.B.; Benediktsson, J.A.J. Land-cover mapping by Markov modeling of spatial–contextual information in very-high-resolution remote sensing images. *Proc. IEEE* **2012**, *101*, 631–651. [CrossRef]
4. Yue, L.; Shen, H.; Li, J.; Yuan, Q.; Zhang, H.; Zhang, L.J. Image super-resolution: The techniques, applications, and future. *Signal Process.* **2016**, *128*, 389–408. [CrossRef]
5. Zhu, Y.; Chen, X.; Yuan, W.; Chu, Z.; Wong, K.-y.; Lei, D.; Yu, Y.J. A waveguide metasurface based quasi-far-field transverse-electric superlens. *Opto-Electron. Adv.* **2021**, *4*, 210013. [CrossRef]
6. Xiong, Z.; Melzer, J.E.; Garan, J.; McLeod, E.J. Optimized sensing of sparse and small targets using lens-free holographic microscopy. *Opt. Express* **2018**, *26*, 25676–25692. [CrossRef]
7. Dorr, F.J. Satellite image multi-frame super resolution using 3D wide-activation neural networks. *Remote. Sens.* **2020**, *12*, 3812. [CrossRef]
8. Xu, J.; Liang, Y.; Liu, J.; Huang, Z.J. Multi-frame super-resolution of Gaofen-4 remote sensing images. *Sensors* **2017**, *17*, 2142. [CrossRef]
9. Bishara, W.; Su, T.-W.; Coskun, A.F.; Ozcan, A.J. Lensfree on-chip microscopy over a wide field-of-view using pixel super-resolution. *Opt. Express* **2010**, *18*, 11181–11191. [CrossRef]
10. Fortin, J.; Chevrette, P. *Realization of a Fast Microscanning Device for Infrared Focal Plane Arrays;* SPIE: Bellingham, WA, USA, 1996; Volume 2743.
11. Wiltse, J.; Miller, J. Imagery improvements in staring infrared imagers by employing subpixel microscan. *J. Optical. Eng.* **2005**, *44*, 056401. [CrossRef]
12. Zhang, X.; Huang, W.; Xu, M.; Jia, S.; Xu, X.; Li, F.; Zheng, Y.J. Super-resolution imaging for infrared micro-scanning optical system. *Opt. Express* **2019**, *27*, 7719–7737. [CrossRef] [PubMed]
13. Yang, C.; Everitt, J.H.; Murden, D. Evaluating high resolution SPOT 5 satellite imagery for crop identification. *Comput. Electron. Agric.* **2011**, *75*, 347–354. [CrossRef]
14. Reulke, R.; Becker, S.; Haala, N.; Tempelmann, U.J. Determination and improvement of spatial resolution of the CCD-line-scanner system ADS40. *ISPRS J. Photogramm. Remote. Sens.* **2006**, *60*, 81–90. [CrossRef]
15. Tang, X.; Hu, F.; Wang, M.; Pan, J.; Jin, S.; Lu, G. Inner FoV Stitching of Spaceborne TDI CCD Images Based on Sensor Geometry and Projection Plane in Object Space. *Remote. Sens.* **2014**, *6*, 6386–6406. [CrossRef]
16. Datsenko, D.; Elad, M.J.M. Example-based single document image super-resolution: A global MAP approach with outlier rejection. *Syst. Signal Process.* **2007**, *18*, 103–121. [CrossRef]
17. Kim, K.I.; Kwon, Y. Example-based learning for single-image super-resolution. In Proceedings of the Joint Pattern Recognition Symposium, Munich, Germany, 10–13 June 2008; pp. 456–465.
18. Keren, D.; Peleg, S.; Brada, R. Image sequence enhancement using sub-pixel displacements. In Proceedings of the Computer Vision and Pattern Recognition, Computer Society Conference, Ann Arbor, MI, USA, 5–9 June 1988.
19. Alam, M.S.; Bognar, J.G.; Hardie, R.C.; Yasuda, B.J.J. Infrared image registration and high-resolution reconstruction using multiple translationally shifted aliased video frames. *IEEE Trans. Instrum. Meas.* **2002**, *49*, 915–923. [CrossRef]
20. Irani, M.; Peleg, S.J.I. Super resolution from image Sequences. In Proceedings of the 10th International Conference on Pattern Recognition, Atlantic City, NJ, USA, 16–21 June 1990.
21. Xie, W.; Zhang, F.; Chen, H.; Qin, Q. Blind Super-Resolution Image Reconstruction Based on POCS Model. In Proceedings of the International Conference on Measuring Technology & Mechatronics Automation, Zhangjiajie, China, 11–12 April 2009.
22. Stark, H.; Oskoui, P. High-resolution image recovery from image-plane arrays, using convex projections. *JOSA A* **1989**, *6*, 1715. [CrossRef]
23. Tom, B.C.; Katsaggelos, A.K. Reconstruction of A High-Resolution Image by Simultaneous Registration, Restoration, and Interpolation Of Low-Resolution Images. In Proceedings of the Image Processing, International Conference, Washington, DC, USA, 23–26 October 1995.
24. Schultz, R.R.; Stevenson, R.L.J. Extraction of high-resolution frames from video sequences. *IEEE Trans. Image Process* **1996**, *5*, 996–1011. [CrossRef] [PubMed]
25. Wronski, B.; Garcia-Dorado, I.; Ernst, M.; Kelly, D.; Krainin, M.; Liang, C.-K.; Levoy, M.; Milanfar, P.J. Handheld multi-frame super-resolution. *ACM Trans. Graph.* **2019**, *38*, 1–18. [CrossRef]

26. Lucas, B.D.; Kanade, T. An iterative image registration technique with an application to stereo vision. In Proceedings of the IJCAI'81: 7th International Joint Conference on Artificial Intelligence, Vancouver, BC, Canada, 24–28 August 1981; pp. 674–679.
27. Takeda, H.; Farsiu, S.; Milanfar, P.J. Kernel regression for image processing and reconstruction. *IEEE Trans. Image Process* **2007**, *16*, 349–366. [CrossRef]
28. Takeda, H.; Farsiu, S.; Milanfar, P. Robust kernel regression for restoration and reconstruction of images from sparse noisy data. In Proceedings of the 2006 International Conference on Image Processing, Atlanta, Georgia, 8–11 October 2006; pp. 1257–1260.
29. Chang, E.; Cheung, S.; Pan, D.Y. Color filter array recovery using a threshold-based variable number of gradients. In Proceedings of the Sensors, Cameras, and Applications for Digital Photography, San Jose, CA, USA, 27–28 January 1999; pp. 36–43.
30. Heide, F.; Steinberger, M.; Tsai, Y.-T.; Rouf, M.; Pająk, D.; Reddy, D.; Gallo, O.; Liu, J.; Heidrich, W.; Egiazarian, K.J. Flexisp: A flexible camera image processing framework. *ACM Trans. Graph.* **2014**, *33*, 1–13. [CrossRef]
31. Tan, H.; Zeng, X.; Lai, S.; Liu, Y.; Zhang, M. Joint demosaicing and denoising of noisy bayer images with ADMM. In Proceedings of the 2017 IEEE International Conference on Image Processing (ICIP), Beijing, China, 17–20 September 2017; pp. 2951–2955.

photonics

MDPI

Article

Comparative Analysis of the Effect of Cutting Piezoelectric Ceramics on Optically Oriented Compensation Capability

Bo Li [1,2] and Dapeng Tian [1,2,*]

1 The Key Laboratory of Airborne Optical Imaging and Measurement, Changchun Institute of Optics, Fine Mechanics and Physics, Chinese Academy of Sciences, Changchun 130033, China; libo221@mails.ucas.ac.cn
2 University of Chinese Academy of Sciences, Beijing 100049, China
* Correspondence: d.tian@ciomp.ac.cn

Abstract: In a photoelectric tracking system, a deformable mirror can be used to adjust the transmission path of a light beam to achieve accurate measurement and tracking of the target position. The single-piezoelectric-sheet deformable mirror is a commonly used wavefront correction device with outstanding advantages, such as its simple structure, low cost, and easy preparation. However, it usually has a large cross-linking value and a weak ability to correct higher-order wavefront aberrations. To solve this problem, a novel single-piezoelectric-sheet deformable mirror driving structure is proposed in this paper, in which the coupling between the driving units is weakened by laser cutting the piezoelectric ceramics along the electrode gaps to reduce the cross-linking value of the deformable mirror. In order to verify the effectiveness of this driving structure, this paper takes the 85-unit single-piezoelectric-sheet deformable mirror as an example and simulates the influence function for different cutting depth conditions. The simulation results show that the cross-linking value between neighboring drive units decreases significantly as the piezoelectric-ceramic-laser cutting depth of the drive electrode gap increases. When the piezoelectric ceramic was laser cut to a depth of half of the overall thickness, the cross-linking value decreased by 6.8%.

Keywords: piezoelectric deformable mirror; adaptive optics; cross-linking value; photoelectric imaging

1. Introduction

The photoelectric tracking system is a technology based on optical technology to achieve the function of the target tracking and positioning system, which usually includes cameras and photoelectric sensors to monitor the position and motion information of target objects [1]. It is widely used in aerospace, navigation, military, and robotics. Photoelectric tracking systems can obtain real-time information about the motion of the target and control it as needed in order to achieve accurate tracking and positioning of the target. In some cases, photoelectric tracking systems need to correct wavefront aberrations in the optical system in order to achieve the system to maintain high-precision tracking of the target [2].

Adaptive optics is a technique applied to optical systems to correct aberrations by adjusting the shape or characteristics of optical elements in real time to obtain better image quality. This technique was initially developed to solve the aberration problem in ground-based telescopes under the influence of atmospheric turbulence [3]. Adaptive optics systems utilize sensors to monitor the propagation of light waves in real time and correct these aberrations with controllers and compensation devices. Deformable mirrors were first proposed to solve the problem of aberrations in ground-based telescopes under the effect of atmospheric turbulence. As one of the core devices of adaptive optics systems, it has good application prospects in both military and civilian fields, and optimizing its parameter design has a crucial impact on the overall performance [4,5]. Atmospheric turbulence can lead to phase differences in the incident light wavefront, thus reducing the resolution and imaging quality of the telescope. By adding a deformable mirror to the

Citation: Li, B.; Tian, D. Comparative Analysis of the Effect of Cutting Piezoelectric Ceramics on Optically Oriented Compensation Capability. *Photonics* **2023**, *10*, 1136. https://doi.org/10.3390/photonics10101136

Received: 7 September 2023
Revised: 8 October 2023
Accepted: 9 October 2023
Published: 11 October 2023

optical system of the telescope, the shape of the mirrors can be adjusted in real time to offset the phase difference caused by atmospheric turbulence, thus improving the observing capability of the telescope [6]. According to the different working principles and materials, deformable mirrors can be categorized into various types, such as electrostatic-driven [7], electromagnetic-driven [8], and piezoelectric-driven [9]. Among these, the piezoelectric deformable mirror is a common type. It utilizes the piezoelectric effect, in which certain crystals change shape under the action of an electric field. It makes use of the piezoelectric effect, which usually has the advantages of a compact and lightweight structure, fast response speed, high deformation accuracy, and wide application band [10].

The piezoelectric deformable mirror is made of piezoelectric crystal material, and the curvature and shape of the mirror are adjusted by applying a voltage to change the shape of the crystal [11,12]. The piezoelectric deformable mirror has significant advantages over other types of mirrors [13]. First of all, it has a fast response speed, which can quickly respond to external excitation and achieve high-speed shape adjustment, suitable for fast optical systems. Secondly, it has high shape adjustment accuracy, which can achieve sub-nanometer deformation and is suitable for high-precision optical systems. Thirdly, its power consumption is low; a piezoelectric deformable mirror only needs to apply a very small voltage when working, so its power consumption is relatively low [14]. Fourthly, it is highly reliable, with a simple, lightweight structure and no internal mechanical parts or wearing parts. Finally, the precision of piezoelectric deformable mirrors can reach sub-micron level, which is finer than other principle mirrors. Moreover, due to the special nature of the piezoelectric effect, piezoelectric deformable mirrors can realize a wide range of surface deformations, which can satisfy the demand of adjusting different wavelengths [15]. However, there are some drawbacks; the displacement range of piezoelectric deformable mirrors is small and not suitable for large displacement applications. The performance of piezoelectric materials is greatly affected by temperature [16]. Its working temperature range is strict, and the mechanical and electrical properties of piezoelectric materials change at different temperatures, causing instability, so temperature compensation is needed [17]. The larger cross-linking values are due to the fact that the piezoelectric deformable mirrors are designed to obtain higher mechanical strength and stability and to improve the rigidity and stability of the material. The structural design of the piezoelectric ceramics has not been updated, resulting in larger cross-linking values for piezoelectric deformable mirrors compared to other types, which is reflected in this paper.

Based on the content of our knowledge, other studies usually only consider the effect of the overall thickness of the piezoelectric ceramic, the shape of the electrodes, the way the electrodes are arranged, and the way the edges are supported on the deformable mirror performance [18–20]. Ningbo University prepared a 19-unit 50 mm caliber deformable mirror prototype in [21]. A two-wafer deformable mirror with 20 electrodes on a 63 mm aperture was designed in reference [22]. Piezoelectric-stack deformable mirrors with 61 piezoelectric actuators and a 60 mm aperture and dual-wafer deformable mirrors with 48 electrodes and a 50 mm aperture have been established in reference [23]. To this end, a simple and low-cost 85-unit silicon-based single-piezoelectric-sheet deformable mirror is proposed in this paper, in which the processed thickness of the piezoelectric ceramic is adjusted by laser cutting, and the electrode shape is ensured to remain unchanged. The processing of the thickness does not exceed 50% of the piezoelectric material's thickness (Laser cutting the PZT sheet deeper than 50% caused serious carbonation and made the actuators fragile.). The deflection and cross-linking value of the deformable mirror are taken as two indicators affecting the calibration performance.

Later in this paper, there is a significant decrease in the cross-link value when cutting the non-electrode region of the piezoelectric material. This is due to the fact that the cutting process destroys the structure of the piezoelectric ceramic material and reduces its cross-linking properties. It is possible to add degrees of freedom to the single-piezoelectric-sheet deformable mirror to achieve a faster deformation speed; reduce the inertia effect, making the response faster and better adapted to external changes; reduce the thickness of the piezoceramics by cutting, which reduces the size and the weight; and reduce the thickness of the piezoceramics to increase their intrinsic frequency, thus improving the frequency response range of the deformable mirror.

The remainder of this article is organized as follows. Section 2 describes the background, underlying theory, and modeling of piezoelectric deformable mirrors. Section 3 introduces and presents the analysis of the proposed method in this study. Section 4 describes the analysis of the simulation results of the proposed method. Finally, Section 5 concludes this article.

2. Background and Modeling of Piezoelectric Deformable Mirrors

This section introduces background on piezoelectric deformable mirrors and their modeling. It also describes some basic concepts of piezoelectric deformable mirrors, such as influence functions and cross-linking values.

2.1. Piezoelectric Materials and the Piezoelectric Effect

The piezoelectric material chosen for the piezoelectric deformable mirror designed in this article is PZT-5H. PZT-5H is a commonly used lead zirconate titanate (PZT) ceramic material with good piezoelectric properties. It has a high piezoelectric coefficient and piezoelectric stress constant, which can produce a large deformation response when an electric field is applied. It also has high mechanical strength and toughness to withstand large stresses and strains, resulting in better stability and durability during deformable mirror operation.

The piezoelectric effect refers to the fact that when an external force is applied to the surface of a piezoelectric material, the material separates the positive and negative charges, resulting in an imbalance in the internal distribution of charges, which leads to the formation of potential differences and voltages [24]. The schematic diagram is shown in Figure 1a. The inverse piezoelectric effect is the inverse process of the positive piezoelectric effect. When an external electric field is applied to a piezoelectric material, the material deforms, resulting in mechanical stress. The schematic diagram is shown in Figure 1b. PDM utilizes the inverse piezoelectric effect, whereby an applied electric field creates mechanical stress through the piezoelectric material, which facilitates the displacement of the mirror.

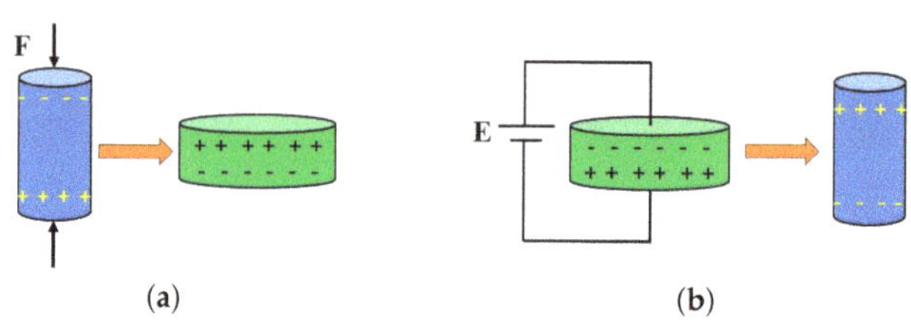

Figure 1. Piezoelectric effect schematic diagram: (**a**) Piezoelectric effect schematic diagram. (**b**) Inverse piezoelectric effect schematic.

2.2. Piezoelectric Deformable Mirror Structure

The initial model of the 85-unit PDM is shown in Figure 2a and consists of a 500 μm thick layer of monocrystalline silicon material and a 300 μm thick layer of the piezoelectric material—lead zirconate titanate (PZT-5H)—with a silver-coated layer on the surface of the optics, which is generally in the range from about 80 nm to 150 nm in thickness.

The diameter of the deformable mirror is 50 cm, and the effective calibrated aperture is 46 cm. In this paper, we investigate the effect of different thicknesses of the piezoelectric layer on the optical system. The inner surface of the piezoelectric layer is covered with a metal film as a common electrode, while the outer surface is covered with a patterned array of electrodes. The operating voltage range was −150~+150 V. We kept the thickness of the electrode region constant and cut the non-electrode region by laser cutting and other methods. In addition, we coated the outer surface of the wafer with a metallic or dielectric reflective layer. We chose monocrystalline silicon material for the deformable mirrors' optical mirror compared to conventional glass materials because of its excellent mechanical properties, compatibility with micromachining processes, and the easy availability and low cost of commercially polished silicon wafers. These wafers can achieve thicknesses of 200~1000 μm and diameters of 50~100 mm, and the polished surfaces have good optical properties.

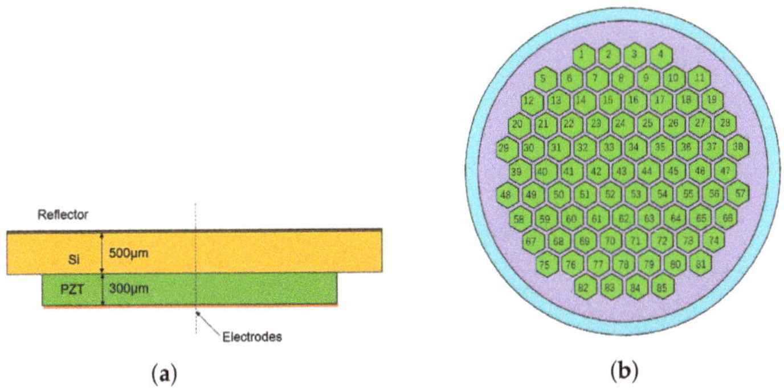

Figure 2. Schematic diagram of 85-unit piezoelectric deformable mirror. (**a**) Cross-sectional view of deformable mirror. (**b**) Distribution of the 85 electrodes.

2.3. Cross-Linking Value

Cross-linking value is a parameter of the strength and stiffness of the surface of a deformable mirror. It describes the properties of the deformable mirror material, including its stiffness, elasticity, and deformability [25]. The driver cross-linking value, which is the ratio of the deformation δ of the energized driver in a single influence function to the deformation δ of the drivers in adjacent positions. The detailed schematic, shown in Figure 3, is as follows

$$\omega = \frac{\delta_2}{\delta_1} \tag{1}$$

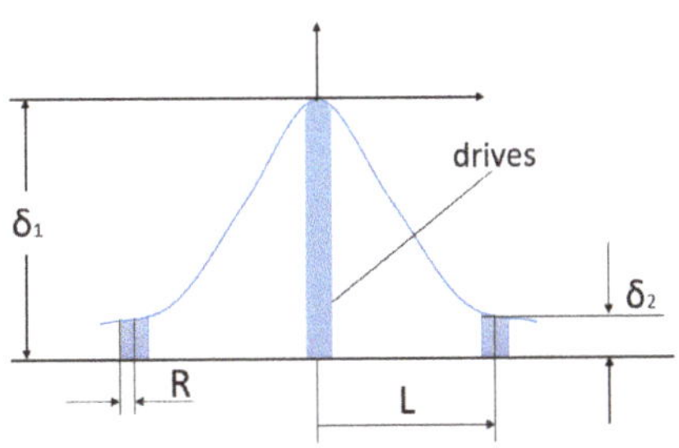

Figure 3. Figure of the method for calculating the cross-linking values of deformable mirrors.

The larger the change in cross-linking value, the smoother the wavefront change, and the smaller the cross-linking value, the steeper the wavefront change. Different cross-linking values can also seriously affect the ability of the deformable mirror to correct the phase of the wavefront. The definitions and calculations of the cross-linking value in this article are described later.

3. Design of Electrode Gap Cutting for Piezoelectric Deformable Mirrors

This section focuses on how to select the ortho-hexagonal pattern among several common piezoelectric deformable mirror electrode patterns, the design of the cutting region, and the method of calculating the cross-linking value based on the piezoelectric deformable mirror structures covered in this paper.

3.1. Design of Electrode Pattern and Selection of Cutting Area

The dominant electrode patterns today are square hexagons or circles. Comparisons have shown that more actuators can be mounted on a hexagonal electrode than on a circular electrode for the same area with the same diameter of mirror or piezoelectric ceramic material. Figure 4 illustrates this point. The individual hexagonal electrodes in (a,b) have the same area as the circular electrodes, but 85 are arranged in (a), while only 69 are arranged in (b). Additionally, the piezoelectric material has the same area, so (a) has a higher fill ratio than (b), which is more valuable in practical applications.

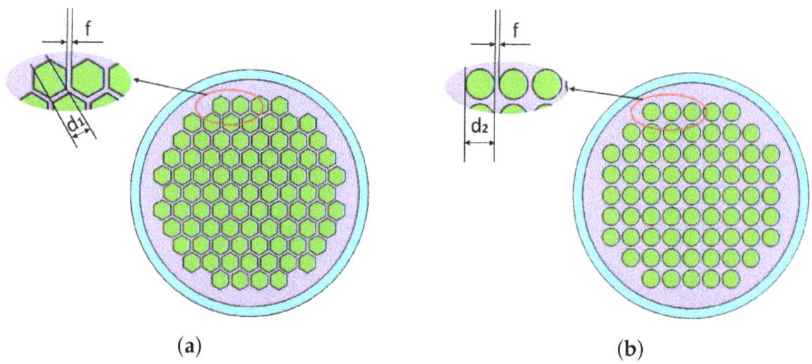

Figure 4. Figures describing deformable mirrors with different electrode shapes. (**a**) Distribution of electrode shapes in an ortho-hexagonal shape. (**b**) Circular electrode shape distribution diagram. d_1 = 2 mm, d_2 = 3.638 mm, f = 0.69 mm.

As shown in Figure 4, at equal actuator areas and electrode spacing, our piezoelectric deformable mirror with circular mirror aperture can accommodate 85 hexagonal actuators in a hexagonal lattice (Figure 4a) but only 69 circular actuators in a square lattice (Figure 4b). We chose the hexagonal arrangement because it provides a dense actuator placement and achieves a high fill ratio.

$$3 \times \sin\frac{\pi}{3} \times {d_1}^2 = \pi\left(\frac{d_2}{2}\right)^2 \tag{2}$$

An increase in the number of actuators is beneficial to the performance of a piezoelectric deformable mirror. It provides a greater deformation range, higher resolution, and faster response. Therefore, electrodes with regular hexagonal shapes are used in this study.

In other studies exploring the effect of the thickness of piezoelectric deformable mirrors on their performance, usually, only the thickness of the whole electrode has been considered for the study. In contrast, in this paper, a more innovative cutting region is considered, where the piezoelectric ceramic material is cut at different depths except for the electrode region, such as the purple area in Figure 4a.

In order to investigate the effect of the cutting depth of the piezoelectric ceramic electrode gap region on the performance of the piezoelectric deformable mirror, four groups of programs were designed, as shown in Figure 5. The cutting depths of the cut electrode gap region were 0 μm, 37.5 μm, 75 μm, and 150 μm, respectively (12.5%, 25%, and 50% of the total thickness of 300 μm). The deformable mirror's deformable ability changes were observed by designing different depths of cut.

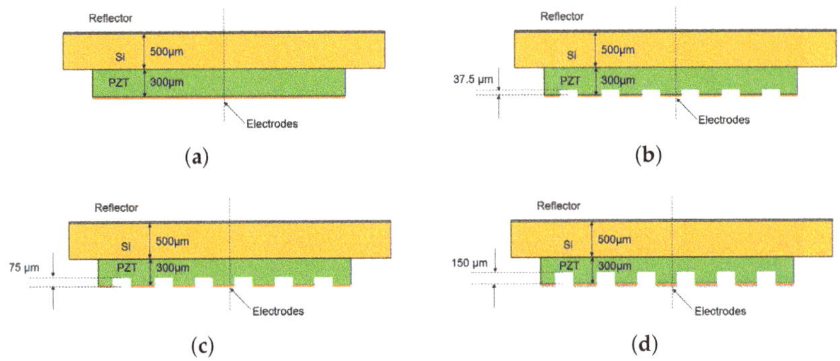

Figure 5. Schematic diagram of cutting depth: (**a**) 0 μm depth of cut; (**b**) 37.5 μm depth of cut; (**c**) 75 μm depth of cut; (**d**) 150 μm depth of cut.

3.2. Proposed Method of Calculating the Cross-Linking Value for This Structure

For the structure of the 85-unit piezoelectric deformable mirror proposed in this paper, a method for calculating the cross-linking value is proposed. If the electrode sheet used to calculate the cross-linking value is located at the edge, the displacements of the center points of several electrode regions nearest to its center point are added and averaged. If the electrode sheet for calculating the cross-linking value is not located at the edge, the calculation parameters are taken from the several electrode regions closest to its center point, which is usually six drivers in this structure. As shown in Figure 6, when analyzing electrode No. 43, which is located at the center, we need to consider electrodes No. 33, 34, 42, 44, 52, and 53, which are at the closest distance around it. When analyzing electrode 1 at the edge, we only need to consider electrodes No. 2, 6, and 7 because they are the only three electrodes within the closest distance. This is calculated by averaging the displacements of electrodes No. 33, 34, 42, 44, 52, and 53 and comparing them with the displacement of electrode No. 43. The displacements of electrodes No. 2, 6, and 7 are averaged and compared with the displacement of electrode No. 1.

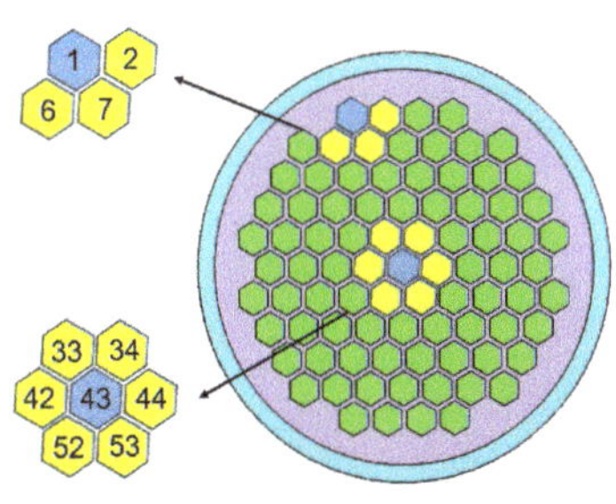

Figure 6. Figure of illustration of the calculation of the cross-linking values of this structure.

3.3. Simulation Parameter Setting

The designed 85-unit piezoelectric deformable mirror structure was imported into finite-element simulation software. In this study, 50 V was applied individually and separately in the order of electrode numbering, and fixed constraints were applied to the mirror edges. The mechanical restriction tooling for the free actuator electrodes is shown in Figure 7. The base is not included in the simulation and the perimeter of the piezoelectric sheet deformation mirror is glued to the tooling with epoxy adhesive. When the piezoelectric sheet is energized, there is no effect on the deformation of the tooling.

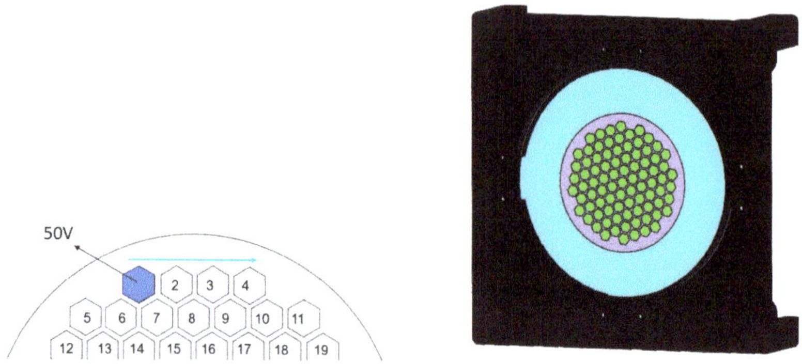

(**a**) (**b**)

Figure 7. Figure of a description of the condition setting for piezoelectric deformable mirrors in the simulation. (**a**) Schematic diagram of the sequence of applying 50 V to the electrodes. (**b**) Schematic diagram for applying fixed constraints to mirror edges.

The piezoelectric deformable mirror is modeled using finite-element simulation software, applying electrostatic field and solid mechanics to establish the relationship between the stated voltage and the amount of deformation through the piezoelectric effect. In the finite-element model, the diameter and thickness of the mirror are set to be 50 cm and 500 μm, respectively; the diameter of the piezoelectric material is 46 cm, the thickness of the piezoelectric material is 300 μm, the electrode material is silver with a thickness of about 1–3 μm, and a free tetrahedral mesh is delineated for the overall structure. The material parameters of the piezoelectric deformable mirror are shown in Table 1:

Table 1. Deformable mirror material parameters.

	Unit	**Si**	**PZT-5H**	**Ag**
Young's modulus	[E/GPa]	170	68	83
Poisson's ratio	[v]	0.28	0.3	0.37
Piezoelectric coefficient	[d31/(pm·V^{-1})]	-	−450	-

4. Results

This section shows the results of applying various different constraints to the piezoelectric deformable mirror and simulating it in finite-element software. It contains the amount of mirror deformation, the influence function curve, the cross-linking value, and the linearity index between input and output.

First, we place a fixed constraint on the edge of the mirror and then apply 50 V to all electrodes simultaneously. The displacement results are shown in Figure 8. It can be seen that at any cutting depth, the maximum value of the entire mirror displacement is located in the most central area of the mirror. This is due to the fixed constraint we place on the specular edge, so the position variant closer to the edge is smaller. From the overall point of view, the maximum deformation of the mirror gradually decreases as the cutting depth deepens. This phenomenon occurs because the volume of the piezoelectric material decreases, and the charge generated by the inverse piezoelectric effect decreases.

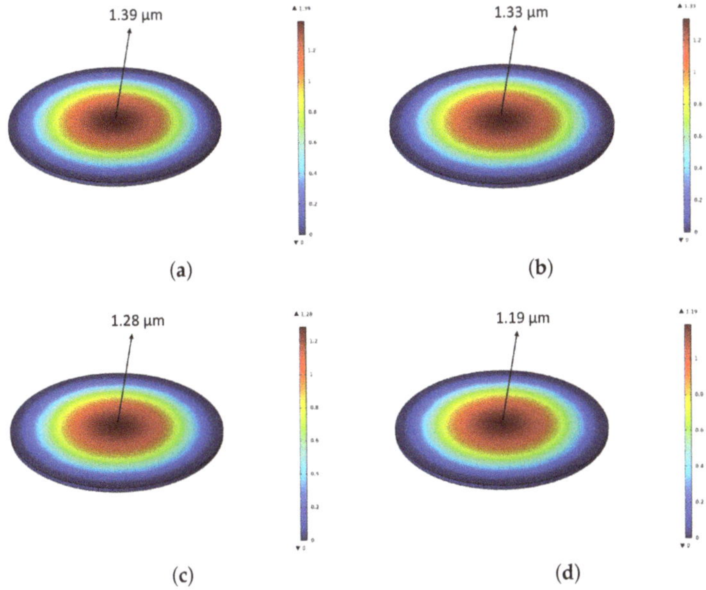

Figure 8. Figures of simulated deformation of all electrodes electrified with 50 V simulated deformable mirror. (**a**) Figure is the simulation of deformable mirror shape variable of 0 μm depth of cut. (**b**) Figure is the simulation of deformable mirror shape variable of 37.5 μm depth of cut. (**c**) Figure is the simulation of deformable mirror shape variable of 75 μm depth of cut. (**d**) Figure is the simulation of deformable mirror shape variable of 150 μm depth of cut.

We applied a voltage value of 50 V to each of the individual electrodes by applying them in the order of the electrode serial number and taking the center point of each positive hexagonal electrode as the coordinates for counting the displacement. Using the mirror surface as the zero reference surface, the resulting histogram of the shape variable of each electrode region is obtained, as shown in Figure 9.

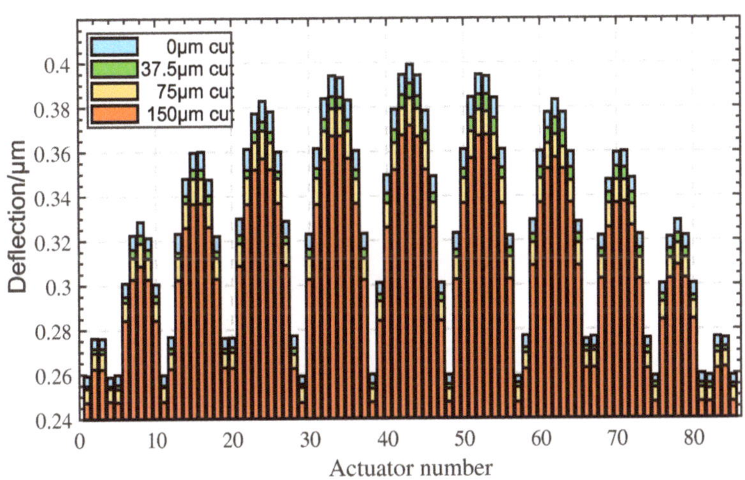

Figure 9. Figure of the amount of deformation produced at the center of a single electrode when a voltage of 50 V is applied to that electrode, with the actuator number in the horizontal coordinate and the amount of deflection in the vertical coordinate.

The overall graph of deflection shows a sinusoidal-like function curve, with the overall trend increasing the closer we become to electrode 43. The reason for this phenomenon is that the closer electrode No. 43 is, the closer its position on the piezoelectric material is to the center, and the less the effect of the edge fixation constraint is applied. In the process of converging to electrode No. 43, there will also be the image of a small sinusoidal wave. For examples, see No. 0–4, No. 5–11, No. 12–19. . . . The reason is the same as above: there is a significant decrease in the deflection at the position close to the edge.

In the structural design of the piezoelectric deformable mirror, the electrodes are placed in a circular array. Therefore, when plotting the influence function curves to infer the wavefront characteristics of the entire surface shape, we selected five electrodes on the circumferential diameter. However, electrode No. 47 is close to the edge, and, within the deformable mirror with a small diameter, the error is large, so it is rounded off. The remaining four electrodes can be used as representatives of the rest of the electrodes, and, with them, the wavefront phase of the whole mirror can be fitted by a specific algorithm. So, No. 43, 44, 45, and 46 are chosen as the main research objects of this paper. The detailed description is in Figure 10.

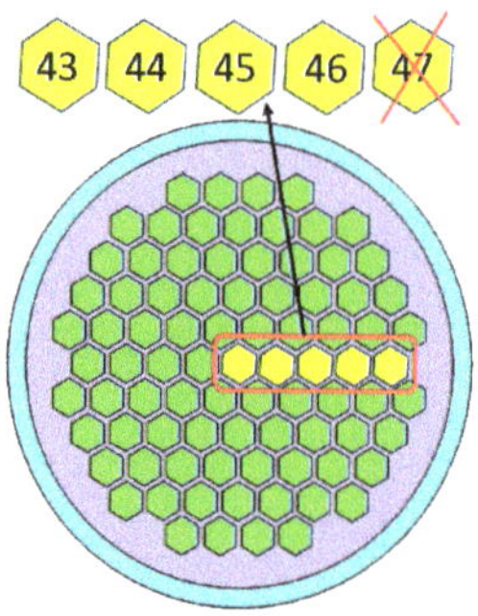

Figure 10. Figure of illustration on how to select electrodes for plotting the influence function.

The results were obtained by applying 50 V to each of these four electrodes individually in the finite-element software and then extracting the cross-section data. Figure 11 shows the cross-section lines of the actuator influence function based on the four design options (corresponding to the numbering in Figure 1b). Electrodes No. 43, 44, 45, and 46 are analyzed and numbered from the center region to the boundary regions where the actuator can locally deform when a voltage is applied. The difference in the deformation of each actuator is mainly affected by the electrode position, in which the deformable mirror of the centermost actuator No. 43 has a maximum deformation of 0.399 μm without cutting, and 0.371 μm with a cutting amount of 150 μm, while the deformable mirror of the outermost actuator No. 46 has a deformation of 0.349 μm without cutting within the effective aperture diameter and 0.349 μm without cutting. The maximum deformation of the deformable mirror surface of the outermost actuator No. 46 was 0.349 μm when no cutting was performed in the effective aperture, and the maximum deformation of the deformable mirror surface was 0.326 μm when the cutting amount was 150 μm; the relationship between the maximum deformation and the number of electrodes is shown in Table 2. As can be seen from Figure 11, the peak value of the influence function curve of each actuator decreases with the increase in the cutting depth. When the cutting depth is the same, the peak value of the influence function is larger and closer to the center position. The cross-linking values of the deformable mirrors reflect the influence of adjacent actuators, which will be analyzed later.

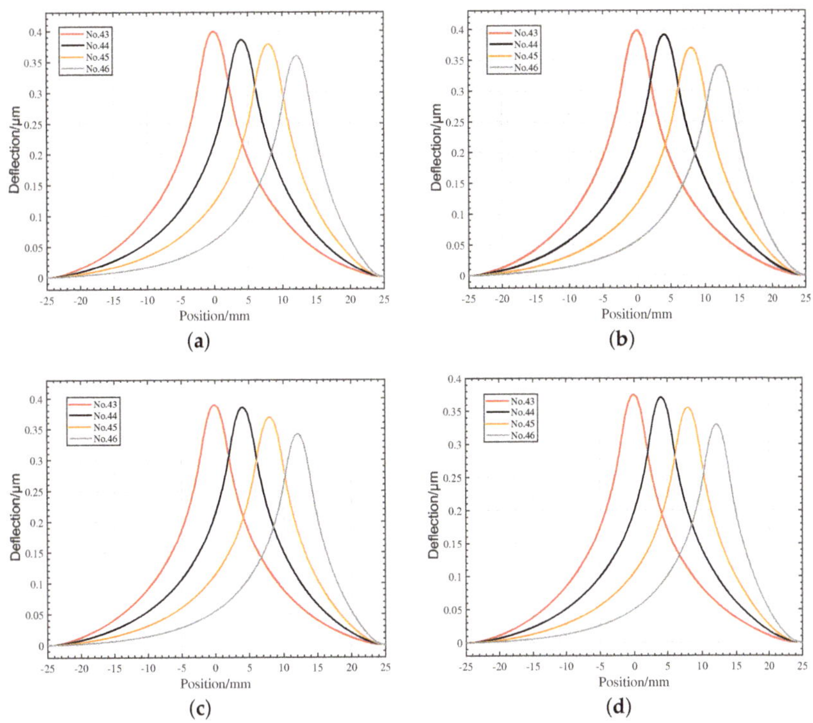

Figure 11. Figures of the influence function curves for specific electrodes at different depths of cut: (**a**) 0 μm depth of cut; (**b**) 37.5 μm depth of cut; (**c**) 75 μm depth of cut; (**d**) 150 μm depth of cut.

Table 2. Maximum deformation of electrodes No. 43, 44, 45, and 46 analyzed by influence function curves for four cutting depths.

	0 μm	37.5 μm	75 μm	150 μm
No. 43	0.399	0.390	0.383	0.371
No. 44	0.394	0.385	0.379	0.367
No. 45	0.379	0.369	0.363	0.351
No. 46	0.349	0.341	0.336	0.326

The maximum deformation versus the number of electrodes is shown in Table 2, which lists the maximum deformation induced by electrodes No. 43, 44, 45, and 46 at an on-voltage of 50 V when electrode gap cuttings of 0 μm, 37.5 μm, 75 μm, and 150 μm were performed. Taking electrode No. 43 as an example, the maximum deformation shows a decreasing trend when the depth of the cut is gradually increased from 0.399 μm, 0.390 μm, and 0.383 μm to the final 0.371 μm. It can be concluded that the maximum deformation decreases with the increase in the depth of cut.

In this section, the cross-linking values are also calculated and analyzed, and the results are shown in Figure 12. It can be concluded that a deeper cutting depth makes the cross-linking values decrease. This is analyzed with the most minor cutting depth of 150 μm, where the smallest cross-linking value is 35.14% at electrode No. 1. In contrast, the largest cross-linking value is 52.84% at electrode No. 43 (the most central position). An overall trend can also be seen in that, for a certain depth of cut, and the closer the electrode position is to the central region, the greater the cross-linking value.

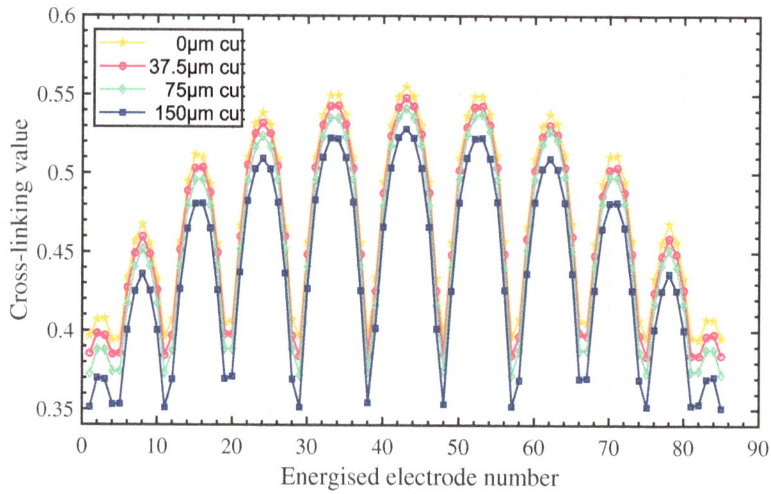

Figure 12. Cross-linking value of deformable mirror.

It can be seen in Figure 12 that the general pattern is very similar to Figure 9. The reason is also the same, it is due to the position of the electrode rows that lead to this result. We can see from the figure that cutting only the electrode gap can lead to a significant decrease in the cross-linking value. Therefore, cutting the electrode gap should be considered to achieve the purpose of obtaining the best cross-linking value when designing the deformation mirror when the electrode shape and electrode distribution cannot be changed.

5. Conclusions

In this paper, based on 85-unit piezoelectric deformable mirrors, four electrode gap cutting depths were designed. The advantages and disadvantages of different electrode modes were compared, and the effects of the piezoelectric ceramic actuator gap cutting depths on the distribution of the actuator displacement capability and cross-linking values were investigated. The results show that the displacement capability and cross-link value decrease with the increase in the cutting depth and increase with the decrease in the depth. In general, the linear relationship between the input voltage and output deformation is better, and the response is faster when the cross-linking value decreases. Therefore, the cutting piezoelectric deformable mirror can also meet the demand in terms of its compensation ability. The deformable mirror model established by the finite-element method can effectively predict the effect of the cutting depth on the performance of the deformable mirror's electrode gap region and give the magnitude of the actuator displacement and cross-linking value under the simulation conditions, which are guides to the manufacturing process and the use environment of actual piezoelectric deformable mirrors and lay the foundation for the design and manufacture of more-unit deformable mirrors. In the future, we plan to consider deformable mirrors with more electrode shapes, divide the depth level of the cut into more detailed levels, and search for literature to use different cutting methods. Moreover, we will search for a better optical bonding adhesive to adhere piezoelectric materials and mirrors and, subsequently, plan to compensate the piezoelectric hysteresis by control means to further enhance the performance of deformable mirrors.

Author Contributions: Conceptualization, B.L. and D.T.; methodology, B.L. and D.T.; software, B.L.; validation, B.L. and D.T.; formal analysis, B.L.; investigation, B.L. and D.T.; resources, B.L. and D.T.; data curation, B.L.; writing—original draft preparation, B.L.; writing—review and editing, B.L. and D.T.; visualization, D.T.; supervision, D.T.; project administration, B.L. and D.T.; funding acquisition, D.T. All authors have read and agreed to the published version of the manuscript.

Funding: This research was supported by the CAS Project for Young Scientists in Basic Research (YSBR-066); The Innovation Fund Project of the National Defense Science and Technology of the Chinese Academy of Sciences (CXJJ-21S014); the National Science Foundation of China (T2122001); The Youth Innovation Promotion Association of the Chinese Academy of Sciences (2023230); and Capital construction funds within the budget of Jilin Province (2023C030-6).

Institutional Review Board Statement: Not applicable.

Informed Consent Statement: Not applicable.

Data Availability Statement: The data that support the findings of this study are available from the corresponding author upon reasonable request.

Acknowledgments: The authors thank Jian Chen for his supervisory work on this paper. The authors thank the reviewers for their valuable suggestions.

Conflicts of Interest: The authors declare no conflict of interest.

References

1. Nie, G.; Bodda, S.; Sandhu, H.; Han, K.; Gupta, A. Computer-Vision-Based Vibration Tracking Using a Digital Camera: A Sparse-Optical-Flow-Based Target Tracking Method. *Sensors* **2022**, *22*, 6869. [CrossRef] [PubMed]
2. Mai, V.; Kim, H. Beaconless angle-of-arrival tracking with improved receiver sensitivity and tracking precision for free-space optical communications. *Opt. Commun.* **2023**, *527*, 128963. [CrossRef]
3. Hippler, S. Adaptive optics for extremely large telescopes. *J. Astron. Instrum.* **2019**, *8*, 1950001. [CrossRef]
4. Hardy, J.W. *Adaptive Optics for Astronomical Telescopes*; Oxford Optical and Imaging Sciences: Oxford, UK, 1998; Volume 16.
5. Tyson, R. *Topics in Adaptive Optics*; BoD–Books on Demand: Norderstedt, Germany, 2012.
6. Ellerbroek, B.; Vogel, C. Simulations of closed-loop wavefront reconstruction for multiconjugate adaptive optics on giant telescopes. *Astron. Adapt. Opt. Syst. Appl.* **2003**, *5169*, 206–217.
7. Bifano, T. MEMS deformable mirrors. *Nat. Photonics* **2011**, *5*, 21–23. [CrossRef]
8. Toporovsky, V.; Kudryashov, A.; Samarkin, V.; Panich, A.; Sokallo, A.; Malykhin, A.; Skrylev, A.; Sheldakova, J. Small-aperture stacked-array deformable mirror made of the piezoceramic combs. In Proceedings of the Laser Resonators, Microresonators, and Beam Control XXIII, Online, 6–12 March 2021; Volume 11672, pp. 183–188.
9. Lukin, V.; Antoshkin, L.; Bolbasova, L.; Botygina, N.; Emaleev, O.; Kanev, F.; Konyaev, P.; Kopylov, E.; Lavrinov, V.; Lavrinova, L.; et al. The history of the development and genesis of works on adaptive optics at the Institute of atmospheric optics. *Atmos. Ocean. Opt.* **2020**, *33*, 85–103. [CrossRef]
10. Kudryashov, A.; Rukosuev, A.; Nikitin, A.; Galaktionov, I.; Sheldakova, J. Real-time 1.5 kHz adaptive optical system to correct for atmospheric turbulence. *Opt. Express* **2020**, *28*, 37546–37552. [CrossRef] [PubMed]
11. Han, X.; Ma, J.; Bao, K.; Cui, Y.; Chu, J. Piezoelectric deformable mirror driven by unimorph actuator arrays on multi-spatial layers. *Opt. Express* **2023**, *31*, 13374–13383. [CrossRef] [PubMed]
12. Samarkin, V.; Sheldakova, J.; Toporovsky, V.; Rukosuev, A.; Kudryashov, A. High-spatial resolution stacked-actuator deformable mirror for correction of atmospheric wavefront aberrations. *Appl. Opt.* **2021**, *60*, 6719–6724. [CrossRef] [PubMed]
13. Toporovsky, V.; Kudryashov, A.; Samarkin, V.; Rukosuev, A.; Nikitin, A.; Sheldakova, Y.; Otrubyannikova, O. Cooled stacked-actuator deformable mirror for compensation for phase fluctuations in a turbulent atmosphere. *Atmos. Ocean. Opt.* **2020**, *33*, 584–590. [CrossRef]
14. Hall, S.; Regis, J.; Renteria, A.; Chavez, L.; Delfin, L.; Vargas, S.; Haberman, M.; Espalin, D.; Wicker, R.; Lin, Y. Paste extrusion 3D printing and characterization of lead zirconate titanate piezoelectric ceramics. *Ceram. Int.* **2021**, *47*, 22042–22048. [CrossRef]
15. Hu, S.; Xu, B.; Zhang, X.; Hou, J.; Wu, J.; Jiang, W. Double-deformable-mirror adaptive optics system for phase compensation. *Appl. Opt.* **2006**, *45*, 2638–2642. [CrossRef] [PubMed]
16. Toporovsky, V.; Kudryashov, A.; Skvortsov, A.; Rukosuev, A.; Samarkin, V.; Galaktionov, I. State-of-the-Art Technologies in Piezoelectric Deformable Mirror Design. *Photonics* **2022**, *9*, 321. [CrossRef]
17. Rousset, G.; Lacombe, F.; Puget, P.; Hubin, N.; Gendron, E.; Fusco, T.; Arsenault, R.; Charton, J.; Feautrier, P.; Gigan, P.; et al. NAOS—The first AO system of the VLT: On-sky performance. In Proceedings of the Adaptive Optical System Technologies II, Waikoloa, HI, USA, 22–28 August 2002; Volume 4839, pp. 140–149.
18. Xie, L.; Zhong, Z.; Zhang, B. Influence of partially failed actuators of power failure on correction ability of deformable mirrors. *Acta Opt. Sin.* **2021**, *41*, 0223001. [CrossRef]
19. Furieri, T.; Bonora, S. Large field of view wavefront correction with deformable lenses. In Proceedings of the Free-Space Laser Communications XXXV, San Francisco, CA, USA, 28 January–3 February 2023; Volume 12413, pp. 323–327.
20. Jiang, L.; Hu, L.; Hu, Q.; Xu, X.; Wu, J.; Yu, L.; Huang, Y. Sixty-Nine-Element Voice Coil Deformable Mirror for Visible Light Communication. *Photonics* **2023**, *10*, 322. [CrossRef]
21. Zhong, G.; Han, X.; Zhang, X.; Lou, J.; Ma, J. Fabrication and Characterization of Deformable Mirror Driven by Piezoelectric Unimorph Actuator Array. *Acta Photonica Sin.* **2022**, *51*, 0151125.

22. Sheldakova, J.; Toporovsky, V.; Galaktionov, I.; Nikitin, A.; Rukosuev, A.; Samarkin, V.; Kudryashov, A. Flat-top beam formation with miniature bimorph deformable mirror. In Proceedings of the Laser Beam Shaping XX, Online, 24 August–4 September 2020; Electr Network, AUG 24-SEP 04, Volume 11486.
23. Galaktionov, I.; Sheldakova, J.; Toporovsky, V.; Kudryashov, A. Numerical analysis of efficiency of light focusing through a moderately scattering medium with the use of deformable mirrors. In Proceedings of the Laser Beam Shaping XXII, San Diego, CA, USA, 21–26 August 2022; Volume 12218.
24. Gowda, H.; Wallrabe, U.; Wapler, M. Higher order wavefront correction and axial scanning in a single fast and compact piezo-driven adaptive lens. *Opt. Express* **2023**, *31*, 23393–23405. [CrossRef] [PubMed]
25. Greenwood, D. Bandwidth specification for adaptive optics systems. *J. Opt. Soc. Am.* **1977**, *67*, 390–393. [CrossRef]

photonics

Communication

Disturbance-Observer-Based LQR Tracking Control for Electro-Optical System

Chao Liu [1,2,3,4], Yao Mao [1,2,3,*] and Xiaoxia Qiu [1,2]

1 Institute of Optics and Electronics, Chinese Academy of Sciences, Chengdu 610209, China; liuchao191@mails.ucas.ac.cn (C.L.); qiuxiaoxia19@mails.ucas.ac.cn (X.Q.)
2 University of Chinese Academy of Sciences, Beijing 100049, China
3 Key Laboratory of Optical Engineering, Chinese Academy of Sciences, Chengdu 610209, China
4 The School of Electronics, Electrical and Communication Engineering, Beijing 101408, China
* Correspondence: maoyao@ioe.ac.cn

Abstract: To improve the dynamic property and the disturbance suppression ability of an electro-optical tracking system, this paper presents a disturbance-observer-based LQR tracking control method. The disturbance-observer-based robust controller is composed of three parts: one is the LQR tracking controller, one is the reference model controller and the other is a compensatory controller designed with the output of the disturbance observer. The uncertainty and disturbances are considered in the controller design. By Lyapunov stability theory and linear matrix inequality (LMI) technique, the sufficient conditions for observer gain and controller gain of the tracking reference model of the electro-optical system are given. Simulation and experimental results show that the proposed method in this paper not only improved the disturbance suppression ability of the electro-optical tracking system but also improved the dynamic property of the electro-optical tracking system, such as rise time, settling time and system overshoot. Specially, compared with other methods in this paper, the tracking accuracy and the disturbance suppression ability of the proposed method are about two to three times higher. The method presented in this paper has important reference value in the field of electro-optical system applications. But, with the development of electro-optical system applications, the tracking accuracy and disturbance suppression ability of the proposed method cannot meet the actual requirements of an electro-optical system. The next step of this paper will consider a variety of practical requirements, such as the controller saturation problem and tracking reference target with strong maneuverability, and further optimize the proposed method.

Keywords: disturbance-observer-based control (DOBC); LQR; LMI technique; Lyapunov stability theory; electro-optical tracking platform

Citation: Liu, C.; Mao, Y.; Qiu, X. Disturbance-Observer-Based LQR Tracking Control for Electro-Optical System. *Photonics* **2023**, *10*, 900. https://doi.org/10.3390/photonics10080900

Received: 27 June 2023
Revised: 21 July 2023
Accepted: 27 July 2023
Published: 3 August 2023

1. Introduction

The electro-optical tracking platform is a complex and high-precision directional tracking system integrating optical, mechanical and electrical properties. It is widely used in long-distance laser communication, quantum communication, inertial measurement unit and other fields [1–4]. The electro-optical tracking platform is mainly used to realize real-time precision tracking and measuring of moving targets. However, it is often affected by external disturbances and internal uncertainties in engineering control applications. These disturbances seriously affect the stability performance and control effect of the system and may even cause instability of the closed-loop system. Therefore, many researchers are devoted to dealing with the disturbance and internal uncertainty of electro-optical tracking systems [5–7]. In general, the aforementioned disturbance suppression methods of electro-optical tracking platforms can be classified into the following two categories. The first category is a multiloop feedback control system composed of high-sampling-rate inertial sensors, such as the micro-electro-mechanical system (MEMS) accelerometers, fiber optical gyroscopes (FOG) and high-resolution position detectors. The disturbance suppression capabilities of the multiloop feedback control system is the superposition of the effects of each loop, but this method is insufficient for disturbance suppression capacity or dealing with internal uncertainty, and it can only provide basic disturbance suppression [8,9]. More seriously, when suffering from strong disturbances, the controlled variables might have too large fluctuations, which could even lead to instability of the closed-loop system. The second category mainly uses the direct feedforward method based on measurement to suppress disturbance. This method requires accurate identification of disturbance transfer characteristics outside the system. However, it is hard or even impossible to measure the disturbances in many actual processes, including the inertial uncertainty. Therefore, it is of practical interest to improve the disturbance rejection ability of the stable control platform to be able to observe and compensate for the disturbance source [10].

Based on the above situation and to further improve the disturbance suppression ability of the system, disturbance-observer-based control (DOBC) is introduced into the electro-optical tracking system in this paper. And this method, based on DOBC, does not require accurate model information [11,12]. In practical applications, the electro-optical tracking system requires motion tracking of the position, velocity or acceleration curve of a given time series with a certain precision. Meanwhile, the electro-optical tracking system must also meet certain control performance indicators, such as minimum tracking time and minimum cost. In this way, the system can track the specified trajectory faster, more accurately and more effectively. As we know, there is little research on the optimal tracking control of electro-optical tracking systems subject to external disturbances. At present, various optimal control methods are popular in the control field, including linear quadratic regulator optimal control (LQR), adaptive dynamic programming control [13,14], etc., to achieve ideal dynamic and steady-state performance.

The LQR is a well-known design technique in modern optimal control theory and has been widely used in many applications [15,16]. In contrast with pole placement, the desired performance objectives are directly addressed by minimizing a quadratic function of the state and control input. The resulting optimal control law has many excellent properties, including closed-loop stability. Furthermore, the trade-off between state regulatory requirements and control energy consumption in the LQR can be controlled by choosing the weighting matrices $\boldsymbol{Q}$ and $\boldsymbol{R}$ [17–19]. However, the solution to the LQR problem depends on solving the Riccati equation. Before solving the Riccati equation, designers often need to determine some undetermined parameters in advance. The selection of these parameters will not only affect the quality of the conclusion but also affect the solvability of the problem, which brings great conservatism to the solution of the problem. Meanwhile, there are still some problems in solving the Riccati equation itself. At present, there are many methods for solving the Riccati equation, but most of them are iterative methods, and the convergence of these methods cannot be guaranteed.

In view of the above problems, linear matrix inequality (LMI) technology can be well solved [20,21]. One advantage of using LMI is that it makes it easy to include other specifications for controller design [22,23]. Therefore, various design specifications can be rewritten into the LMI, and the resulting LMI constraints can be efficiently solved using newly developed convex optimization algorithms.

In this paper, a LQR-DOB tracking control method to achieve the optimal tracking of the desired trajectory under the condition of modeling error and uncertain disturbance is proposed. In summary, the contribution of this paper is as follows:

1. This paper proposes the LQR-DOB tracking control method, which solves the uncertainty of the model and the instability of the system caused by uncertain disturbance;
2. Using standard techniques, the DOB gain and LQR controller gain of the tracking reference model design is reduced to a convex constraint problem, which can be efficiently solved with the LMI approach;
3. The stability constraint of the electro-optical tracking closed-loop system is considered by using Lyapunov theory in the LMI framework;

4. Compared with other control methods, the disturbance suppression ability and dynamic response performance of the system, such as rise time, settling time and system overshoot, have been significantly improved under the proposed method.

The rest of this paper is organized as follows. In Section 2, the electro-optical tracking platform is modeled. In Section 3, the LQR-DOB tracking controller is designed and analyzed. In Section 4, the simulated and experimental results are presented. In Section 5, the direction of future work is pointed out. Finally, Section 6 concludes the paper.

2. Modeling of The Electro-Optical Tracking Platform

The main structure of the electro-optical tracking stable platform is shown in Figure 1a. A detector such as PSD receives the beacon of light reflected by the tip-tilt mirror and sends the position error signal to the controller. The controller calculates the correction angle of the mirror, and then through the D/A converter, the output of the controller drives the motors connected to the mirror. The aim is to stabilize the light at the center of the detector by rapidly deflecting the mirror under the influence of the disturbance.

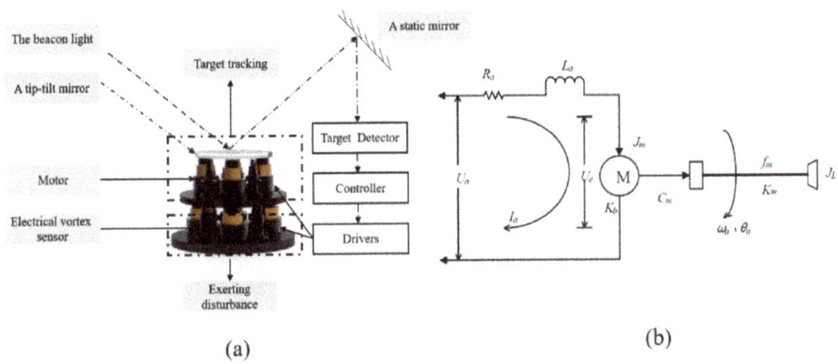

Figure 1. (**a**) The schematic of the electro-optical tracking system. (**b**) The physical model structure of the plant.

Mathematical modeling is the foundation of control. In Figure 1b, using the potential plus the torque balance equation, we obtain

$$\begin{cases} U_a = R_a I_a(s) + L_a s I_a(s) + K_b s \theta_a(s) \\ C_m I_a = (J_L s^2 + f_m s + K_m)\theta_a(s) \end{cases}, \tag{1}$$

where $U_a, I_a, R_a, L_a, K_b, C_m, f_m, K_m$ are the motor voltage, current, resistance, inductor, back electromotive force coefficient, torque coefficient, viscous friction and spring stiffness, respectively. Meanwhile, J_L, θ_a are the load inertia and the relative position angle of the motor-driven tilt mirror, respectively. Then, the controlled system plant can be modeled as

$$G(s) = \frac{\theta_a(s)}{U_a(s)} = \frac{C_m}{(J_L s^2 + f_m s + K_m)(L_a s + R_a) + K_b C_m s}. \tag{2}$$

Moreover, it can also be factorized with the typical resonance element and inertia element, which is

$$G(s) = \frac{\theta_a(s)}{U_a(s)} = \frac{K}{(s^2 + as + b)} \frac{1}{(Ts + 1)}, \tag{3}$$

where $a = 2\zeta_{ol}\omega_{ol}, b = \omega_{ol}^2$. ζ_{ol}, ω_{ol} are the damping ratio and natural frequency of the open-loop system, respectively. K is the system open-loop gain. And T is the parasitic time constant.

Since the inertia element in the controlled plant only affects the characteristics of the high-frequency part of the electro-optical tracking platform, the frequency characteristics from the voltage input U_a to the angle output θ_a can be approximated to a typical resonance element. Therefore, the general form of the controlled system object for low and intermediate frequencies can be expressed as

$$G(s) = \frac{\theta_a(s)}{U_a(s)} = \frac{K}{s^2 + as + b}, \tag{4}$$

where the meanings of a, b and K are consistent with those in Equation (3). Convert the controlled system object in Equation (4) into state-space equation form as

$$\begin{cases} \begin{bmatrix} \dot{x}_1(t) & \dot{x}_2(t) \end{bmatrix}^T = \boldsymbol{A_1}\begin{bmatrix} x_1(t) & x_2(t) \end{bmatrix}^T + \boldsymbol{B_1}u(t) \\ y = \boldsymbol{C_1}\begin{bmatrix} x_1(t) & x_2(t) \end{bmatrix}^T + \boldsymbol{D_1}u(t) \end{cases}, \tag{5}$$

where $\boldsymbol{A_1} = \begin{bmatrix} 0 & 1 \\ -b & -a \end{bmatrix}, \boldsymbol{B_1} = \begin{bmatrix} 0 \\ K \end{bmatrix}, \boldsymbol{C_1} = \begin{bmatrix} 1 & 0 \end{bmatrix}, \boldsymbol{D_1} = 0, x_1(t) = y, x_2(t) = v$; y, v represent the position and speed of the system, respectively. However, in the actual working environment, the electro-optical tracking platform will not only be affected by external interference but its characteristics will also change with the change in attitude and load. Therefore, the electro-optical tracking system in Equation (5) can be converted into

$$\begin{cases} \dot{\boldsymbol{x}}(\boldsymbol{t}) = (\boldsymbol{A_1} + \boldsymbol{\Delta A_1})\boldsymbol{x}(\boldsymbol{t}) + \boldsymbol{B_1}[u(t) + d(t)] + \boldsymbol{D_1}w(t) \\ \boldsymbol{z}(\boldsymbol{t}) = \boldsymbol{C_1}\boldsymbol{x}(\boldsymbol{t}) \end{cases}, \tag{6}$$

where $\boldsymbol{x}(\boldsymbol{t})$ denotes system state variable; $u(t)$ stands for the control input; $\boldsymbol{z}(\boldsymbol{t})$ is the controlled output; $\boldsymbol{\Delta A_1}$ denotes the parameter uncertainty; and $d(t)$ and $w(t)$ are the disturbances, where $w(t)$ is square integrable on $[\ 0, \ \ +\infty\ )$.

3. The LQR-DOB Tracking Controller

In this section, the LMI-LQR-DOB tracking controller is designed for the electro-optical tracking system with uncertainty and disturbance. The main objective of this work is to design a controller ensuring that the electro-optical tracking system can track the reference signal generated by the following model

$$\begin{cases} \dot{\boldsymbol{x}}_{\boldsymbol{r}}(\boldsymbol{t}) = \boldsymbol{A_r}\boldsymbol{x_r}(\boldsymbol{t}) + \boldsymbol{B_r}r(t) \\ z_r(\boldsymbol{t}) = \boldsymbol{C_1}\boldsymbol{x_r}(\boldsymbol{t}) \end{cases}, \tag{7}$$

where $\boldsymbol{x_r}(\boldsymbol{t})$ denotes the state vector of the reference system, and $r(t)$ is the bounded reference input. $\boldsymbol{A_r}$, $\boldsymbol{B_r}$, $\boldsymbol{C_1}$ are known constant matrices, and $\boldsymbol{A_r}$ is Hurwitz.

The following assumptions, lemmas and definition are adopted throughout this work.

Assumption 1 ([24]). *The system satisfies the controllable and observable condition, that is,* $(\boldsymbol{A_1}, \boldsymbol{B_1})$ *is controllable and* $(\boldsymbol{A_1}, \sqrt{\boldsymbol{Q}})$ *is observable.*

Assumption 2 ([24]). *There exist two matrices* $\boldsymbol{K_1}$,$\boldsymbol{K_2}$ *such that* $\boldsymbol{A_r}$, $\boldsymbol{B_r}$ *in the reference model in Equation (7) satisfies* $\boldsymbol{A_r} = \boldsymbol{A_1} + \boldsymbol{B_1}\boldsymbol{K_1}$ *and* $\boldsymbol{B_r} = \boldsymbol{B_1}\boldsymbol{K_2}$.

Assumption 3 ([25]). *The uncertainty* ΔA_1 *can be expressed as* $\Delta A_1 = D_2F_1(t)E_1$, *where* D_2, E_1 *are known constant matrices, and* $F_1(t)$ *is an unknown matrix satisfying* $\|F_1(t)\| \leq 1$.

Lemma 1 ([26]). *Assume that* $\boldsymbol{X}$ *and* $\boldsymbol{Y}$ *are vectors or matrices with appropriate dimension. The following inequality*

$$\boldsymbol{X^TY} + \boldsymbol{Y^TX} \leq \alpha\boldsymbol{X^TX} + \alpha^{-1}\boldsymbol{Y^TY}, \tag{8}$$

holds for any constant $\alpha > 0$.

Lemma 2 ([26]). *Assume that $\mathbf{H_1}$ and $\mathbf{H_2}$ are symmetric matrices, $\mathbf{S_1}$ and $\mathbf{S_2}$ are vectors or matrices with appropriate dimension and $\mathbf{F^T F} \leq \mathbf{I}$. The following inequality*

$$\begin{bmatrix} H_1 & S_1^T F S_2 \\ * & H_2 \end{bmatrix} \leq \begin{bmatrix} H_1 + \varepsilon S_1^T S_1 & 0 \\ 0 & H_2 + \frac{1}{\varepsilon} S_2^T S_2 \end{bmatrix}, \tag{9}$$

holds for any constant $\varepsilon > 0$.

Notations: The symmetric term is denoted as $*$, i.e.,$\begin{bmatrix} X & Y \\ Y^T & Z \end{bmatrix} = \begin{bmatrix} X & Y \\ * & Z \end{bmatrix}$.

Proof. Premultiplying and postmultiplying simultaneously by $\begin{pmatrix} x_1 & x_2 \end{pmatrix}$ and $\begin{pmatrix} x_1 & x_2 \end{pmatrix}^T$ with $\begin{bmatrix} H_1 & S_1^T F S_2 \\ * & H_2 \end{bmatrix}$ yields

$$x_1^T(t) H_1 x_1(t) + x_1^T(t) S_1^T F S_2 x_2(t) + x_2^T(t) S_2^T F^T S_1 x_1(t) + x_2^T(t) H_2 x_2(t). \tag{10}$$

□

According to Lemma 2 and Equation (10), for $\mathbf{F^T F} \leq \mathbf{I}$, we have

$$\begin{array}{l} x_1^T(t) H_1 x_1(t) + x_1^T(t) S_1^T F S_2 x_2(t) + x_2^T(t) S_2^T F^T S_1 x_1(t) + x_2^T(t) H_2 x_2(t) \\ \leq x_1^T(t)(H_1 + \alpha S_1^T S_1) x_1(t) + x_2^T(t)(H_2 + \alpha^{-1} S_2^T S_2) x_2(t) \\ = \left(x_1^T(t) x_2^T(t)\right) \begin{pmatrix} H_1 + \alpha S_1^T S_1 & 0 \\ 0 & H_2 + \alpha^{-1} S_2^T S_2 \end{pmatrix} \begin{pmatrix} x_1(t) \\ x_2(t) \end{pmatrix} \end{array}. \tag{11}$$

Combining Equations (10) and (11), we have Equation (9).

Lemma 3. *Schur complement [25]: For a given symmetric matrix $\mathbf{S} = \begin{bmatrix} S_{11} & S_{12} \\ S_{21} & S_{22} \end{bmatrix}$, where $\mathbf{S_{11}}$ is $r \times r$ dimensional and $\mathbf{S_{22}}$ is $(n-r) \times (n-r)$ dimensional. The following three conditions are equivalent:*

(i) $\mathbf{S < 0}$;

(ii) $\mathbf{S_{11} < 0, S_{22} - S_{12}^T S_{11}^{-1} S_{12} < 0}$;

(iii) $\mathbf{S_{22} < 0, S_{11} - S_{12} S_{22}^{-1} S_{12}^T < 0}$.

Defining the tracking error as $e_x(t) = x(t) - x_r(t)$ and invoking equations in Equations (6) and (7), we have

$$\dot{e}_x(t) = (\mathbf{A_1} + \mathbf{\Delta A_1})\mathbf{x(t)} + \mathbf{B_1}[u(t) + d(t)] - \mathbf{A_r x_r(t)} - \mathbf{B_r} r(t) + \mathbf{D_1} w(t). \tag{12}$$

The controller $u(t)$ in Figure 2 is designed as

$$u(t) = u_f(t) + u_l(t) - \hat{d}(t), \tag{13}$$

where $u_f(t)$ is the reference model matching controller, $u_l(t)$ is an LQR tracking controller and $\hat{d}(t)$ is the estimation of disturbance $d(t)$.

The reference model matching controller is given by

$$u_f(t) = K_1 x_r(t) + K_2 r(t), \tag{14}$$

where K_1, K_2 are the gain matrices satisfying Assumption 2. Substituting Equations (13) and (14) into Equation (12), we obtain

$$\dot{e}_x(t) = (\mathbf{A_1} + \mathbf{\Delta A_1}) e_x(t) + \mathbf{B_1 u_l} + \mathbf{B_1} e_d(t) + \mathbf{\Delta A_1 x_r(t)} + \mathbf{D_1} w(t), \tag{15}$$

where $e_d(t) = d(t) - \hat{d}(t)$ is the disturbance error vector.

The LQR tracking controller $u_l(t)$ is designed for the following error system:

$$\dot{e}_x(t) = (\boldsymbol{A_1} + \Delta \boldsymbol{A_1})e_x(t) + \boldsymbol{B_1}u_l. \tag{16}$$

For the error tracking system in Equation (16) above, we consider an auxiliary function as

$$J(t) = \int_0^{\infty} \left[e_x^{\mathrm{T}}(t)\left(\boldsymbol{Q} + \boldsymbol{K}^{\mathrm{T}}\boldsymbol{R}\boldsymbol{K}\right)e_x(t)\right]dt, \tag{17}$$

which is selected to design the LQR tracking controller $u_l(t)$, where $\boldsymbol{Q}$ is the semipositive definite state weighting matrix, $\boldsymbol{R}$ is the positive definite control weighting matrix and $\boldsymbol{K}$ is the gain of the LQR tracking controller.

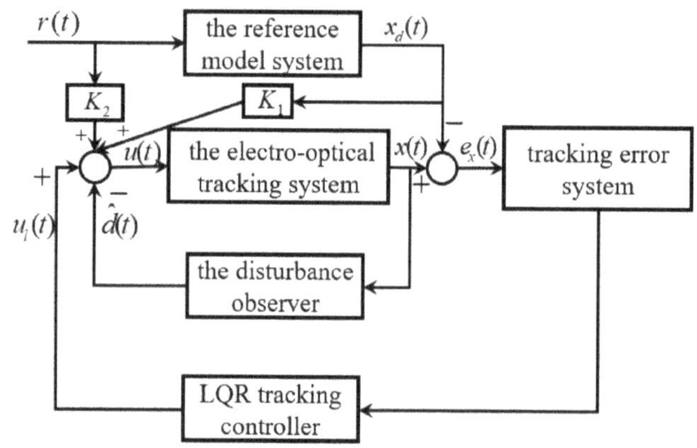

Figure 2. Block diagram of LQR-DOB tracking control.

The LQR tracking controller $u_l(t)$ is designed as

$$u_l(t) = \boldsymbol{K}e_x(t). \tag{18}$$

The above LQR tracking controller design problem can be expressed as the following optimization problem by LMI technology:

$$\min J = \int_0^{\infty} \left[e_x^{\mathrm{T}}(t)\left(\boldsymbol{Q} + \boldsymbol{K}^{\mathrm{T}}\boldsymbol{R}\boldsymbol{K}\right)e_x(t)\right]dt < \gamma, \tag{19}$$

where γ is the upper bound of the LQR performance index. Under the condition that Assumption 1 is satisfied, the above LQR tracking controller design problem is transformed into the following inequality relationship:

$$(\boldsymbol{A_1} + \Delta\boldsymbol{A_1})\boldsymbol{x} + \boldsymbol{x}(\boldsymbol{A_1} + \Delta\boldsymbol{A_1})^T + \boldsymbol{B_1}\boldsymbol{W} + (\boldsymbol{B_1}\boldsymbol{W})^{\mathrm{T}} + x_0x_0^{\mathrm{T}} < 0, \tag{20}$$

$$\mathrm{trace}(\sqrt{\boldsymbol{Q}}\boldsymbol{X}(\sqrt{\boldsymbol{Q}})^T) + \mathrm{trace}(\boldsymbol{Y}) < \gamma, \tag{21}$$

$$\begin{bmatrix} -\boldsymbol{Y} & \sqrt{\boldsymbol{R}}\boldsymbol{W} \\ (\sqrt{\boldsymbol{R}}\boldsymbol{W})^T & -\boldsymbol{X} \end{bmatrix} < 0, \tag{22}$$

where $\boldsymbol{X} \in S^n$, S^n is the set of symmetric positive definite matrices; $\boldsymbol{Y} \in S^r$, S^r is also the set of symmetric positive definite matrices; $\boldsymbol{W} \in R^{r\times n}$, $R^{r\times n}$ is the set of $r \times n$ matrices; and x_0 is the initial value of state variable x, and the trace operator is defined as $trace(\boldsymbol{S}) = \sum_{i=1}^{n} s_{ii}$ with $\boldsymbol{S} = (s_{ij})_{n\times n}$.

By substituting the uncertainty ΔA_1 in Assumption 3 into Equation (20) and using Lemma 3: (Schur Complement), we can further convert Equation (20) to

$$\begin{bmatrix} A_1X + XA_1^T + B_1W + (B_1W)^T + \alpha D_2D_2^T + x_0x_0^T & XE_1^T \\ * & -\alpha I \end{bmatrix} < 0. \tag{23}$$

Combining Equations (21)–(23), the gain of LQR tracking controller can be determined by setting

$$K = WX^{-1}. \tag{24}$$

Then, we design the disturbance observer as

$$\begin{cases} \hat{d}(t) = \sigma(t) + Lx(t) \\ \dot{\sigma}(t) = -L[A_1x(t) + B_1u(t) + B_1\hat{d}(t)] \end{cases}, \tag{25}$$

where $\hat{d}(t)$ is the estimation of $d(t)$, $\sigma(t)$ denotes the auxiliary variable of the designed observer and L is the disturbance observer gain. The disturbance error system has the following form:

$$\begin{aligned} \dot{e}_d(t) &= \dot{d}(t) - \dot{\sigma}(t) - L\dot{x}(t) = -LB_1e_d(t) - L\Delta A_1x(t) - LD_1w(t) + \dot{d}(t) \\ &= -LB_1e_d(t) - L\Delta A_1e_x(t) - LD_1w(t) + \dot{d}(t) - L\Delta A_1x_r(t). \end{aligned}, \tag{26}$$

Combining Equations (7), (12) and (19), we have

$$\begin{cases} \dot{e}(t) = A_{e1}e(t) + D_{e1}w_1(t) \\ e_e(t) = C_{e1}e(t) \end{cases}, \tag{27}$$

where $e^T(t) = \begin{bmatrix} e_x^T(t) & e_d^T(t) & x_r^T(t) \end{bmatrix}, w_1^T(t) = \begin{bmatrix} w^T(t) & \dot{d}^T(t) & r^T(t) \end{bmatrix}$, $A_{e1} = \begin{bmatrix} A_1 + B_1K + \Delta A_1 & B_1 & \Delta A_1 \\ -L\Delta A_1 & -LB_1 & -L\Delta A_1 \\ 0 & 0 & A_r \end{bmatrix}, D_{e1} = \begin{bmatrix} D & 0 & 0 \\ -LD_1 & I & 0 \\ 0 & 0 & B_r \end{bmatrix}$, $C_{e1} = \begin{bmatrix} C_1 & 0 & C_1 \end{bmatrix}$.

Now, a Lyapunov function is chosen as

$$V(t) = e^T(t)\tilde{P}e(t), \tag{28}$$

where $\tilde{P} = diag(P_1, P_2, P_3)$ with $P_i > 0 (i = 1, 2, 3)$. The derivative of $V(t)$ along the closed-loop system Equation (27) is

$$\dot{V}(t) = e^T(t)\left(\tilde{P}A_{e1} + {A_{e1}}^T\tilde{P}\right)e(t) + e^T(t)\tilde{P}D_{e1}\omega_1(t) + \omega_1^T(t){D_{e1}}^T\tilde{P}e(t). \tag{29}$$

Then, introducing the auxiliary function as

$$J_1(t) = V(t) - \int_0^t w_1^T(s)w_1(s)ds. \tag{30}$$

The initial condition $x(t)$ is assumed to be zero. By using the fact that $V(0) = 0$ and the Equation (30), the term $J_1(t)$ becomes

$$J_1(t) = \int_0^t \left[\dot{V}(s) - w_1^T(s)w_1(s)\right]ds = \int_0^t \tilde{e}^T(s)\Phi\tilde{e}(s)ds, \tag{31}$$

where $\tilde{e}^T(t) = \begin{bmatrix} e^T(t) & w_1^T(t) \end{bmatrix}, \Phi = \begin{bmatrix} A_{e1} + {A_{e1}}^T\tilde{P} + {C_{e1}}^TC_{e1} & \tilde{P}D_{e1} \\ * & -I \end{bmatrix}$.

Using the Lemma 1 and Lemma 2, we can obtain $\tilde{\boldsymbol{\Phi}} \leq \tilde{\Lambda}$, and the term $\tilde{\Lambda}$ has the following form:

$$\tilde{\Lambda} = \begin{bmatrix} \tilde{\Lambda}_{11} & P_1B_1 & 0 & P_1D_1 & 0 & 0 \\ * & \tilde{\Lambda}_{22} & 0 & -P_2LD_1 & P_2 & 0 \\ * & * & \tilde{\Lambda}_{33} & 0 & 0 & P_3B_1 \\ * & * & * & -I & 0 & 0 \\ * & * & * & * & -I & 0 \\ * & * & * & * & * & -I \end{bmatrix}, \tag{32}$$

with

$$\begin{array}{l} \tilde{\Lambda}_{11} = P_1(A_1 + B_1K_3) + (A_1 + B_1K_3)^{\mathrm{T}}P_1 + \alpha_1 P_1 D_2 D_2^{\mathrm{T}} P_1 \\ +\alpha_1^{-1} E_1^{\mathrm{T}} E_1 + \alpha_2 P_1 D_2 D_2^{\mathrm{T}} P_1 + \alpha_3^{-1} E_1^{\mathrm{T}} E_1 \\ \tilde{\Lambda}_{22} = -P_2 L B_1 + (LB_1)^{\mathrm{T}} P_2 + \alpha_4^{-1} P_2 L D_2 D_2^{\mathrm{T}} L^{\mathrm{T}} P_2 + \alpha_3 P_2 L D_2 D_2^{\mathrm{T}} L^{\mathrm{T}} P_2 \\ \tilde{\Lambda}_{33} = P_3 A_r + A_r^{\mathrm{T}} P_3 + \alpha_2^{-1} E_1^{\mathrm{T}} E_1 + \alpha_4 E_1^{\mathrm{T}} E_1 \\ \alpha_i (i = 1,2,3,4) > 0 \end{array}.$$

If $\tilde{\Lambda} < \mathbf{0}$ holds, we have $\tilde{\boldsymbol{\Phi}} < \mathbf{0}$, i.e., $J_1(t) < \mathbf{0}$. Defining $\tilde{L} = P_2L$ and applying Lemmas 2 and 3 to the inequality $\tilde{\Lambda} < \mathbf{0}$, we obtain

$$\tilde{\Lambda}'' = \begin{bmatrix} \tilde{\Lambda}''_{11} & P_1B_1 & 0 & P_1D_1 & 0 & 0 & \tilde{\Lambda}''_{17} & 0 \\ * & \tilde{\Lambda}''_{22} & 0 & -P_2LD_1 & P_2 & 0 & 0 & \tilde{\Lambda}''_{28} \\ * & * & \tilde{\Lambda}''_{33} & 0 & 0 & P_3B_r & 0 & 0 \\ * & * & * & -I & 0 & 0 & 0 & 0 \\ * & * & * & * & -I & 0 & 0 & 0 \\ * & * & * & * & * & -I & 0 & 0 \\ * & * & * & * & * & * & -\tilde{\Lambda}''_{77} & 0 \\ * & * & * & * & * & * & * & -\tilde{\Lambda}''_{88} \end{bmatrix}, \tag{33}$$

with

$$\begin{array}{l} \tilde{\Lambda}''_{11} = P_1(A_1 + B_1K_3) + (A_1 + B_1K_3)^{\mathrm{T}}P_1 + \alpha_1^{-1} E_1^{\mathrm{T}} E_1 + \alpha_3^{-1} E_1^{\mathrm{T}} E_1 \\ \tilde{\Lambda}''_{22} = -P_2 L B_1 + (LB_1)^{\mathrm{T}} P_2 \\ \tilde{\Lambda}''_{33} = P_3 A_r + A_r^{\mathrm{T}} P_3 + \alpha_2^{-1} E_1^{\mathrm{T}} E_1 + \alpha_4 E_1^{\mathrm{T}} E_1 \\ \tilde{\Lambda}''_{17} = \begin{pmatrix} P_1D_2 & P_1D_2 \end{pmatrix} \\ \tilde{\Lambda}''_{28} = \begin{pmatrix} \tilde{L}D_2 & \tilde{L}D_2 \end{pmatrix} \\ \tilde{\Lambda}''_{77} = diag(\alpha_1^{-1} I, \alpha_2^{-1} I) \\ \tilde{\Lambda}''_{88} = diag(\alpha_3^{-1} I, \alpha_4 I) \end{array}.$$

Using the fact that $e_e^T(t)e_e(t) = e^T(t)C_e^T C_e e(t)$ and $J_1(t) < \mathbf{0}$ and invoking Equation (28), if $C_e^T C_e < \gamma^2 \tilde{P}$ holds, we have

$$e_e^T(t)e_e(t) < \gamma^2 \int_0^t w_1^T(s)w_1(s)ds. \tag{34}$$

If $w_1(t) = 0$ and $\tilde{\boldsymbol{\Phi}} < \mathbf{0}$, we have

$$\dot{V}(t) = e^{\mathrm{T}}(t)\left(\tilde{P}A_e + A_e^{\mathrm{T}}\tilde{P}\right)e(t) < 0. \tag{35}$$

Therefore, the augmented closed-loop system Equation (27) is asymptotically stable under the following conditions:

$$\tilde{\Lambda}'' < \mathbf{0}, C_e^T C_e < \gamma^2 \tilde{P}, \tag{36}$$

where $C_e^T C_e < \gamma^2 \tilde{P}$ can be further simplified to

$$\begin{bmatrix} C_1^T C_1 - \gamma^2 P_1 & 0 & C_1^T C_1 \\ 0 & -\gamma^2 P_2 & 0 \\ C_1^T C_1 & 0 & C_1^T C_1 - \gamma^2 P_3 \end{bmatrix} < 0. \tag{37}$$

Finally, the disturbance observer gain L is obtained as $L = P_2^{-1}\tilde{L}$ by solving the Equations (33) and (37).

In view of the above discussion, the design process of the DOB-based LQR tracking controller is summarized as follows for easy reference:

- Step 1: According to the controlled object in Equation (39) and the reference model system in Equation (40), the gain matrix in the reference model matching controller K_1, K_2 by Assumption 2 is calculated;
- Step 2: Set the LQR weighting matrix $\boldsymbol{Q}, \boldsymbol{R}$; give the prescribed upper bound of LQR performance index γ and the value of input signal $r(t)$; and determine the values of other parameters $D_1, E_1, F_1, \omega(t), d$, etc;
- Step 3: Compute the gain K of the LQR tracking controller by combining Equations (21)–(23);
- Step 4: Compute the gain L of the disturbance observer by combining Equations (33) and (37).

At this point, the LQR-DOB tracking controller design of the electro-optical tracking system with uncertainty and disturbance is completed. Specifically, the disturbance observer is designed as Equation (25) to estimate the disturbances; the reference model matching controller is designed as Equation (14) to track the electro-optical tracking system in Equation (6) with $L_2 - L_\infty$ performance; and the LQR tracking controller is designed as Equation (18).

4. Simulation Analysis and Experimental Verification

4.1. Simulation Analysis

The position transfer function of the controlled object obtained by the electro-optical tracking system through experimental fitting is

$$G(s) = \frac{207}{s^2 + 25.78s + 1151.2}. \tag{38}$$

Convert the above controlled object in Equation (38) into the state-space equation:

$$\begin{aligned} \begin{bmatrix} \dot{x}_1(t) \\ \dot{x}_2(t) \end{bmatrix} &= \begin{bmatrix} 0 & 1 \\ -1151.2 & -25.78 \end{bmatrix} \begin{bmatrix} x_1(t) \\ x_2(t) \end{bmatrix} + \begin{bmatrix} 0 \\ 207 \end{bmatrix} u(t) \\ y(t) &= \begin{bmatrix} 1 & 0 \end{bmatrix} \begin{bmatrix} x_1(t) \\ x_2(t) \end{bmatrix} \end{aligned}, \tag{39}$$

where $x_1(t)$ and $x_2(t)$, respectively, represent the position and speed of the system.

The tracked reference track signal in this paper is generated by the following reference model system, which is shown as

$$\begin{aligned} \begin{bmatrix} \dot{x}_{r1}(t) \\ \dot{x}_{r2}(t) \end{bmatrix} &= \begin{bmatrix} 0 & 1 \\ -4 & -2.828 \end{bmatrix} \begin{bmatrix} x_{r1}(t) \\ x_{r2}(t) \end{bmatrix} + \begin{bmatrix} 0 \\ 1 \end{bmatrix} u(t) \\ y(t) &= \begin{bmatrix} 1 & 0 \end{bmatrix} \begin{bmatrix} x_{r1}(t) \\ x_{r2}(t) \end{bmatrix} \end{aligned}. \tag{40}$$

According to the condition in Assumption 2 and through the above controlled object in Equation (39) and the reference model system in Equation (40), the gain matrix in the reference model matching controller can be obtained as

$$K_1 = \begin{bmatrix} 5.542 & 0.111 \end{bmatrix}, K_2 = 0.0048. \tag{41}$$

Other parameters are given as $D_1 = 0.01\begin{bmatrix} 1 & 1 \end{bmatrix}^T, E_1 = 0.01\begin{bmatrix} 1 & 1 \end{bmatrix}, F_1 = \sin(t)$, $\omega(t) = e^{-5t}, d = \sin(t), Q = \begin{bmatrix} 10 & 0 \\ 0 & 10000 \end{bmatrix}, R = 0.1$, and the upper bound of the LQR performance index $\gamma = 10$ and $r(t) = 1$ is a step signal.

By solving the LMIs in Equations (21)–(24), the gain of the LQR tracking controller is obtained as

$$K = \begin{bmatrix} -0.0596 & -0.003 \end{bmatrix}. \tag{42}$$

By solving the LMIs in Equations (33) and (37), the disturbance observer gain $\boldsymbol{L}$ is obtained as

$$L = \begin{bmatrix} 0.0072 & 0.2924 \end{bmatrix}. \tag{43}$$

In the process of solving the disturbance observer (DOBC) by Equations (33) and (37), the $\tilde{P}$ parameter is shown as

$$P_1 = \begin{bmatrix} 30.2459 & 0.5843 \\ 0.5843 & 0.0377 \end{bmatrix}, P_2 = 31.1002, P_3 = \begin{bmatrix} 17.3446 & 1.1595 \\ 1.1595 & 3.7649 \end{bmatrix}. \tag{44}$$

The eigenvalue of Equation (44) is

$$eig(P_1) = \begin{bmatrix} 0.0264 & 30.2572 \end{bmatrix}^T, eig(P_2) = 31.1002, eig(P_3) = \begin{bmatrix} 3.666 & 17.4429 \end{bmatrix}^T. \tag{45}$$

It can be seen from Equation (45) that the eigenvalue of matrix $\tilde{P}$ is greater than zero, which satisfies the conditions for the LMI method to solve the above inequality relations.

Figure 3 shows the response comparison diagram of the system tracking reference position under sinusoidal disturbance $sin(t)$. The premise parameters such as the LQR weighting matrix $\boldsymbol{Q}, \boldsymbol{R}$, the prescribed upper bound of LQR performance index γ, the value of input signal $r(t)$ and the values of $D_1, E_1, F_1, \omega(t), d$, etc., of all control methods in Figure 3 are guaranteed to be consistent. It can be seen that compared with LQR + DOB with the H_∞ control method, the method proposed in this paper significantly improves the dynamic properties of the system, such as rise time and settling time. The improvement of the dynamic properties of the system is mainly due to the good frequency response characteristics of the LQR tracking controller. Meanwhile, it can be seen that the disturbance observer with $L_2 - L_\infty$ performance index and the model reference tracking controller aim to enhance the robustness and disturbance suppression ability of the system. In addition, it can also be seen in Figure 3 that the gain parameters of DOB observer and controller adjusted by the proposed method are valid. In other words, the LQR tracking control method based on disturbance observer can realize the optimal tracking control of the electro-optical tracking system under the modeling error and uncertain disturbance. This has important practical reference and application value for electro-optical tracking systems.

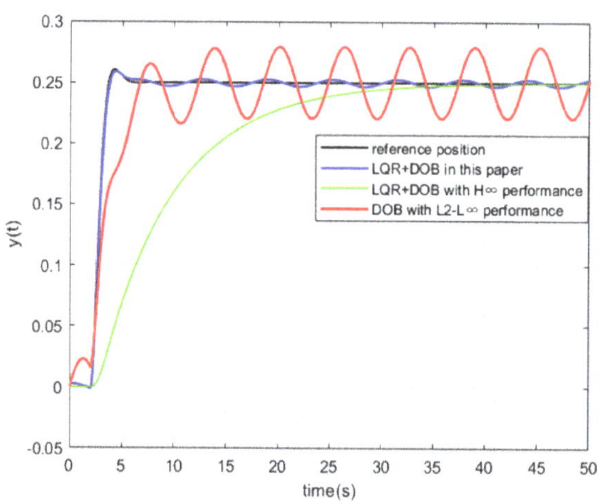

Figure 3. The response comparison diagram of the system tracking reference position under sinusoidal disturbance.

The performance indexes of the tracking reference position of the proposed method in this paper, the LQR + DOB with H_∞ performance control method, and the DOB with $L_2 - L_\infty$ performance control method, such as settling time (T_s) and rise time (T_r), are presented in Table 1 for comparison.

Table 1. Position tracking performance measures.

Method	T_s(s)	T_r(s)
LQR + DOB in this paper	2.97	1.4
LQR + DOB with H_∞ performance	27.36	17.5
DOB with $L_2 - L_\infty$ performance	7.56	4.1

Figure 4 shows the response comparison diagram of the system tracking reference speed under sinusoidal disturbance $sin(t)$. The same conclusion can be drawn from Figure 4 as from Figure 3. Compared with the LQR + DOB with H_∞ control method and the DOB with $L_2 - L_\infty$ control method, the method proposed in this paper significantly improves the dynamic property and disturbance suppression ability of the system. The performance indexes of the tracking reference speed, such as settling time (T_s) and rise time (T_r), are presented in Table 2 for comparison.

Table 2. Speed tracking performance measures.

Method	T_s(s)	T_r(s)
LQR + DOB in this paper	5	2
LQR + DOB with H_∞ performance	34	4.3
DOB with $L_2 - L_\infty$ performance	10.7	2.2

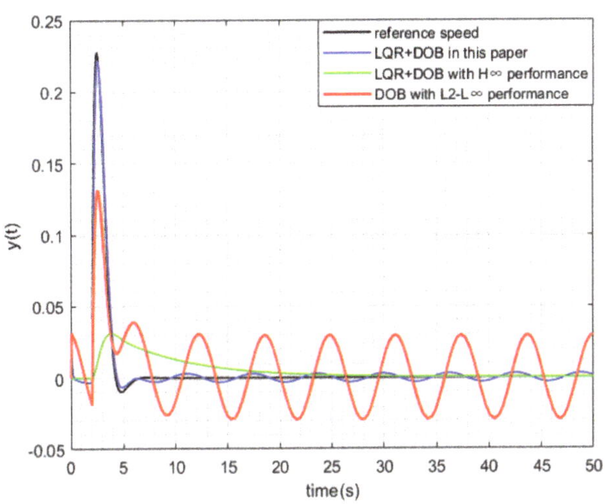

Figure 4. The response comparison diagram of the system tracking reference speed under sinusoidal disturbance.

To sum up, the LQR tracking control method based on disturbance observer can realize the optimal tracking control of the electro-optical tracking system under the modeling error and uncertain disturbance. The LQR tracking controller improves the dynamic response of the system. The model reference tracking controller enhances the robustness of the system. And the DOB with $L_2 - L_\infty$ performance improves the disturbance suppression ability of the electro-optical tracking system. Figure 5 shows the comparison of the disturbance observer under different methods. It can be seen that the DOB under the proposed method can observe the disturbance in real time to compensate. And the disturbance observation progress of the system is relatively high.

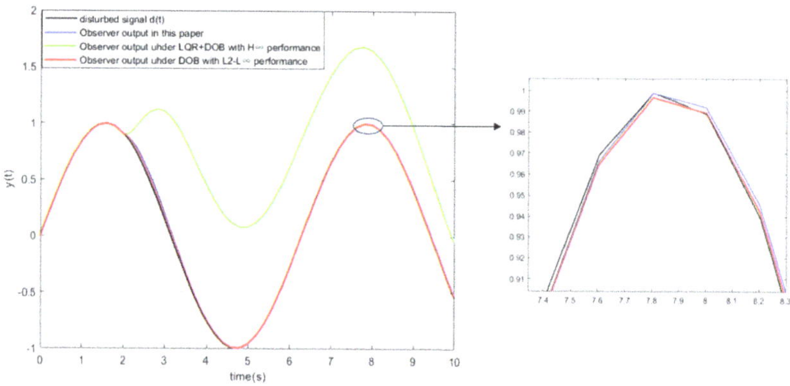

Figure 5. The comparison of disturbance observer under different methods.

4.2. Experimental Verification

To verify the improvement of the dynamic response performance and disturbance suppression ability of the proposed method on the stability control platform, we used the experimental devices shown in Figure 6 for verification.

The electro-optical tracking experimental platform is a two-axis system. This experiment aims at one axis due to the symmetry of the two axes. As shown in Figure 6, the laser light is used to simulate the beacon of light. An apparatus constructed by two superimposed tip-tilt mirror platforms is used to verify the previous analysis. One is used to stabilize the light, and the other is to simulate disturbance, which is measured by position sensors. The electro-optical tracking platform is mounted on the disturbance platform. And both platforms are driven by the voice coil motors. The mirror reflects the laser light into the PSD, which detects the stabilization error at the sampling rate of 5 kHz. In the electro-optical stable tracking system, two main problems need to be solved: one is how to ensure the stability of the optical axis, and the other is the target tracking technology. Stability is a prerequisite for tracking. Therefore, better disturbance suppression ability of the electro-optical platform is conducive to improving the tracking accuracy of the system. The main purpose of this experiment is to verify that the proposed method can significantly improve the disturbance suppression ability and tracking performance of the electro-optical tracking system.

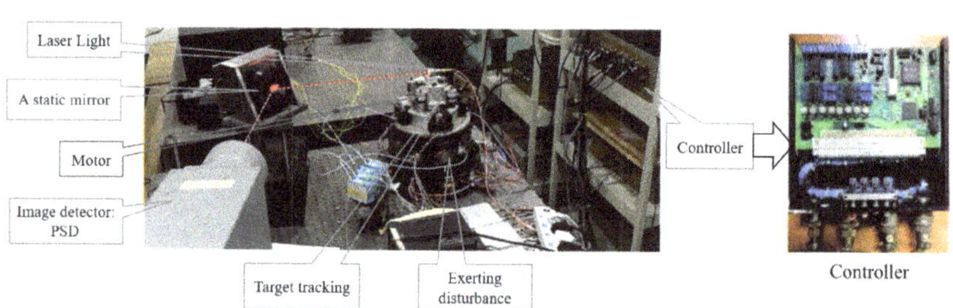

Figure 6. The electro-optical tracking experimental platform.

The disturbed platform is locked when the stable platform is scanned for open-loop position. The characteristic of the electro-optical controlled plant is shown in Figure 7 by inputting the sweep signal to the system. The transfer function of the controlled object obtained by the system identification is as shown in Equation (38). The stability test is to drive the signal to the disturbed platform in the closed loop of the stable platform position and compare the position signal output by the stable platform PSD with that of the disturbed platform.

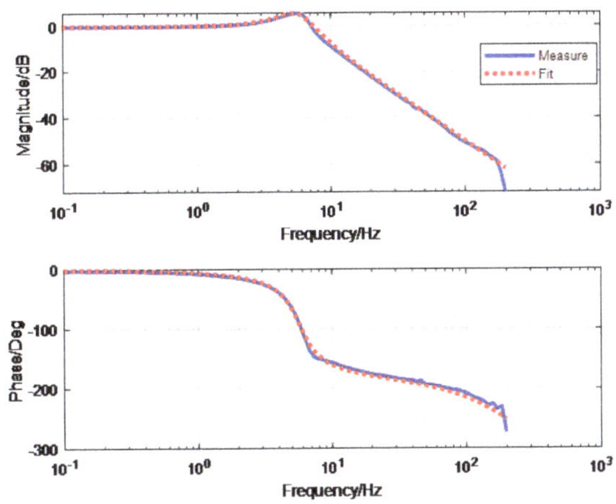

Figure 7. The characteristic of the electro-optical controlled plant.

Firstly, the LQR-DOB tracking control method in this paper is applied to the electro-optical tracking experimental platform. And the disturbance $10sin(t)$ is applied to the disturbance platform. The disturbance input of the electro-optical tracking experimental platform is the value measured by the sensor on the disturbance platform. When the electro-optical tracking platform completes the tracking of the specified target, we simulate the internal disturbance of the electro-optical tracking platform by changing the load on the stable platform. Then, we put a small iron on a stable platform and continue to observe the tracking accuracy and disturbance suppression effect of our control method.

Secondly, the LQR + DOB with H_∞ control method and the DOB with $L_2 - L_\infty$ control method are also applied to the electro-optical tracking experimental platform. In addition, the operation of external disturbance and internal disturbance in the experiment is consistent with the above.

Figure 8 shows the tracking position comparison of the system under different methods. Based on the experimental results, it can be seen that the method proposed in this paper can significantly improve the disturbance suppression ability of the system and dynamic property, such as rise time, settling time and system overshoot. Meanwhile, we can also see that the experimental results are consistent with the above simulation results. The method presented in this paper is effective in the electro-optical tracking system.

Figure 9 shows the tracking speed comparison of the system under different methods. Similarly, compared with the LQR + DOB with H_∞ control method and the DOB with $L_2 - L_\infty$ control method, the method proposed in this paper significantly improves the dynamic property and disturbance suppression ability of the system.

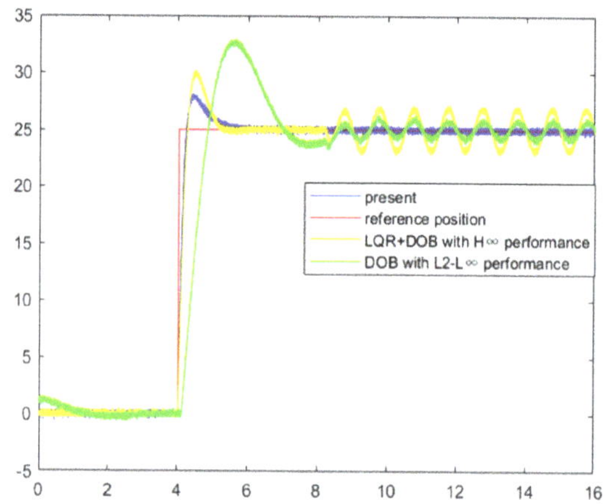

Figure 8. The tracking position comparison of the system under different methods.

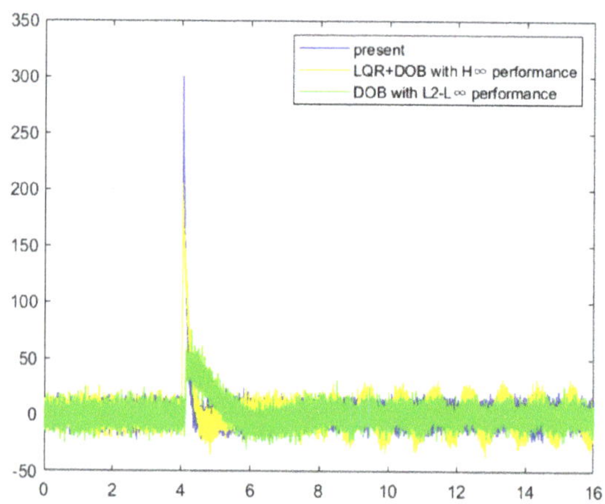

Figure 9. The tracking speed comparison of the system under different methods.

5. Discussion

The LQR-DOB tracking control method proposed in this paper solves the problem of system instability caused by model uncertainty and uncertain disturbance in an electro-optical tracking system. With the increased maneuverability of tracking target, the corresponding control strategy needs to be further studied to achieve the purpose of tracking faster reference signals. From the perspective of control theory, the higher type of control loop has the advantage of tracking faster signals. The design of the high-type control loop has been challenging in academia and industry; that is, it is very difficult to set controller parameters in the high-type control loop. In our future work, high-type control combined with LQR optimal control is introduced into the electro-optical tracking system to improve the disturbance suppression ability, tracking ability and tracking accuracy of the system. In addition, the nonlinear model of the electro-optical tracking system in practical applications can more accurately reflect the characteristics of the system object. Therefore, our future work will focus on designing a nonlinear controller with high-type control combined with optimal control to improve the dynamic response performance of the system and restrain internal and external disturbances. This has a very important application value for electro-optical tracking systems.

6. Conclusions

This paper presents an LQR-DOB tracking control method to solve the problems of modeling error and uncertain disturbance in an electro-optical tracking control system. Using standard techniques, the DOB gain and controller gain of the tracking reference model design is reduced to a convex constraint problem, which can be efficiently solved with the LMI approach. Meanwhile, the stability constraint of the electro-optical tracking closed-loop system is considered by using Lyapunov theory in this framework. Compared with the LQR + DOB with H_∞ control method and the DOB with $L_2 - L_\infty$ control method under the same disturbance condition, the method proposed in this paper can significantly improve the dynamic properties of the system, such as rise time, settling time and system overshoot. The improvement of the dynamic properties of the system is mainly due to the good frequency response characteristics of the LQR tracking controller. Meanwhile, the disturbance observer with $L_2 - L_\infty$ performance index and the model reference tracking controller aim to enhance the robustness and disturbance suppression ability of the system. Specifically, compared with the other methods in this paper, the tracking accuracy and the disturbance suppression ability of the proposed method is about two to three times higher.

However, with the increase in target tracking maneuverability in the electro-optical tracking system, the tracking accuracy and disturbance suppression ability of the system under the proposed method are reduced. To meet the needs of the practical applications of electro-optical tracking systems, the next work of our paper is to further optimize the method in this paper and further solve the problem that the tracking accuracy and disturbance suppression ability of the system decline under the premise of strong tracking target mobility. Meanwhile, many practical constraints, such as controller saturation, will be considered in the next work of this paper. In general, the method proposed in this paper has important reference value for electro-optical tracking systems.

Author Contributions: Conceptualization, C.L., Y.M. and X.Q.; methodology, C.L.; software, C.L. and X.Q.; validation, C.L. and Y.M.; formal analysis, X.Q.; investigation, Y.M.; resources, Y.M.; data curation, X.Q.; writing—original draft, C.L.; writing—review and editing, C.L. and Y.M.; visualization, C.L. and X.Q.; supervision, Y.M.; project administration, Y.M.; funding acquisition, Y.M. All authors have read and agreed to the published version of the manuscript.

Funding: The work described in this article was supported by the National Natural Science Foundation of China (No.61905253).

Informed Consent Statement: Informed consent was obtained from all subjects involved in the study.

Data Availability Statement: The data presented in this study are available from the corresponding author upon reasonable request.

Acknowledgments: The authors sincerely thank the three anonymous reviewers for their valuable comments and suggestions during the peer review process.

Conflicts of Interest: The authors declare no potential conflict of interest with respect to the research, authorship and/or publication of this article.

References

1. Zhou, X.; Li, X. Trajectory tracking control for electro-optical tracking system using ESO based fractional-order sliding mode control. *IEEE Access* **2021**, *9*, 45891–45902. [CrossRef]
2. Zhou, X.; Li, X. Trajectory tracking control for electro-optical tracking system based on fractional-order sliding mode controller with super-twisting extended state observer. *ISA Trans.* **2021**, *117*, 85–095. [CrossRef]
3. Zhao, T.; Tong, W.; Mao, Y. Hybrid Nonsingleton Fuzzy Strong Tracking Kalman Filtering for High Precision Photoelectric Tracking System. *IEEE Trans. Ind. Inform.* **2022**, *19*, 2395–2408. [CrossRef]
4. Duan, Q.; Mao, Y.; Zhang, H.; Xue, W. Add-on integration module-based proportional-integration-derivative control for higher precision electro-optical tracking system. *Trans. Inst. Meas. Control* **2021**, *43*, 1347–1362. [CrossRef]
5. Wang, J.; Guo, Y.; Kong, L.; Rao, C. Experimental demonstration of LQG control with disturbance mitigation on multiple modes in adaptive optics system. *Optik* **2020**, *202*, 163594. [CrossRef]

6. Shi, Q.; Wang, H.; Cheng, H.; Wang, M.; Cheng, F.; Liu, F. Fuzzy disturbance observer-based modified variable gains super-twisting sliding mode composite control for a two-axis airborne optoelectronic stabilized platform. *Optik* **2021**, *228*, 166132. [CrossRef]
7. Wang, J.; Song, Y.; Jiang, H.; Dong, K.; Liu, Y. Prototype development of multi-target tracking system for space multi-node laser communication network. *Optik* **2023**, *274*, 170552. [CrossRef]
8. Deng, J.; Xue, W.; Zhou, X.; Mao, Y. On Dual Compensation to Disturbances and Uncertainties for Inertially Stabilized Platforms. *Int. J. Control Autom. Syst.* **2022**, *20*, 1521–1534. [CrossRef]
9. Zhang X, Wang J. Multi-loop Feedback Analysis of Distributed Power Inverters in Microgrids. In Proceedings of the 2023 IEEE International Conference on Control, Electronics and Computer Technology (ICCECT), Jilin, China, 28–30 April 2023; IEEE: Piscataway, NJ, USA, 2023; pp. 1431–1434.
10. Deng, J.; Xue, W.; Liang, W.; Zhou, X.; Mao, Y. On adjustable and lossless suppression to disturbances and uncertainties for nonminimum-phase laser pointing system. *ISA Trans.* **2023**, *136*, 727–741. [CrossRef]
11. Toor, W.A.; Ashraf, M. Performance Analysis of Learning-Based Disturbance Observer for Pulsed Superconducting Cavity Field Control. *Electronics* **2023**, *12*, 1556. [CrossRef]
12. Lubing, X.; Yangfei, Y.; Yang, Y.;Weixing, Z. DOB Tracking Control Algorithm for Unmanned Aerial Vehicles with TS Disturbance Modeling. In Proceedings of the 2021 40th Chinese Control Conference (CCC), Shanghai, China, 26–28 July 2023; IEEE: Piscataway, NJ, USA, 2021; pp. 503–508.
13. Grunwald, R.; Jurke, M.; Bock, M.; Liebmann, M.; Bruno, B.P.; Gowda, H.; Wallrabe, U. High-flexibility control of structured light with combined adaptive optical systems. *Photonics* **2022**, *9*, 42. [CrossRef]
14. Pochechuev, M.S.; Fedotov, I.V.; Solotenkov, M.A.; Andreeva, M.S.; Lanin, A.A.; Fedotov, A.B.; Zheltikov, A.M. Adaptive Wave-Front Shaping and Beam Focusing through Fiber Bundles for High-Resolution Bioimaging. *Photonics* **2021**, *9*, 21. [CrossRef]
15. Abbasi, A.; Moshayedi, A.J. Trajectory tracking of two-wheeled mobile robots, using LQR optimal control method, based on computational model of KHEPERA IV. *J. Simul. Anal. Nov. Technol. Mech. Eng.* **2018**, *10*, 41–50.
16. Liu, C.; Duan, Q.; Zhang, C.; Luo, Y.; Qiu, X.; Xia, W.; Li, H.; Mao, Y. Extending the LQR to the design of PID type-ii and type-iii control loops. *IET Control Theory Appl.* **2023**, *17*, 713–743. [CrossRef]
17. Okasha, M.; Kralev, J.; Islam, M. Design and Experimental Comparison of PID, LQR and MPC Stabilizing Controllers for Parrot Mambo Mini-Drone. *Aerospace* **2022**, *9*, 298. [CrossRef]
18. Elkhatem, A.S.; Engin, S.N. Robust LQR and LQR-PI control strategies based on adaptive weighting matrix selection for a UAV position and attitude tracking control. *Alex. Eng. J.* **2022**, *61*, 6275–6292. [CrossRef]
19. Bilgic, H.H.; Sen, M.A.; Yapici, A.; Yavuz, H.; Kalyoncu, M. Meta-heuristic tuning of the LQR weighting matrices using various objective functions on an experimental flexible arm under the effects of disturbance. *Arab. J. Sci. Eng.* **2021**, *46*, 7323–7336. [CrossRef]
20. Németh, Z.; Kuczmann, M. Linear-Matrix-Inequality-Based Controller and Observer Design for Induction Machine. *Electronics* **2022**, *11*, 3894. [CrossRef]
21. Kim, Y.; Tran, T.V.; Kim, K.H. LMI-based model predictive current control for an LCL-filtered grid-connected inverter under unexpected grid and system uncertainties. *Electronics* **2022**, *11*, 731. [CrossRef]
22. Gritli, H.; Zemouche, A.; Belghith, S. On LMI conditions to design robust static output feedback controller for continuous-time linear systems subject to norm-bounded uncertainties. *Int. J. Syst. Sci.* **2021**, *52*, 12–46. [CrossRef]
23. Feng, Z.Y.; Guo, H.; She, J.; Xu, L. Weighted sensitivity design of multivariable PID controllers via a new iterative LMI approach. *J. Process Control* **2022**, *110*, 24–34. [CrossRef]
24. Wang, X.; Chen, X.; Wen, L. The LQR baseline with adaptive augmentation rejection of unmatched input disturbance. *Int. J. Control Autom. Syst.* **2017**, *15*, 1302–1313. [CrossRef]
25. Duan, G.R.; Yu, H.H. *LMIs in Control Systems: Analysis, Design and Applications*; CRC Press: Boca Raton, FL, USA, 2013.
26. Li, Y.; Sun, H.; Zong, G.; Hou, L. Anti-disturbance control for time-varying delay Markovian jump nonlinear systems with multiple disturbances. *Int. J. Syst. Sci.* **2017**, *48*, 3186–3200. [CrossRef]

photonics

MDPI

Article

Joint-Transceiver Equalization Technique over a 1.4 km Multi-Mode Fiber Using Optical MIMO Technique in IM/DD Systems

Jasmeet Singh [1,2,*] and Andreas Ahrens [1]

1 Department of Electrical Engineering and Computer Science, Hochschule Wismar, Philipp-Müller-Straße 14, 23966 Wismar, Germany; andreas.ahrens@hs-wismar.de
2 ETSIST, Universidad Politécnica de Madrid, Campus Sur, Calle Nikola Tesla s/n, 28031 Madrid, Spain
* Correspondence: jasmeet.singh@hs-wismar.com

Abstract: In optical fiber communication, recent advances in multiple-input and multiple-output (MIMO) systems using space-division multiplexing have helped achieve higher spectral efficiency and data rates. Propagating higher-order modulation formats over MIMO systems further strengthens the capacity of the transmission link. In the optical-MIMO system, the dispersion impairments originating from a 1.4 km long multi-mode fiber (MMF) are mitigated using the proposed joint-transceiver equalization technique. A numerical convex optimization algorithm is used to compute and optimize the pre- and post-equalization (PPE) coefficients jointly restricted by cost and power budgets. The potential of the proposed joint-PPE technique is tested on an MMF link, which is severely degraded by dispersion compared to a single-mode fiber channel. From the experimental results, the average optical received power gain necessary to reach 10^{-4} bit-error rate is improved by nearly 2.5 dB using the joint-PPE compared to the post-equalization only based on the minimum mean-squared error principle. When the efficiency of the conventional zero-forcing (ZF) principle-based PPE and the joint-PPE is compared, the joint-PPE scheme outperforms the ZF-PPE by approximately 1.5 dB. The enhancement in the transmission quality is observed with experimentally measured eye diagrams using the joint-PPE scheme. Under the analyzed scenarios, computer simulation also confirms the hypothesis, which establishes the effectiveness of the proposed joint-transceiver equalization over the conventional ZF-PPE scheme. Moreover, the simulated performance benefits of the joint-PPE are evaluated using the singular value decomposition (SVD) technique. Improvement of ≈3.86 dB in the average optical received power gain required to reach 10^{-4} bit-error rate is witnessed with the PAM-4 format. Overall, the joint-transceiver equalization technique is proven to be beneficial in optical MIMO systems.

Keywords: optical fiber communication; multi-mode fiber; optical MIMO technique; pre- and post-equalization; space-division multiplexing

Citation: Singh, J.; Ahrens, A. Joint-Transceiver Equalization Technique over a 1.4 km Multi-Mode Fiber Using Optical MIMO Technique in IM/DD Systems. *Photonics* **2023**, *10*, 696. https://doi.org/10.3390/photonics10060696

Received: 4 May 2023
Revised: 8 June 2023
Accepted: 13 June 2023
Published: 19 June 2023

1. Introduction

In recent years, phenomenal growth in the demand for bandwidth has been witnessed due to the progress in cloud computing, high-definition video streaming, and the internet of things. The surge in the increasing data requirements has inspired the expansion of telecommunication networks by developing better modulation and coding techniques. Optical networks have provided support to match the high-bandwidth requirements from short-haul to the long-haul transmission links.

Single-mode fibers (SMFs) are preferred for long-distance optical systems due to their low dispersion properties and low signal attenuation. Despite the availability of degrees of freedom such as time, wavelength, or polarization, the capacity of an SMF is rapidly reaching Shannon's limit [1–3]. Furthermore, the small core diameter of an SMF causes sub-micron alignment tolerances. Multi-mode fibers (MMFs) consist of a larger core (50 µm or 62.5 µm) and hence have higher alignment margins. Due to the larger core diameter, the MMF core can support multiple parallel data channels using up to 100 transverse spatial mode groups [4–6]. The orthogonality between different mode groups is only preserved for a short distance. In an intensity modulation/direct detection (IM/DD) system for short-reach optical links, an MMF is preferred over an SMF due to higher spectral efficiency, cost-effectiveness and simple implementation. However, MMFs adversely affect the available bandwidth distance product as a result of intermodal dispersion and mode-dependent losses such as mode scrambling [6–8].

The technique for creating multiple parallel data channels using the transverse spatial extent of an MMF is known as space-division multiplexing (SDM) [9]. Offset light-launching conditions excite two or more parallel data streams in the SDM technique. The considerable enhancement in the data rates and the mitigation of mode scrambling are handled by incorporating the multiple-input and multiple-output (MIMO) signal-processing technique [6,7]. Moreover, the spectral efficiency of the transmission link is boosted using the higher-order modulation (HOM) formats, where the constellation size is more than two. In particular, in an IM/DD system, the pulse-amplitude modulation (PAM) formats are preferred over subcarrier modulation (SCM) [10], carrier-less amplitude modulation (CAP) [11], or discrete multitone (DMT) [12,13] in terms of simple implementation and better sensitivity regarding the received optical power. However, an MMF channel is severely impaired by intermodal dispersion compared to an SMF link [8]. Dispersion compensation mechanisms include physical components, such as dispersion-shifted fibers or fiber Bragg gratings, and digital mitigation using equalization.

In an IM/DD system, when a trade-off between complexity and cost-effectiveness is considered, digital dispersion compensation mechanisms, especially linear MIMO equalizers, are preferred. On the other hand, the performance of non-linear MIMO equalizers surpasses most optimal linear MIMO equalizers based on the minimum mean-squared error (MMSE) principle [13–17]. However, a linear MMSE-based equalizer is selected because of comprehensive implementation, simple realization and analytical traceability [3,18]. Additionally, a further boost in the system performance is evident in the presence of a transceiver equalization compared to the equalization applied either at the transmitter or the receiver [14,15,19]. According to the literature, the zero-forcing (ZF) principle is commonly adapted for the design of the pre-equalizer. Moreover, a separate equalization principle is utilized to calculate the post-equalization coefficients [16,17,20,21]. Additionally, the performance evaluation of these pre- and post-equalization (PPE) schemes is validated either on the single-input and single-output (SISO) system model or over an SMF fiber [16,17,21,22]. Although the PPE schemes are reported in [14,15,19,23], the experimental comparison with conventional PPE schemes using a longer MMF and higher data rate needs to be investigated. Moreover, addressing the lower convergence rate to achieve the optimal design of PPE coefficients is crucial [19,23]. This research gap limits the feasible use of a pre-equalizer on the transmitter side due to high complexity. In MIMO signal processing, the singular value decomposition (SVD) technique is well established, and it has been proven to be advantageous as the MIMO channel is transformed into several parallel transmission layers [24–27]. The advantages of an SDM-based MIMO model deploying an MMF channel with HOM formats have not been characterized in depth. Therefore, this work focuses on low-cost SDM-based IM/DD systems for short-distance transmission.

Against this background, the novelty of this work is the enhancement of the transmission quality and the bit-error rate (BER) performance using the proposed transceiver–joint equalization scheme over an optical MIMO system with an MMF transmission link. This novel joint-PPE scheme mitigates the impairments originating from a 1.4 km long MMF channel. The performance benefits of the joint-PPE scheme are yielded by optimally designing both of the PPE coefficients simultaneously using a numerical optimization solver. The performance analysis of the transceiver–joint equalization is also conducted with a SVD-based MIMO system incorporating the HOM format. The analytical model-based simulations and the experimental setup verify the benefits of the provided joint-PPE scheme.

Additionally, the supremacy of the joint-PPE scheme is established over the conventional ZF-PPE and MMSE-based post-equalization only (PE-only).

Hereinafter, this paper is organized as follows: Section 1 describes the system model in the electrical domain. An analytical description of the provided system model and the quality criteria are also included in this section. Furthermore, the optimization and design of the joint-PPE scheme are presented in the Section 3, and the solver using the interior-point algorithm is explained in detail. In the Section 4, the simulation and experimental results are discussed and compared with other equalization techniques. Finally, the conclusion is outlined in the Section 5. The literature with a similar overlap in this area is reported in the Section 6.

2. System Model

The proposed transceiver–joint equalization technique in a MIMO system is analyzed using an electrical system model with $n_{\mathrm{T}} = n_{\mathrm{R}} = 2$, where n_{T} and n_{R} represent number of transmitters and receivers, respectively. In Figure 1, a $(n_{\mathrm{R}} \times n_{\mathrm{T}})$ MIMO transmission model with $n_{\mathrm{R}} = n_{\mathrm{T}} = 2$ is segregated into three parts, namely the transmitter, the channel, and the receiver. The symbols utilized in the system model are described in Table 1. The MIMO transmitter consists of a serial-to-parallel (S/P) converter and a pre-equalizer. Using the S/P converter, encoded serial binary data are transformed into n_{T} parallel data streams, where n_{T} represents the total number of MIMO inputs. The parallel symbol streams $s_{\mu}(k)$, with $\mu = 1, 2, \ldots, n_{\mathrm{T}}$ and $k \in \mathbb{N}$ are mapped on to M_{μ}-ary PAM constellation. The bit-to-symbol mapping is conducted using Gray coding as it improves the system performance. The transmit symbol space of $s_{\mu}(k) \in \{s_{\mu,0}, s_{\mu,1}, \ldots, s_{\mu,m}\}$, where $m = [0, 1, \ldots, M_{\mu} - 1]$, consists of $s_{\mu,m}$ symbols in accordance to the chosen M_{μ}-ary PAM constellation. It is worth noting that the symbols in $s_{\mu}(k)$ are independent and identically distributed. The amplitude of the symbols is defined as

$$s_{\mu,m} = \left(\frac{m}{M_{\mu} - 1}\right) \cdot P_{\mathrm{s},\mu}^{(\max)} \quad \text{, where } M_{\mu} \neq 1 \ . \tag{1}$$

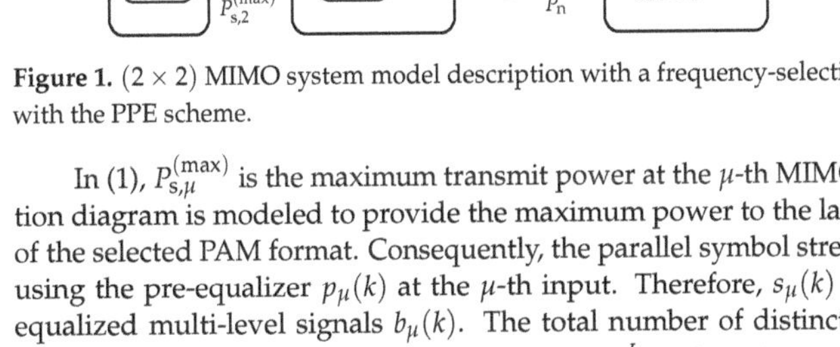

Figure 1. (2×2) MIMO system model description with a frequency-selective channel incorporated with the PPE scheme.

In (1), $P_{\mathrm{s},\mu}^{(\max)}$ is the maximum transmit power at the μ-th MIMO input. The constellation diagram is modeled to provide the maximum power to the last symbol irrespective of the selected PAM format. Consequently, the parallel symbol streams are pre-equalized using the pre-equalizer $p_{\mu}(k)$ at the μ-th input. Therefore, $s_{\mu}(k)$ is converted into pre-equalized multi-level signals $b_{\mu}(k)$. The total number of distinct pre-equalized signal levels N_{μ} at the μ-th input is defined as $N_{\mu} = 2^{L_{\mathrm{P}}}$, where L_{P} represents the number of pre-equalizer filter taps. The resulting pre-equalized signal $b_{\mu}(k) = [b_{\mu,0}, b_{\mu,1}, \ldots, b_{\mu,N_{\mu}-1}]$ complies with the power constraint, which is given as

$$s_{\mu,M_{\mu}-1} = b_{\mu,N_{\mu}-1} = \mathrm{q} \cdot P_{\mathrm{s},\mu}^{(\max)} \ , \tag{2}$$

where q is a conversion factor for converting an electrical signal into an optical signal. Although implementing the pre-equalization, a constant peak-to-average power ratio is also maintained in (2).

Table 1. List of symbols and their respective description.

Symbol	Description
n_T	Number of MIMO inputs
n_R	Number of MIMO outputs
μ	MIMO input index, $\mu = 1, 2, \ldots, n_\mathrm{T}$
ν	MIMO output index, $\nu = 1, 2, \ldots, n_\mathrm{R}$
$s_\mu(k)$	Input symbol sequence at μ-th MIMO input
M_μ	Modulation format index at μ-th MIMO input
$s_{\mu,m}$	Gray coded input symbols
$P_{\mathrm{s},\mu}^{(\max)}$	Maximum transmit power at μ-th MIMO input
$p_\mu(k)$	Pre-equalizer filter coefficients
$b_\mu(k)$	Pre-equalized symbol stream
L_p	Length of the pre-equalizer filter
N_μ	Number of distinct pre-equalizer levels
$h_{\nu\mu}(k)$	Channel taps
$w_\mu(k)$	IM/DD noise mechanisms
$r_\nu(k)$	Received signal
$f_{\nu\mu}(k)$	Post-equalization filter
$\boldsymbol{H}$	Frequency-selective channel matrix with $(n_\mathrm{R} \times n_\mathrm{T})$ dimensions
$\boldsymbol{r}$	Received signal matrix with $(n_\mathrm{R} \times 1)$ dimensions
$\boldsymbol{b}$	Pre-equalized transmitted signal with $(n_\mathrm{T} \times 1)$ dimensions
$\boldsymbol{w}$	IM/DD noise mechanism with $(n_\mathrm{R} \times 1)$ dimensions
$\boldsymbol{P}$	Pre-equalizer with $(n_\mathrm{T} \times n_\mathrm{T})$ dimensions
$\mathbf{s}$	Input signal with $(n_\mathrm{T} \times 1)$ dimensions
$\boldsymbol{F}$	Post-equalizer with $(n_\mathrm{R} \times n_\mathrm{R})$ dimensions
θ_μ	Noise-weighting factor
L_f	Length of the post-equalizer filter
P_n	Noise power
ϱ_μ	Transmission quality
U_A	Half-vertical eye-opening
$P_\mathrm{BER}^{(\mu)}$	BER at μ-th MIMO input
$\mathrm{erfc}(\cdot)$	Complimentary error operator
$P_\mathrm{BER}^{(s_{\mu,m})}$	Symbol-specific BER
$\boldsymbol{S}$	Unitary matrix with $(n_\mathrm{R} \times n_\mathrm{R})$ dimensions
H	Conjugate transpose operator (Hermitian)
$\boldsymbol{V}$	Eigen matrix with $(n_\mathrm{R} \times n_\mathrm{T})$ dimensions
$\boldsymbol{D}$	Unitary matrix with $(n_\mathrm{T} \times n_\mathrm{T})$ dimensions
$h_{m,i}^{(\mu)}$	Hamming distance

Subsequently, the multi-level pre-equalized signal is propagated through the MIMO channel, which is an MMF link. The transmitted signal is deteriorated with the inter-symbol interference (ISI) and the inter-channel interference (ICI) originating from the MMF channel $h_{\nu\mu}(k)$, where $\nu = 1, 2, \ldots, n_\mathrm{R}$ and the number of MIMO outputs is denoted by n_R. The temporal pulse broadening and chirping occur due to the frequency-selective behavior of the MMF channel $h_{\nu\mu}(k)$. Moreover, the indirect components of the channel $h_{\nu\mu}(k)$, when $\nu \neq \mu$, are responsible for the crosstalk between the MIMO layers. An MMF channel supports several guided modes and provides easier light-launching conditions into the fiber due to the larger core diameter. Hence, the parallel data streams are generated by offset light-launching conditions to achieve a higher spectral efficiency. This technique is known as SDM using an MMF [6]. In the given system model, the first layer is created by launching the light into the center of the MMF core. Similarly, the second MIMO layer is formed by a 15 μm offset launching condition into the fiber core, as shown in Figure 2.

When traveling through the MMF channel, the transmitted signal is degraded due to ISI and ICI, originating from the channel. Afterwards, the IM/DD noise impairments $w_\mu(k)$ further degrade the transmitted signal $b_\mu(k)$ [26]. At the MIMO receiver, the post-processing of the received signal $r_\nu(k)$ with $\nu = 1, 2, \ldots, n_\mathrm{R}$ is conducted using a post-equalizer $f_{\nu\mu}(k)$. The residual interferences are mitigated with the help of an MMSE-based post-equalizer. When designing the post-equalizer, the MMSE principle is preferred over the ZF principle as the optimized coefficients of an MMSE equalizer contribute less to noise amplification.

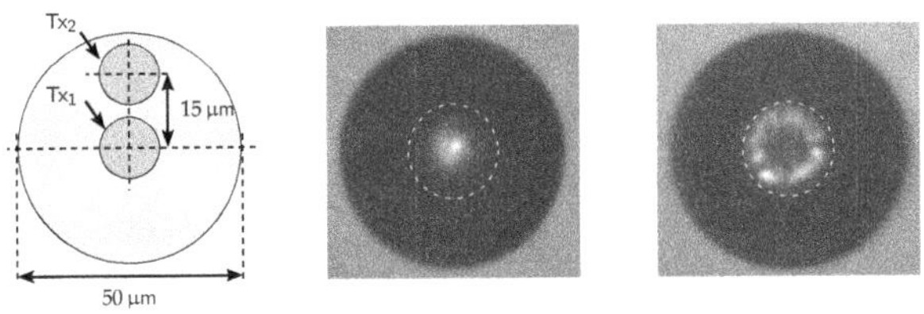

(a) Light launching conditions at the MMF core. (b) Centric launching at the first transmitter Tx_1. (c) Eccentric launching at the second transmitter Tx_2.

Figure 2. In (**a**), the transmitter's configuration for launching the light into an MMF core is depicted. Measured mean power distribution with (**b**) centric (offset: 0 µm) and (**c**) eccentric (offset: 15 µm) light-launching conditions into an MMF core, which is represented by the white dashed circle.

2.1. *Analytical Description of the System Model*

A frequency-selective channel-based $(n_\mathrm{T} \times n_\mathrm{R})$ MIMO system with $n_\mathrm{R} = n_\mathrm{T} = 2$, as shown in Figure 1, is analytically represented as

$$\boldsymbol{r} = \boldsymbol{H} \cdot \boldsymbol{b} + \boldsymbol{w} \ . \tag{3}$$

From (3), the pre-equalized transmitted signal is represented as $\boldsymbol{b}$ with $(n_\mathrm{T} \times 1)$ dimensions. The $(n_\mathrm{R} \times n_\mathrm{T})$ channel matrix $\boldsymbol{H}$ describes the block-oriented system model, and is defined as

$$\boldsymbol{H} = \begin{bmatrix} \boldsymbol{H}_{11} & \cdots & \boldsymbol{H}_{1\,n_\mathrm{T}} \\ \vdots & \ddots & \vdots \\ \boldsymbol{H}_{n_\mathrm{R}\,1} & \cdots & \boldsymbol{H}_{n_\mathrm{R}\,n_\mathrm{T}} \end{bmatrix} , \tag{4}$$

where $\boldsymbol{H}_{n_\mathrm{R}\,n_\mathrm{T}}$ is the SISO channel matrix, which characterizes the influence of the channel over the pre-equalized transmitted signal $\boldsymbol{b}$. The $(n_\mathrm{R} \times 1)$ noise vector $\boldsymbol{w}$ pertains to the noise in an IM/DD system. Finally, the received signal is represented as $\boldsymbol{r}$ with $(n_\mathrm{R} \times 1)$ dimensions. The transmitted signal $\boldsymbol{b}$ is decomposed as

$$\boldsymbol{r} = \boldsymbol{H} \cdot \boldsymbol{P} \cdot \boldsymbol{s} + \boldsymbol{w} \ , \tag{5}$$

where $\boldsymbol{P}$ is the pre-equalizer with a size of $(n_\mathrm{T} \times n_\mathrm{T})$, and $\boldsymbol{s}$ represents the input vector of $(n_\mathrm{T} \times 1)$. For a limited increase in the complexity, the pre-equalizer $\boldsymbol{P}$ is considered to be a diagonal matrix. Ultimately, the post-equalization is executed to mitigate the residual interference. The final expression of the given system model with PPE is formulated as

$$\begin{aligned} \boldsymbol{y} &= \boldsymbol{F} \cdot \boldsymbol{r} \ , \\ \boldsymbol{y} &= \underbrace{\boldsymbol{F} \cdot \boldsymbol{H} \cdot \boldsymbol{P}}_{\text{Term 1}} \cdot \boldsymbol{s} + \underbrace{\boldsymbol{F} \cdot \boldsymbol{w}}_{\text{Term 2}} \ , \end{aligned} \tag{6}$$

with $\boldsymbol{F}$ denoting the post-equalizer with $(n_\mathrm{R} \times n_\mathrm{R})$ dimensions. In (6), the first term implies that the channel influence can be mitigated with the aid of PPE. Similarly, the noise amplification due to post-equalization is evident in (6). The system performance is adversely degraded due to noise enhancement from post-equalization. Hence, a well-designed pre-equalizer will be capable of compensating for the channel along with negligible noise amplification. When a trade-off between complexity and cost-effectiveness is considered, a pre-equalizer with a few taps is preferred in conjunction to a post-equalizer.

The quantification of the noise amplification due to the post-equalization filter taps is expressed with the noise weighting factor (NWF) θ_μ and is given as

$$\theta_\mu = \sum_{\nu=1}^{n_R} \sum_{k=0}^{L_f-1} |f_{\mu\nu}[k]|^2 \qquad \text{for} \quad \mu = 1, 2, \ldots, n_T \ , \tag{7}$$

where L_f is the number of post-equalization taps and $f_{\mu\nu}[k]$ represents the post-equalizer taps for μ-th input and ν-th output. From (7), it is evident that the amplitude of the post-equalization taps is responsible for the noise amplification.

2.2. Quality Criteria

The BER is considered to be the key indicator to evaluate the performance of an MMF channel. The transmission quality ϱ_μ of the μ-th MIMO input is defined as

$$\varrho_\mu = \frac{U_{A,\mu}^2}{\widetilde{P}_{n,\mu}} \ , \tag{8}$$

where $U_{A,\mu}$ represents the half-vertical eye-opening of the μ-th input and the noise power after the post-equalization is defined with $\widetilde{P}_{n,\mu}$. From (7) and (8), the relationship between θ_μ and $\widetilde{P}_{n,\mu}$ is given by

$$\widetilde{P}_{n,\mu} = \theta_\mu \cdot P_n \ . \tag{9}$$

By substituting (9) into (8), the resulting transmission quality at the μ-th input is

$$\varrho_\mu = \frac{U_{A,\mu}^2}{\widetilde{P}_{n,\mu}} = \frac{U_{A,\mu}^2}{\theta_\mu \cdot P_n} \ . \tag{10}$$

In (10), the noise amplification factor is inversely proportional to the transmission quality. Typically, for an SDM-based MIMO system, the BER performance $P_{BER}^{(\mu)}$ of the μ-th input is

$$P_{BER}^{(\mu)} = \frac{1}{\log_2(M_\mu)} \left(\frac{M_\mu - 1}{M_\mu}\right) \text{erfc}\left(\sqrt{\frac{\varrho_\mu}{2}}\right) \ , \tag{11}$$

with erfc$(\cdot)$ denoting the complimentary error function, the constellation size and the transmission quality at the μ-th input is represented by M_μ and ϱ_μ, respectively.

By combining (10) with (11), the final BER performance is evaluated as

$$P_{BER}^{(\mu)} = \frac{1}{\log_2(M_\mu)} \left(\frac{M_\mu - 1}{M_\mu}\right) \text{erfc}\left(\sqrt{\frac{U_{A,\mu}^2}{2 \cdot \theta_\mu \cdot P_n}}\right) \ . \tag{12}$$

In (11), when an error is encountered during the transmission, it is assumed that the error will only be limited to its neighboring symbol, i.e., only a one-bit error due to the use of Gray coding. However, more-than-one-bit errors can occur at low received power due to the high noise variance. Second, the noise variance of each symbol is considered to be identical in (11). In contrast, the noise variances are different in the shot noise dominant region. Hence, the area integration method is adapted to accurately estimate the BER incorporating probability density function (PDF). The symbol-specific BER $P_{BER}^{(s_{\mu,m})}$ is calculated as

$$P_{BER}^{(s_{\mu,m})} = \frac{1}{\log_2(M_\mu)} \sum_{i=0}^{M_\mu-1} h_{m,i}^{(\mu)} \cdot \int_{R_i} p(y_{\nu,m})\, \mathrm{d}y \ , \tag{13}$$

where $h_{m,i}$ represents the Hamming distance between the symbols $s_{\mu,m}$ and $s_{\mu,i}$ at μ-th input. The PDF of the received symbol $y_{v,m}$ is described as $p(y_{v,m})$ and R_i is the decision region. The overall BER is defined as

$$P_{\mathrm{BER}}^{(\mu)} = \frac{1}{M_\mu} \sum_{m=0}^{M_\mu - 1} P_{\mathrm{BER}}^{(s_{\mu,m})} \ . \tag{14}$$

In summary, the system performance will be degraded due to the noise amplification originating from the extensive utilization of the post-equalizer. Theoretically, this deterioration in the system performance is shown in (12). From an analytical perspective, it is worth investigating, in order to practically estimate the usefulness of shifting the post-equalizer taps to the transmitter side with the given system model. The challenges of designing the optimal PPE filter and evaluating the system performance are addressed in the following sections.

3. Design and Optimization of Pre- and Post-Equalization

According to the literature, a few-tap pre-equalizer is preferred to compensate for the channel influence without any noise enhancement [16,17,20,22]. The zero-forcing design of a pre-equalizer is chosen predominantly due to the simple implementation. Moreover, the post-equalizer filter taps are separately calculated. By contrast, in the proposed transceiver–joint equalization, the pre-equalizer and the post-equalizer are calculated and jointly optimized to yield performance benefits. The pre-equalizer is characterized as a finite impulse response (FIR) filter, and the role of pre-equalizer taps $p_{i,\mu}$ is shown as

$$b_\mu(k) = \sum_{i=1}^{L_\mathrm{P}} p_{i,\mu} \cdot s_\mu[k - (i-1)], \qquad \mu = 1, \ldots, n_\mathrm{T} \ , \tag{15}$$

where L_P is the length of the pre-equalizer at the μ-th MIMO layer. The structure of a linear FIR pre-equalizer present on each MIMO layer is shown in Figure 3. The main goal is to design and optimize the PPE filter taps to enhance the system performance while maintaining power and cost budgets.

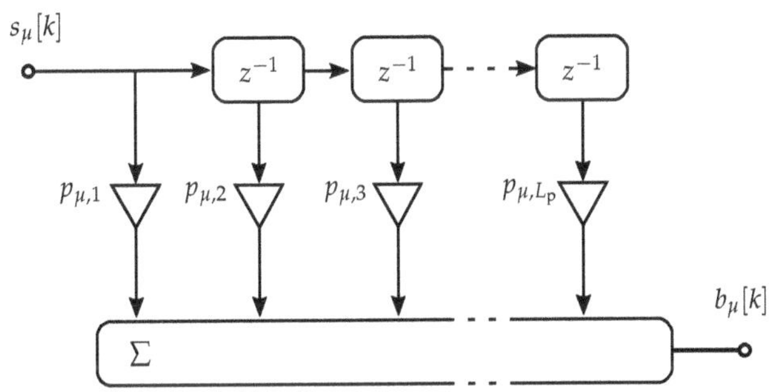

Figure 3. The structure of a linear FIR pre-equalizer filter, where the delay by one symbol is denoted by z^{-1}.

In this work, a numerical convex optimization is applied to calculate and optimize the PPE coefficients. The primal–dual interior-point constrained algorithm is tailored with respect to the analytical solution of the given system model [28]. Hence, the objective function subjected to the constraints is stated as follows

$$\begin{aligned} \underset{p_\mu(k),\, f_{\nu\mu}(k)}{\text{minimize}} \quad & \theta_\mu(h_{\nu\mu}(k)) \qquad\qquad , \nu = 1, \ldots, n_\mathrm{R} \\ \text{subject to} \quad & \boldsymbol{F} \cdot \boldsymbol{H} \cdot \boldsymbol{P} - \boldsymbol{I} = 0 \ , \\ \text{and} \quad & s_{\mu, M_\mu - 1} - \mathrm{q} \cdot P_{\mathrm{s},\mu}^{(\max)} = 0 \ . \end{aligned} \tag{16}$$

The objective function is defined to minimize the NWF on each MIMO layer in accordance with the channel state information. The values of $p_\mu(k)$ and $f_{\nu\mu}(k)$ will converge to their optimal value with each iteration. In (16), the first constraint determines the complete removal of the ISI and ICI with the PPE filter. Additionally, the second constraint assures that the symbol with the highest amplitude is limited to the maximum transmit power of the laser source. In (16), the optimization problem is called the primal problem. In order to solve this problem, the duality optimization is utilized, where the constraints are incorporated with the objective function. The modified objective function of the optimization statement is

$$\min_{p_\mu(k),\, f_{\nu\mu}(k)} \ \theta_\mu(h_{\nu\mu}(k)) + \max_{\alpha, \beta} \left((\boldsymbol{F} \cdot \boldsymbol{H} \cdot \boldsymbol{P} - \boldsymbol{I}) \cdot \alpha + \left(s_{\mu, M_\mu - 1} - \mathrm{q} \cdot P_{\mathrm{s},\mu}^{(\max)} \right) \cdot \beta \right) \ , \tag{17}$$

where α and β represent the scalar slopes of the penalty function. The purpose of the penalty function is to penalize the objective function when the constraints are not followed. Furthermore, the objective function is rewarded when the constraints are obeyed. The objective function (17) is termed as the primal problem, which is converted into the duality problem, as the min and max operators are in the function simultaneously. Thus, the dual problem can be written as

$$\max_{\alpha, \beta} \ \min_{p_\mu(k),\, f_{\nu\mu}(k)} \left(\theta_\mu(h_{\nu\mu}(k)) + (\boldsymbol{F} \cdot \boldsymbol{H} \cdot \boldsymbol{P} - \boldsymbol{I}) \cdot \alpha + \left(s_{\mu, M_\mu - 1} - \mathrm{q} \cdot P_{\mathrm{s},\mu}^{(\max)} \right) \cdot \beta \right) \ . \tag{18}$$

The objective function in (18) can be easily solved with the Lagrangian solver. Hence, in order to find the optimal coefficients of the proposed joint-PPE scheme, the following condition should be satisfied

$$\nabla \left(\theta_\mu(h_{\nu\mu}(k)) + (\boldsymbol{F} \cdot \boldsymbol{H} \cdot \boldsymbol{P} - \boldsymbol{I}) \cdot \alpha + \left(s_{\mu, M_\mu - 1} - \mathrm{q} \cdot P_{\mathrm{s},\mu}^{(\max)} \right) \cdot \beta \right) = 0 \ . \tag{19}$$

To hold the strong duality between the primal and dual problems, there should be zero difference between the primal optimal solution and the dual optimal solution. The strong duality is executed using the Karush–Kuhn–Tucker (KKT) optimality conditions, which are given as

$$\begin{aligned} (\boldsymbol{F} \cdot \boldsymbol{H} \cdot \boldsymbol{P} - \boldsymbol{I}) \cdot \alpha &= 0 \ , \\ \left(s_{\mu, M_\mu - 1} - \mathrm{q} \cdot P_{\mathrm{s},\mu}^{(\max)} \right) \cdot \beta &= 0 \ , \\ \alpha &\geq 0 \ , \\ \beta &\geq 0 \ , \\ \boldsymbol{F} \cdot \boldsymbol{H} \cdot \boldsymbol{P} - \boldsymbol{I} &= 0 \ , \\ s_{\mu, M_\mu - 1} - \mathrm{q} \cdot P_{\mathrm{s},\mu}^{(\max)} &= 0 \ . \end{aligned} \tag{20}$$

For solving such a problem, the KKT conditions are modified, where a variable named t is introduced. This variable t is a positive number to control the degree of perturbation towards the optimal solution. The constraints, mentioned in (20), are adapted with t_n where n is the total number of constraints, and mentioned as

$$\alpha = \frac{t_1}{(\boldsymbol{F} \cdot \boldsymbol{H} \cdot \boldsymbol{P} - \boldsymbol{I})} \ , \qquad \beta = \frac{t_2}{\left(s_{\mu,M_\mu-1} - \mathrm{q} \cdot P_{\mathrm{s},\mu}^{(\mathrm{max})}\right)} \ . \tag{21}$$

The characteristics of t_n are that the zero-duality gap is achieved between the primal and the dual problems when $t_n \rightarrow 0, \forall n$. By combining (19), (20) and (21), the final optimization problem is defined as

$$\nabla\left(\theta_\mu(h_{\nu\mu}(k)) + t_1 \cdot \log(\boldsymbol{F} \cdot \boldsymbol{H} \cdot \boldsymbol{P} - \boldsymbol{I}) + t_2 \cdot \log\left(s_{\mu,M_\mu-1} - \mathrm{q} \cdot P_{\mathrm{s},\mu}^{(\mathrm{max})}\right)\right) = 0 \ . \tag{22}$$

Using the Newton solver, (22) is solved by choosing larger values for t_n, which results in discovering the analytical center of the feasible region. Afterwards, the value of t_n decreases to locate the optimal solution of the given statement.

Ultimately, the jointly designed PPE filter taps are optimized to remove the interferences, given minimal increment in the NWF. From joint-PPE, the post-equalizer will contribute to nominal noise amplification. Therefore, the hypothesis states that the joint-PPE is expected to yield better performance than ZF-based PPE and MMSE-based PE-only.

4. System Model Using SVD

When considering the SVD technique applied to a frequency-selective MIMO channel, the resulting discrete-time block-oriented system is illustrated in Figure 4. The joint-PPE and ZF-PPE schemes are analyzed over the SVD-based (2×2) MIMO system. The advantage of utilizing SVD is that the channel is converted into a diagonal matrix. It implies that the crosstalk components, $\boldsymbol{H}_{n_\mathrm{R}\, n_\mathrm{T}}$ when $n_\mathrm{R} \neq n_\mathrm{T}$, will be completely mitigated. The modified frequency-selective channel $\boldsymbol{H}$ is defined as

$$\boldsymbol{H} = \boldsymbol{S} \cdot \boldsymbol{V} \cdot \boldsymbol{D}^{\mathrm{H}} \ , \tag{23}$$

where $\boldsymbol{V}$ contains the positive square roots of the eigenvalues of $\boldsymbol{H}^{\mathrm{H}}\boldsymbol{H}$ matrix with $(n_\mathrm{R} \times n_\mathrm{T})$ and $\boldsymbol{S}$ denotes the unitary matrix of $(n_\mathrm{R} \times n_\mathrm{R})$ dimensions. The matrix $\boldsymbol{D}^{\mathrm{H}}$ of $(n_\mathrm{T} \times n_\mathrm{T})$ dimension represents the conjugate transpose (Hermitian operation H) of the unitary matrix $\boldsymbol{D}$. The resulting numerical matrix after SVD is the real-valued diagonal matrix $\boldsymbol{V}$, which has the following form

$$\boldsymbol{V} = \begin{bmatrix} V_{11} & \cdots & 0 \\ \vdots & \ddots & \vdots \\ 0 & \cdots & V_{n_\mathrm{R}\, n_\mathrm{T}} \end{bmatrix} . \tag{24}$$

The overall transmission of the system using SVD is described as

$$\boldsymbol{y} = \boldsymbol{F} \cdot \boldsymbol{V} \cdot \boldsymbol{P} \cdot \boldsymbol{s} + \boldsymbol{F} \cdot \tilde{\boldsymbol{w}} \ . \tag{25}$$

The implication of incorporating SVD is distinct eye-openings, which is formulated as

$$U_{\mathrm{A},\mu} = \sqrt{\xi_\mu} \cdot U_{\mathrm{s},\mu} \ , \tag{26}$$

with $U_{s,\mu}$ denoting the half-vertical transmit amplitude at μ-the MIMO input and $\sqrt{\xi_\mu}$ is the positive root of the eigenvalue of $\boldsymbol{H}^{\mathrm{H}}\boldsymbol{H}$. The quality of transmission is calculated as

$$\varrho_\mu = \frac{U^2_{\mathrm{A},\mu}}{\widetilde{P}_{\mathrm{n},\mu}} = \xi_\mu \frac{U^2_{\mathrm{s},\mu}}{\theta_\mu \cdot P_{\mathrm{n}}} \ . \tag{27}$$

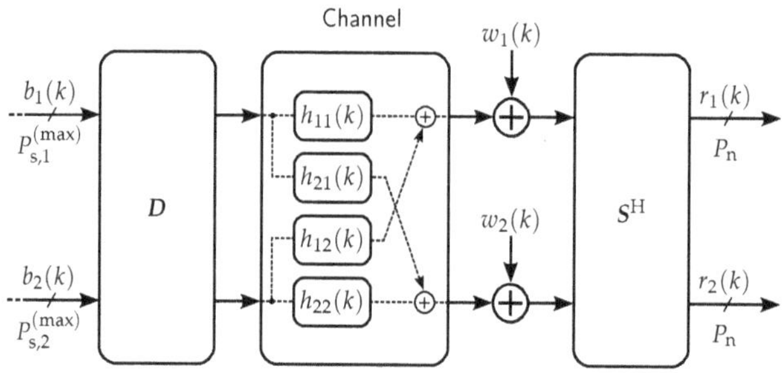

Figure 4. Optical MIMO channel utilizing the SVD technique.

The advantage of incorporating SVD in the system model is transforming the channel $\boldsymbol{H}$ into non-interfering and independent transmission layers with unequal gains. Please note that neither the noise power nor the transmit power are enhanced. The resulting orthogonal system model of the μ-th layer based on SVD is shown in Figure 5.

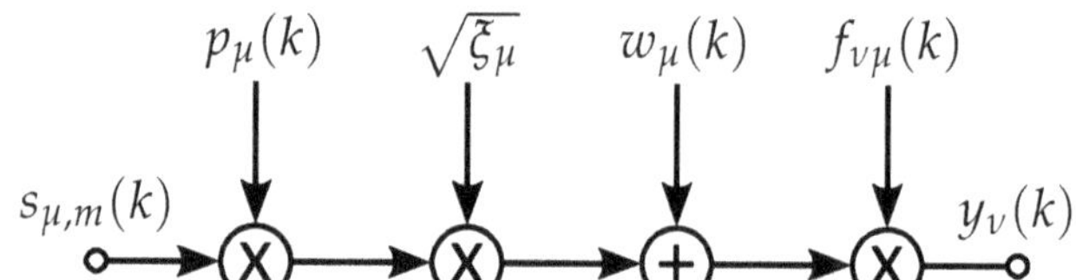

Figure 5. Simplified layer-specific MIMO system model with joint-PPE scheme.

5. Results

In this section, the performance of the proposed joint-PPE scheme is assessed using the given system model with a 1.4 km MMF channel. Moreover, the performance comparison is demonstrated between the ZF-based PPE and the proposed joint-PPE using Monte Carlo simulations and the experimental setup.

5.1. Simulation Results

The discrete-time Monte Carlo simulations are incorporated with a (2×2) MIMO system using a 1.4 km long MMF channel. The simulations are implemented using three sequential steps. In the first step, the MMSE-based PE-only is applied to the provided system model while considering the pre-equalizer as a Dirac delta function. Additionally, the channel state information is computed from the received data using the pilot-based least-square algorithm. The second step involves the design and optimization of the PPE coefficients with the help of the given constrained numerical optimization. In the last step, the optimal joint-PPE coefficients are integrated in the system model, and the BER performance is calculated. Finally, the performance of the joint-PPE, the ZF-PPE and the PE-MMSE (also known as PE-only) are compared. Please note that the ZF-PPE incorporates a ZF-based pre-equalizer with an MMSE-based post-equalizer [29,30].

The primary features of an IM/DD system are simple implementation and cost-effectiveness. Therefore, limiting the complexity of the transmitter is a factor, which should be considered while evaluating the system performance. In order to determine the optimal number of pre-equalizer filter taps, the relationship of the NWF with the pre-equalizer taps is calculated. The results are illustrated in Figure 6, where the optimized pre-equalizer coefficients achieve a significant reduction in the NWF on both the MIMO layers as compared to the PE-MMSE NWF of 7 on the first layer and 58.48 on the second layer. The extent of NWF decrement is directly related to the improvements in the system performance. Clearly, the optimally designed pre-equalizer using the joint-PPE scheme surpasses the ZF-PPE scheme. Hence, the joint-PPE is expected to yield better BER performance. It is noteworthy that the maximum reduction in the NWF is observed with the joint-PPE using $L_\mathrm{p} = 4$ at the first layer. However, the ZF-PPE scheme on the first MIMO layer has attained the maximum reduction in the NWF with $L_\mathrm{p} = 30$. Similarly, the maximum reduction in the NWF on the second MIMO layer is observed at $L_\mathrm{p} = 4$ and $L_\mathrm{p} = 30$ with the joint-PPE and the ZF-PPE schemes, respectively. On the basis of the reduction of the NWF and minimal transmitter's complexity, a pre-equalizer with at most four-taps on each layer is chosen for further performance analysis. Evidently, the joint-PPE requires a considerably smaller number of pre-equalizer taps. Thus, the prerequisites of the IM/DD systems are followed by a few pre-equalizer taps in conjunction with a post-equalizer.

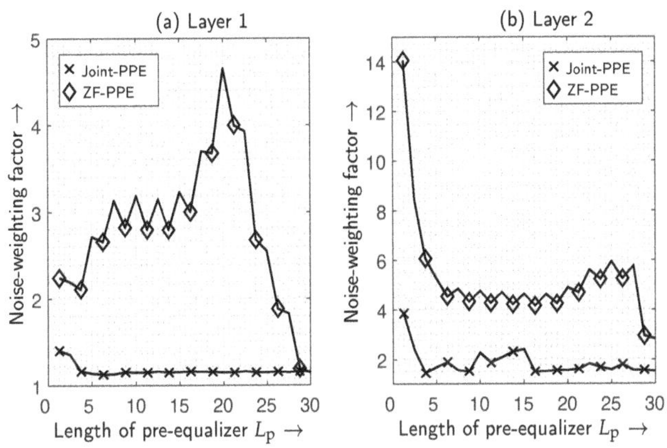

Figure 6. Noise weighting factor with increasing number of pre-equalization coefficients, where a layer-wise comparison between the proposed joint-PPE scheme and the ZF-based PPE is exhibited.

Prior to the system performance evaluation, the measured channel impulse response of a 1.4 km MMF is illustrated in Figure 7. The measured channel state information is obtained using an SDM-based optical MIMO experimental setup with offset launching conditions into an MMF. Afterwards, the estimated channel contains dispersion impairments and mode-dependent losses in accordance with the excited mode groups. This channel information is incorporated in the computer simulation to imitate the SDM technique. From Figure 7, it is clear that both MIMO layers are adversely affected due to crosstalk $h_{\nu\mu}(k)$, when $\nu \neq \mu$. Additionally, a significant amount of ISI is also observed due to differential mode group delays.

The proposed joint-transceiver equalization system performance is compared with the conventional ZF-PPE and PE-MMSE. The BER performances using PAM-2 and PAM-4 transmissions are demonstrated in Figures 8 and 9. The enhancements in the average received power P_r compared to the PE-MMSE are presented in Table 2 at a BER of 10^{-4}. The BER results prove that the joint-PPE outperforms the ZF-PPE and PE-MMSE. Additionally, the BER performance of the joint-PPE can be advanced using the SVD-based joint-PPE. Under the first PAM-2 transmission scenario, a P_r gain SVD-PPE of $\approx$3.52 dB is observed in comparison to PE-MMSE. Similarly, the performance improvement of $\approx$2.54 dB is marked while solely using the joint-PPE. The joint-PPE scheme achieves an identical BER performance with only half of the optimized pre-equalizer taps compared to the ZF-PPE. It is evident by the overlapping BER performances of the joint-PPE with $L_p = 2$ and ZF-PPE with $L_p = 4$. Considering the PAM-4 transmission on both MIMO inputs under the second scenario, a slight improvement is noted in comparison to the first scenario. The joint-PPE in conjunction with SVD yields the best system performance, surpassing PE-MMSE by ≈3.86 dB. The joint-PPE single-handedly achieves a P_r gain of ≈2.73 dB compared to PE-MMSE. An improvement in P_r of ≈1.13 dB indicates that the proposed joint-PPE enhances the system performance to a higher extent than the conventionally deployed ZF-PPE.

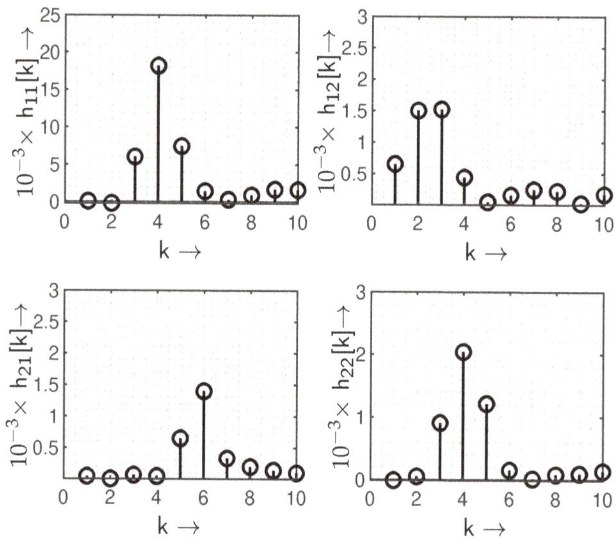

Figure 7. Measured channel impulse response of a 1.4 km long MMF with respect to the symbol rate $f_T = 1/T_s = 5$ GHz per MIMO layer at an operating wavelength of 1550 nm.

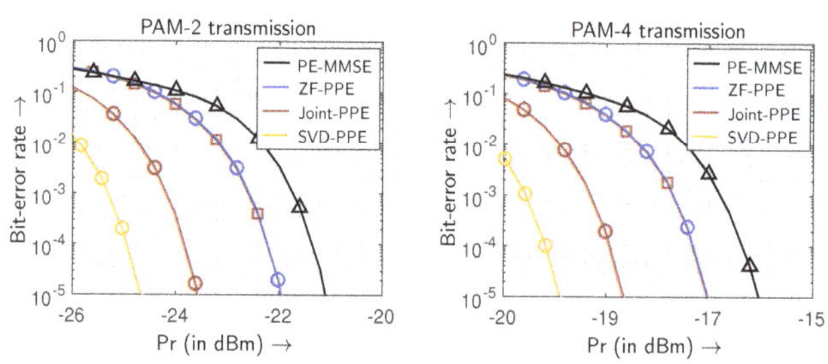

Figure 8. BER dependent on simulated received power P_r for PAM-2 and PAM-4 transmissions. The BER values associated with $L_p = 2$ and $L_p = 4$ are represented with the square and the circle markers, respectively.

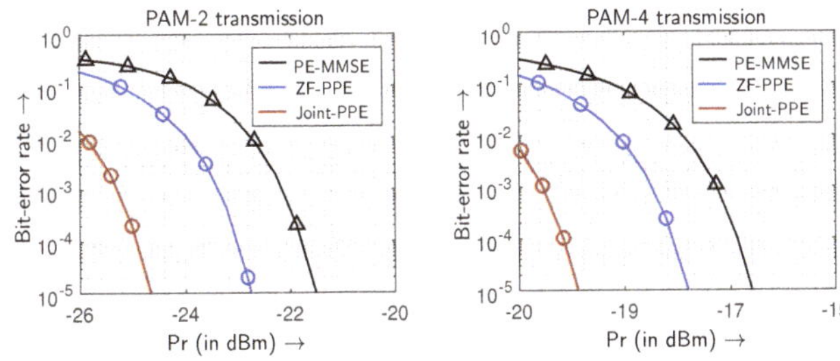

Figure 9. BER dependent on simulated received power P_r for PAM-2 and PAM-4 transmissions for SVD-based optical MIMO system.

Table 2. Overall simulated P_r gains of the (2×2) optical MIMO system at 10^{-4} BER using a 1.4 km long MMF channel at $f_T = 5$ GHz.

Scenario	Modulation Format	Equalization	L_p	L_f	P_r (in dBm) at a BER of 10^{-4}	P_r Gain (in dB)
First scenario	PAM-2	PE-only	-	30	−21.37	-
		ZF-PPE	4	26	−22.19	0.82
		Joint-PPE	4	26	−23.91	2.54
		SVD-PPE	4	26	−24.89	3.52
Second scenario	PAM-4	PE-only	-	30	−16.23	–
		ZF-PPE	4	26	−17.27	1.04
		Joint-PPE	4	26	−18.96	2.73
		SVD-PPE	4	26	−20.09	3.86

In summary, the simulated BER performances and the P_r gains testify to the advantages of using the jointly designed and optimized PPE filter with PAM-2 and PAM-4 formats. Moreover, the hypothesis is established with the analytical proof that, by shifting the post-equalization taps to the transmitter side, the system performance is boosted due to the limited amplification of the noise.

5.2. Experimental Results

A (2×2) optical MIMO experimental setup is developed, which is shown in Figures 10 and 11. Similar to the provided system model, this measurement setup is also divided into three parts: MIMO transmitter, MMF channel and MIMO receiver.

Starting from the transmitter side, a S/P converter is applied to convert a serial data stream into two pseudo-random binary sequences (PRBSs). Afterward, these sequences are mapped on the Gray coded symbols in accordance with the chosen PAM-2 or PAM-4 modulation formats. Additionally, the pilot symbols are also appended to the data symbols for the purpose of channel estimation. By using the above-mentioned optimization statement, the optimal pre-equalizer coefficients are computed using channel information. Consequently, multi-level signals are enabled by applying pre-equalization to the input data streams. In case no channel state information is available, the pre-equalizer acts as a Dirac delta function, and the results for the PE-only are obtained. The signal-generation process is concluded by generating the electrical signals from the input symbols using an arbitrary waveform generator (AWG). The channels of the AWG are operated on the symbol rate of 5 GBaud per MIMO layer. The electrical signals are converted into the optical signals using chirp-free external Mach-Zehnder modulators (MZMs). A lithium-niobate ($LiNbO_3$) substrate is used in the optical circuit of the MZM. The bias of the MZM is manually adjusted, and the quadrature points are stabilized using a temperature controller. The carrier laser sources work on the operating wavelength $\lambda = 1550$ nm. Using variable optical attenuators (VOAs), the overall BER performance is calculated over a distinct range of the average received optical power (P_r). Thereafter, the MMF splice boxes are utilized for the excitation of the selective mode groups by launching the light with a certain offset. In this provided experimental setup, the SDM scheme is enabled by utilizing centric and eccentric launching conditions into a 1.4 km long MMF with the offsets of 0 µm and 15 µm, respectively. The offset light-launching conditions are demonstrated in Figures 2 and 10. Centric launching of the light into the MMF results in the excitation of the fundamental mode group or lower-order mode (LOM) group. Similarly, the high-order mode (HOM) groups are excited with the 15 µm offset launching condition. The last part of the transmitter includes an optical fusion coupler, which is useful in multiplexing the mode groups. In this setup, custom-made fusion couplers are utilized with a slight asymmetric power split ratio as these are well suited for mode multiplexing (MUX) applications [31,32]. The function of a mode MUX is to combine the LOM groups with HOM groups using an x-shaped structure. The LOM groups travel from the center of the fiber core. Thus, these groups are more likely to travel in a transverse direction. However, the HOM groups couple with LOM groups because of their wider electric field distribution. The mode MUX and mode demultiplexer (DMUX) have 45% power-splitting ratio with ≈3 dB insertion loss.

The optical signal, containing the LOM groups and HOM groups, is propagated through the OM4 MMF channel with a length of 1.4 km. The transmitted signal suffers from degradation due to ISI and ICI due to a greater amount of interference present in an MMF compared to an SMF. The measured channel impulse response is demonstrated in Figure 7. The channel state information is obtained by estimating the influence on the appended pilot symbols in the data streams. A least-square pilot estimator algorithm is developed to measure the channel. The significant amount of interference is evident from Figure 7. The challenge of recovering the transmitted data is dealt with by the MIMO receiver.

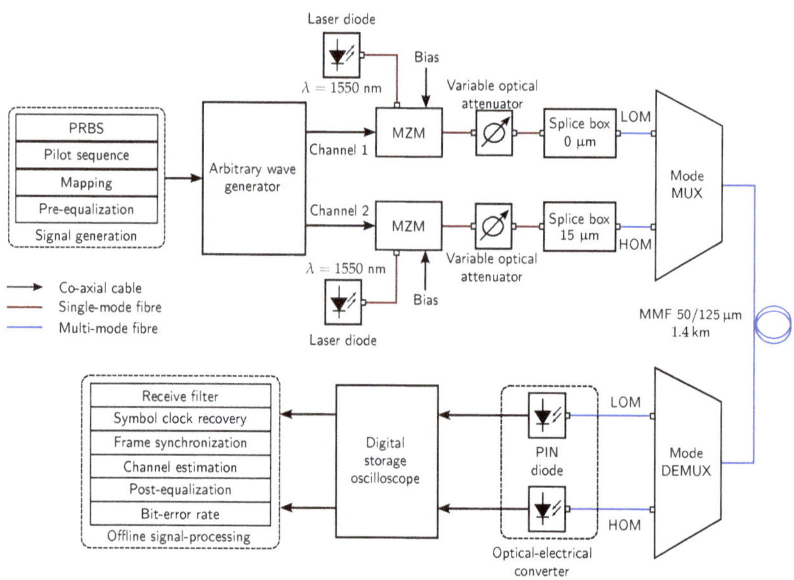

Figure 10. Experimental setup for the joint-PPE scheme using an optical MIMO system with a 1.4 km long MMF channel.

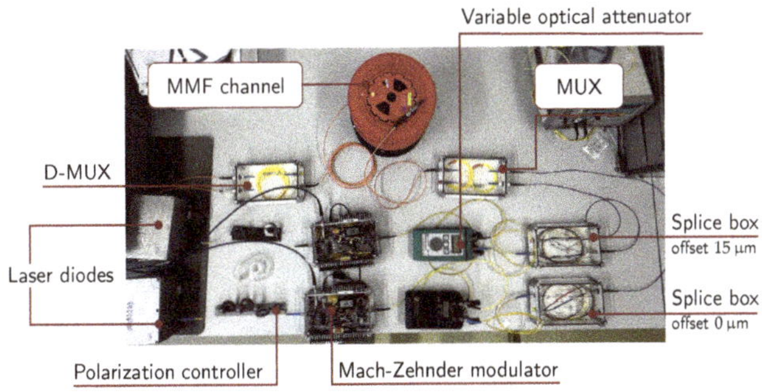

Figure 11. Experimental setup of (2 × 2) optical MIMO system for performance evaluation of the proposed joint-transceiver scheme.

After propagating through the MMF channel, the degraded signal is received by the PIN photo-detector on each layer. These diodes act as optical-to-electrical converters. An inherent increment in capacitance is observed due to the presence of a large detection area. Consequently, the sensitivity of the receiver suffers from this property. An internal trans-impedance amplifier is present, which amplifies the received electrical signals. A digital storage oscilloscope (DSO) is utilized to capture the electrical signals using a resolution of 40 GSa/s at a bandwidth of 13.6 GHz. The stored signals undergo an offline signal-processing chain with one sample per symbol. The first step includes receive filtering, where a rectangular filter similar to the transmit filter response is applied. Afterward, the symbol clock is recovered from the down-sampled filter signal. The next step involves the estimation of the frequency-selective channel. Most importantly, a post-equalizer is used to mitigate all the interference due to the channel. In the end, hard decision decoding is employed to retrieve the original data and subsequently, the performance is evaluated in terms of the BER performance. To obtain reliable and repetitive results, 74 frames are calculated on each MIMO layer for every instance, which adds up to five million bits approximately. The symbol clock recovery is conducted using an open loop timing recovery method for estimating the exact sampling time. Additionally, the beginning of the data frame is computed using the fine-timing acquisition algorithm. In this acquisition process, the cross-correlation between a segment of the received signal and the pilot sequence provides an aid to determine the exact starting of the data frame. Moreover, the forward error correcting decoding is not deployed at the receiver.

Using the experimental setup of a (2 × 2) MIMO system with a 1.4 km long MMF channel, the average optical received power P_r dependent on the overall BER system is illustrated in Figure 12 and in Table 3. The performance comparison is conducted between the PE-MMSE, the ZF-PPE and the joint-PPE schemes. Moreover, the results are also obtained for two different scenarios, which includes the use of PAM-2 and PAM-4 formats. From the BER performances shown in Figure 12, the proposed joint-PPE clearly outperforms the ZF-PPE and the PE-MMSE schemes. When PAM-2 is utilized on both MIMO layers, the joint-PPE scheme achieves the maximum P_r gain of ≈2.58 dB. Moreover, an improvement of ≈1.74 dB in P_r gain is observed by the joint-PPE compared to the ZF-PPE. Similarly, the highest performance improvements of ≈2.68 dB are indicated by the joint-PPE using the PAM-4 format. The P_r gain ≈1.73 dB of the joint-PPE over the conventional ZF-PPE scheme proves the effectiveness of the proposed joint-PPE. The transmission qualities of all three equalization schemes are demonstrated in Figures 13 and 14. The wider eye-opening of the PPE schemes compared to the PE-only verifies that the transmission quality is boosted using a few-tap pre-equalizer. When the transmission qualities of the ZF-PPE and the joint-PPE are measured, better eye-opening is exhibited by the joint-PPE scheme.

Table 3. Measured average optical received power P_r gain of the (2×2) optical MIMO system at 10^{-4} BER using a 1.4 km long MMF channel at $f_T = 5$ GHz.

Scenario	Modulation Format	Equalization	L_p	L_f	P_r (in dBm) at BER of 10^{-4}	P_r Gain (in dB)
First scenario	PAM-2	PE-only	-	30	−7.58	-
		ZF-PPE	4	26	−8.42	0.84
		Joint-PPE	4	26	−10.16	2.58
Second scenario	PAM-4	PE-only	-	30	−2.13	-
		ZF-PPE	4	26	−3.08	0.95
		Joint-PPE	4	26	−4.81	2.68

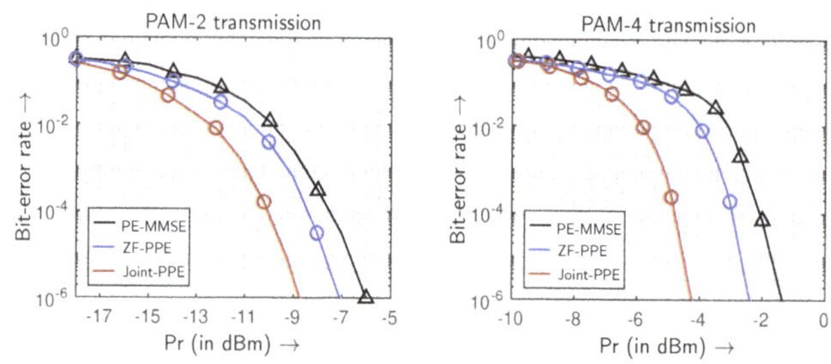

Figure 12. Overall BER performance comparison between the PE-MMSE, ZF-PPE and Joint-PPE with PAM-2 and PAM-4 formats, where four-tap pre-equalizers are utilized in both PPE schemes.

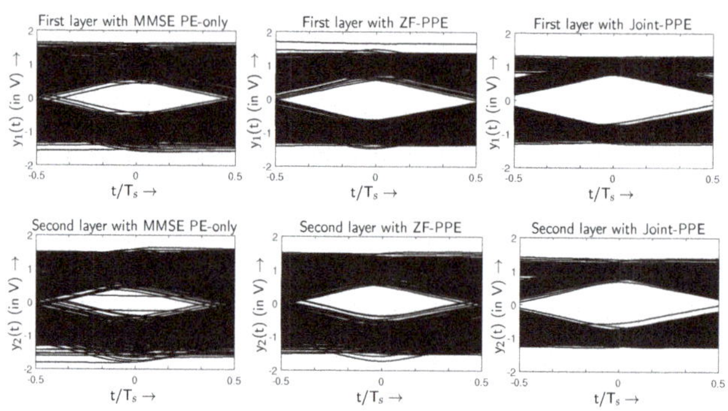

Figure 13. Eye diagrams of the joint-PPE, ZF-PPE and MMSE PE-only using PAM-2 transmission.

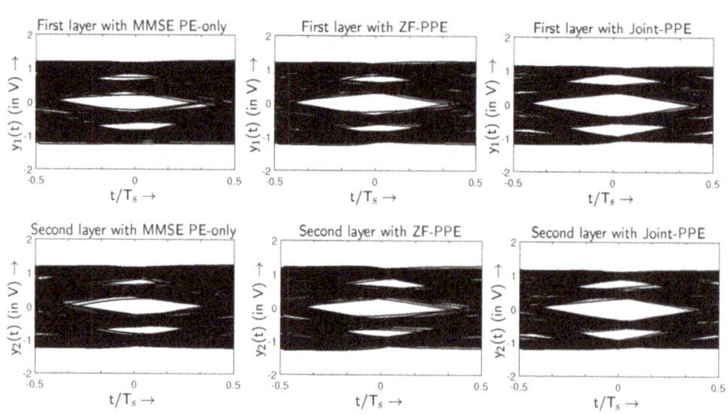

Figure 14. Eye diagrams of the joint-PPE, ZF-PPE and MMSE PE-only using PAM-4 transmission.

6. Discussion

In this work, the effectiveness of the proposed transceiver–joint equalization filter is evaluated and compared with the conventional PPE and the PE-only schemes. The performance evaluation is conducted using the HOM formats over a (2×2) SDM-based optical MIMO system with a 1.4 km MMF transmission link. The average optical received power P_r required to reach a BER of 10^{-4} using the PAM-4 formats is improved by $\approx$2.68 dB with just a four-tap optimally designed pre-equalizer with post-equalization. Similarly, an enhancement of $\approx$2.58 dB in the P_r gain is achieved by the joint-PPE using PAM-2 format. Despite using a longer MMF channel with twice the bit-rate compared to [19], the improvement of $\approx$0.65 dB and $\approx$0.84 dB in the P_r gains at 10^{-4} BER is still accomplished. Additionally, with the improved numerical optimization of the PPE taps, an enhancement of nearly two folds in the required Pr gain at 10^{-4} BER is attained using PAM-2 constellation compared to [23]. In contrary to the existing joint PPE schemes, the efficiency of the proposed transceiver–joint equalization is verified on a critical dispersion-impaired channel with the MIMO configuration. Further research should focus on the channel robustness in a dynamic environment.

7. Conclusions

In conclusion of this work, the performance of a short-reach (2×2) optical MIMO system with PAM-2 and PAM-4 constellations is demonstrated. This work aims to compute the effectiveness of the jointly computed pre- and post-equalizer within the power and cost constraints. The experimental results from this study reveal that the system performance is improved by $\approx$2.5 dB while utilizing the proposed joint-PPE with PAM-2 and PAM-4 over PE-only. Additionally, when a performance evaluation between the conventional PPE scheme and the joint-PPE is conducted, a significant ($\approx$1.5 times) improvement is observed with a constant number of PPE coefficients in both PPE schemes. These results support the hypothesis that the optimally designed PPE coefficients can improve the BER performance of an MMF-based optical system. The findings have implications for increasing the spectral efficiency of critical dispersion-impaired transmission links, and emphasize the need for cost-effectiveness and simple implementation. Future research should focus on optimizing the PPE coefficients with channel robustness. Additionally, the potential of PPE schemes with an MMF link should be explored in the area of quantum communication utilizing quantum key distribution mechanisms [33–35]. In summary, this work adds valuable insight into the optimization and utilization of the joint-PPE with the HOM formats in an optical MIMO system consisting of an MMF link.

Author Contributions: Conceptualization, J.S. and A.A.; methodology, J.S. and A.A.; software, J.S.; validation, A.A.; formal analysis, J.S.; investigation, J.S.; resources, J.S. and A.A.; data curation, J.S.; writing—original draft preparation, J.S.; writing—review and editing, A.A.; visualization, J.S.; supervision, A.A.; project administration, A.A.; All authors have read and agreed to the published version of the manuscript.

Funding: This research received no external funding.

Institutional Review Board Statement: Not applicable.

Informed Consent Statement: Not applicable.

Data Availability Statement: The data that support the findings of this study are available on request from the corresponding author. The data are not publicly available due to privacy or ethical restrictions.

Conflicts of Interest: The authors declare no conflict of interest.

References

1. Bai, N.; Ip, E.; Huang, Y.K.; Mateo, E.; Yaman, F.; Li, M.J.; Bickham, S.; Ten, S.; Liñares, J.; Montero, C.; et al. Mode-division multiplexed transmission with inline few-mode fiber amplifier. *Opt. Express* **2012**, *20*, 2668. [CrossRef] [PubMed]
2. Richardson, D.J.; Fini, J.M.; Nelson, L.E. Space-division multiplexing in optical fibres. *Nat. Photonics* **2013**, *7*, 354–362. [CrossRef]

3. Zhu, Z.; Chen, J.; Zhao, M.; Pang, F.; Zhang, Q.; Ye, N. IM/DD mode division multiplexing transmission enabled by machine learning-based linear and nonlinear MIMO equalization. *Opt. Commun.* **2021**, *488*, 126832. [CrossRef]
4. Chen, H.; van Uden, R.; Okonkwo, C.; Koonen, T. Compact spatial multiplexers for mode division multiplexing. *Opt. Express* **2014**, *22*, 31582. [CrossRef]
5. Sillard, P.; Molin, D.; Bigot-Astruc, M.; Jongh, K.D.; Achten, F.; Velazquez-Benitez, A.M.; Amezcua-Correa, R.; Okonkwo, C.M. Low-Differential-Mode-Group-Delay 9-LP-Mode Fiber. *J. Light. Technol.* **2016**, *34*, 425–430. [CrossRef]
6. Pauwels, J.; der Sande, G.V.; Verschaffelt, G. Space division multiplexing in standard multi-mode optical fibers based on speckle pattern classification. *Sci. Rep.* **2019**, *9*, 17597. [CrossRef]
7. Lei, Y.; Xu, K.; Li, J.; Meng, Z.; Wu, R.; Wan, Z.; Fan, Y.; Zhang, W.; Yin, F.; Dai, Y. Feasibility of Space-Division-Multiplexed Transmission of IEEE 802.11 n/ac-Compliant Wireless MIMO Signals over OM3 Multimode Fiber. *J. Light. Technol.* **2018**, *36*, 2076–2082. [CrossRef]
8. Agrawal, G.P. *Fiber-Optic Communication Systems*; WILEY: Hoboken, NJ, USA, 2021.
9. Velázquez-Benítez, A.M.; Antonio-López, J.E.; Alvarado-Zacarías, J.C.; Fontaine, N.K.; Ryf, R.; Chen, H.; Hernández-Cordero, J.; Sillard, P.; Okonkwo, C.; Leon-Saval, S.G.; et al. Scaling photonic lanterns for space-division multiplexing. *Sci. Rep.* **2018**, *8*, 8897. [CrossRef]
10. Szczerba, K.; Westbergh, P.; Karout, J.; Gustavsson, J.S.; Haglund, Å.; Karlsson, M.; Andrekson, P.A.; Agrell, E.; Larsson, A. 4-PAM for High-Speed Short-Range Optical Communications. *J. Opt. Commun. Netw.* **2012**, *4*, 885. [CrossRef]
11. Olmedo, M.I.; Zuo, T.; Jensen, J.B.; Zhong, Q.; Xu, X.; Popov, S.; Monroy, I.T. Multiband Carrierless Amplitude Phase Modulation for High Capacity Optical Data Links. *J. Light. Technol.* **2014**, *32*, 798–804. [CrossRef]
12. Wei, J.; Stojanovic, N.; Zhang, L.; Calabrò, S.; Rahman, T.; Xie, C.; Charlet, G. Experimental comparison of modulation formats for 200 G/λ IMDD data centre networks. In Proceedings of the 45th European Conference on Optical Communication (ECOC 2019), Dublin, Ireland, 22–26 September 2019; Institution of Engineering and Technology: Hertfordshire, UK, 2019. [CrossRef]
13. Zhang, L.; Wei, J.; Stojanovic, N.; Prodaniuc, C.; Xie, C. Beyond 200-Gb/s DMT Transmission over 2-km SMF Based on A Low-Cost Architecture with Single-Wavelength, Single-DAC/ADC and Single-PD. In Proceedings of the 2018 European Conference on Optical Communication (ECOC), Rome, Italy, 23–27 September 2018. [CrossRef]
14. Rajbhandari, S.; Chun, H.; Faulkner, G.; Haas, H.; Xie, E.; McKendry, J.J.D.; Herrnsdorf, J.; Gu, E.; Dawson, M.D.; OBrien, D. Neural Network-Based Joint Spatial and Temporal Equalization for MIMO-VLC System. *IEEE Photonics Technol. Lett.* **2019**, *31*, 821–824. [CrossRef]
15. Zhai, Z.; Jiang, H.; Fu, M.; Liu, L.; Yi, L.; Hu, W.; Zhuge, Q. An Interpretable Mapping From a Communication System to a Neural Network for Optimal Transceiver-Joint Equalization. *J. Light. Technol.* **2021**, *39*, 5449–5458. [CrossRef]
16. Guo, M.; Qiao, Y.; Tang, X.; Liu, S.; Sun, Z.; Cui, H.; Xu, X.; Rusch, L.A. 112-Gb/s PAM4 with Joint Pre- and Post-Equalization for Data Center Interconnects. In Proceedings of the Asia Communications and Photonics Conference (ACPC) 2019, Chengdu, China, 2–5 November 2019; Optica Publishing Group: Washington, DC, USA, 2019; p. T2G.2.
17. Zou, D.; Chen, Y.; Li, F.; Li, Z.; Sun, Y.; Ding, L.; Li, J.; Yi, X.; Li, L.; Li, Z. Comparison of Bit-Loading DMT and Pre-Equalized DFT-Spread DMT for 2-km Optical Interconnect System. *J. Light. Technol.* **2019**, *37*, 2194–2200. [CrossRef]
18. Iijima, Y.; Yuminaka, Y. Double-Rate Equalization Using Tomlinson-Harashima Precoding for Multi-valued Data Transmission. In Proceedings of the 2016 IEEE 46th International Symposium on Multiple-Valued Logic (ISMVL), Sapporo, Japan, 18–20 May 2016. [CrossRef]
19. Singh, J.; Ahrens, A.; Lochmann, S. Joint Pre- and Post-Equalization with Higher-Order Modulation Formats in SDM-Based Optical MIMO Systems. *Photonics* **2022**, *9*, 876. [CrossRef]
20. Zhang, J.; Yu, J.; Chien, H.C. Single-carrier 400G based on 84-GBaud PDM-8QAM transmission over 2,125 km SSMF enhanced by pre-equalization, LUT and DBP. In Proceedings of the 2017 Optical Fiber Communications Conference and Exhibition (OFC), Los Angeles, CA, USA, 19–23 March 2017; pp. 1–3.
21. Tang, X.; Qiao, Y.; kung Chang, G. Experimental Demonstration of C-band 112-Gb/s PAM4 over 20-km SSMF with Joint Pre- and Post-equalization. In Proceedings of the Optical Fiber Communication Conference (OFC) 2020, San Diego, CA, USA, 8–12 March 2020; Optica Publishing Group: Washington, DC, USA, 2020. [CrossRef]
22. Tang, X.; Qiao, Y.; Chen, Y.W.; Lu, Y.; Chang, G.K. Digital Pre- and Post-Equalization for C-Band 112-Gb/s PAM4 Short-Reach Transport Systems. *J. Light. Technol.* **2020**, *38*, 4683–4690. [CrossRef]
23. Singh, J.; Gotten, M.; Ahrens, A.; Lochmann, S. Joint Pre- and Post-Equalization in Optical MIMO Systems using Multi-Level Signaling. In Proceedings of the 2022 IEEE 95th Vehicular Technology Conference: (VTC2022-Spring), Helsinki, Finland, 19–22 June 2022. [CrossRef]
24. Fazea, Y.; Amphawan, A.; Al-Gumaei, Y.; Al-Samman, A.M.; Al-Rahmi, W.M. Modes power equalization based-singular value decomposition in mode division multiplexing systems for multi-hungry bandwidth applications. *Opt. Fiber Technol.* **2021**, *61*, 102389. [CrossRef]
25. Pankow, J.; Aust, S.; Lochmann, S.; Ahrens, A. Modulation-mode assignment in SVD-assisted optical MIMO multimode fiber links. In Proceedings of the 15th International Conference on Optical Network Design and Modeling—ONDM 2011, Bologna, Italy, 8–10 February 2011; pp. 1–6.
26. Ahrens, A.; Kuhn, V. Modulation-mode and power assignment for MIMO-BICM schemes. In Proceedings of the 2008 International ITG Workshop on Smart Antennas, Darmstadt, Germany, 26–27 February 2008. [CrossRef]

27. Ahrens, A.; Lange, C. Modulation-mode and power assignment in SVD-equalized MIMO systems. *Facta Univ.-Ser. Electron. Energetics* **2008**, *21*, 167–181. [CrossRef]
28. Stephen Boyd, L.V. *Convex Optimization*; Cambridge University Press: Cambridge, UK, 2004.
29. Jiang, Y.; Varanasi, M.K.; Li, J. Performance Analysis of ZF and MMSE Equalizers for MIMO Systems: An In-Depth Study of the High SNR Regime. *IEEE Trans. Inf. Theory* **2011**, *57*, 2008–2026. [CrossRef]
30. Proakis, J.G.; Salehi, M. *Digital Communications*; McGraw-Hill: New York, NY, USA, 2001.
31. Sandmann, A.; Ahrens, A.; Lochmann, S. Experimental Description of Multimode MIMO Channels utilizing Optical Couplers. In Proceedings of the Photonic Networks, 15. ITG Symposium, Leipzig, Germany, 5–6 May 2014; pp. 1–6.
32. Ahrens, A.; Lochmann, S. Optical couplers in multimode MIMO transmission systems measurement results and performance analysis. In Proceedings of the 2013 International Conference on Optical Communication Systems (OPTICS), Reykjavik, Iceland, 29–31 July 2013; pp. 1–6.
33. Xie, Y.M.; Lu, Y.S.; Weng, C.X.; Cao, X.Y.; Jia, Z.Y.; Bao, Y.; Wang, Y.; Fu, Y.; Yin, H.L.; Chen, Z.B. Breaking the Rate-Loss Bound of Quantum Key Distribution with Asynchronous Two-Photon Interference. *PRX Quantum* **2022**, *3*, 020315. [CrossRef]
34. Lucamarini, M.; Yuan, Z.L.; Dynes, J.F.; Shields, A.J. Overcoming the rate–distance limit of quantum key distribution without quantum repeaters. *Nature* **2018**, *557*, 400–403. [CrossRef]
35. Gu, J.; Cao, X.Y.; Fu, Y.; He, Z.W.; Yin, Z.J.; Yin, H.L.; Chen, Z.B. Experimental measurement-device-independent type quantum key distribution with flawed and correlated sources. *Sci. Bull.* **2022**, *67*, 2167–2175. [CrossRef] [PubMed]

photonics

Communication

Deep Compressed Super-Resolution Imaging with DMD Alignment Error Correction

Miao Xu [1], Chao Wang [1], Haodong Shi [1], Qiang Fu [1], Yingchao Li [1], Lianqing Dong [2] and Huilin Jiang [1,*]

[1] School of Opto-Electronic Engineering, Changchun University of Science and Technology, Changchun 130022, China
[2] Beijing Institute of Space Mechanics & Electricity, Beijing 100094, China
* Correspondence: hljiang@cust.edu.cn

Abstract: In the field of compressed imaging, many attempts have been made to use the high-resolution digital micromirror array (DMD) in combination with low-resolution detectors to construct imaging systems by collecting low-resolution compressed data to reconstruct high-resolution images. However, the difficulty of achieving micrometer-level alignment between DMD devices and detectors has resulted in significant reconstruction errors. To address this issue, we proposed a joint input generative adversarial network with an error correction function that simulates the degradation of image quality due to alignment errors, designed an optical imaging system, and incorporated prior imaging system knowledge in the data generation process to improve the training efficiency and reconstruction performance. Our network achieved the ability to reconstruct 4× high-resolution images with different alignment errors and performed outstanding reconstruction in real-world scenes. Compared to existing algorithms, our method had a higher peak signal-to-noise ratio (PSNR) and better visualization results, which demonstrates the feasibility of our approach.

Keywords: super-resolution; deep compressed sensing; alignment error correction

1. Introduction

With the development of single-pixel imaging technology, researchers have been looking for faster imaging speeds and larger pixel arrays. Therefore, scholars have proposed the theory of block-wise compressed imaging (BCI) [1]. This method can be seen as an extension of single-pixel imaging in plane space. Using high-resolution spatial light modulators to block-encode the scene in parallel, compressed data were sampled by low-resolution detectors and finally reconstructed into high-resolution images using a computationally efficient reconstruction algorithm. This achieved the acquisition of high-resolution images based on low-resolution detectors, which greatly reduced the cost of equipment and the amount of transmitted data. This could be applied to infrared detection [2], polarization imaging [3], and target tracking identification [4].

Due to their high resolution, simple control, and fast response, digital micromirror array (DMD) devices have been widely used as spatial light modulators in compressed imaging [5]. Although DMD devices have shown excellent performance in single-pixel imaging, they have also faced some challenges when used in BCI. First, due to the use of off-axis optics in imaging systems, they are susceptible to variations in the field of view, leading to distortions in various points of the image. Second, the manufacturing and calibration errors of the optical system, including aberrations in the lenses and misalignment between DMD and detector pixels, can all degrade the final image quality [6]. Finally, traditional algorithms for block reconstruction also have significant block effect issues.

To address these issues, various methods have been proposed. For example, the leakage of light from DMD and CCD pixels during registration and the effects of specific lenses on the super-resolution imaging results have been analyzed [7]. Mole patterns have

Citation: Xu, M.; Wang, C.; Shi, H.; Fu, Q.; Li, Y.; Dong, L.; Jiang, H. Deep Compressed Super-Resolution Imaging with DMD Alignment Error Correction. *Photonics* **2023**, *10*, 581. https://doi.org/10.3390/photonics10050581

Received: 11 April 2023
Revised: 6 May 2023
Accepted: 12 May 2023
Published: 17 May 2023

been loaded into the optical system, and calibration errors have been corrected based on Mole patterns [8]. Special coding schemes have been designed to reduce the effects of block effects [9]. Nonlinear mappings between measured values and the original object have been established using deep learning networks [10]. A joint input compression imaging network has been proposed, where customized coding modules are used to make the imaging degradation model input of the network [11]. However, traditional CS algorithms often result in slow reconstruction speeds and poor reconstruction quality. While innovative deep learning methods are seldom able to solve the problem of the actual optical system calibration errors that cause degradation.

In this work, we first analyzed the influence of optical system calibration errors on imaging to better simulate the degradation of imaging quality during optical imaging processes; then, a joint input generation adversarial network was proposed, which combined low-resolution degraded images and coding matrices with errors as network inputs to attempt to recover low-resolution images with unknown and complex degradation; finally, an experimental platform was built to validate the feasibility of this method in real-world scenarios, which successfully reconstructed high-resolution images with low sampling rates and reduced the error caused by misalignment.

2. Materials and Methods

2.1. Block-Wise Compressed Imaging System with DMD

As an extension of single-pixel imaging methods in the spatial domain, we designed a BCI system using DMD as a spatial light modulator (SLM). DMD is a reflective digital SLM that is composed of millions of micromirrors on the semiconductor silicon substrate [12]. Each micromirror is independently controlled and modulated by a tilting angle to modulate light. The block-based imaging system consists of five main components: the target, a distant objective, DMD, an imaging objective, and a visible light detector array. The imaging process is shown in Figure 1. Firstly, the telescope objective focuses the scene on the DMD; then, the resulting image is divided into multiple small blocks of the same size and is compressed by a common encoding matrix. Next, the modulated image is focused onto the detector array, where each pixel of the detector collects the intensity information of a target block. By varying the loaded patterns on the DMD, the detector collects a set of low-resolution compressed images. Finally, reconstruction algorithms are used to reconstruct high-resolution images. Compared to single-pixel imaging, BCI can achieve a higher resolution and use fewer sampling times.

In our system, each 4×4 micromirror on the DMD formed a large block and was projected onto one pixel of the detector. The formula for collecting data from each block is as follows:

$$y = \Phi x \tag{1}$$

In our system, $x(16 \times 1)$ is the original information corresponding to each block of the scene, $\Phi(M \times 16)$ is a measurement matrix, each row of Φ corresponds to a single pattern loaded on the DMD, $y(M \times 1)$ is the compressed sampling result, M is the sampling count of the detector and is also the number of patterns loading on the DMD. A parallel combination of the data collected from each block was performed, and the detector ultimately obtained M low-resolution images Y.

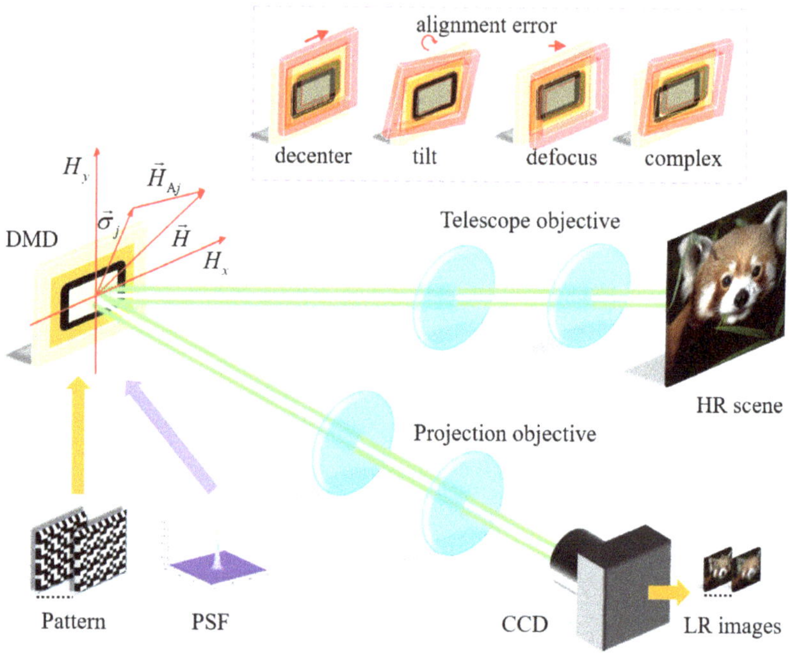

Figure 1. Block-wise compression imaging system diagram.

2.2. *Effect of Alignment Error on Compressed Imaging*

Due to the special structure of the BCI system, DMD devices need to achieve precise pixel-level alignment with the detector. However, it is difficult to achieve precise calibration during actual optical system construction. Alignment errors can bring aberrations to the actual optical system and have a significant impact on the reconstruction results. We needed to analyze the alignment errors and provide guidance for compensation in the reconstruction algorithm.

The classic wave aberration theory was used to analyze the impact of the alignment error on the imaging system. The alignment error caused the wavefront formed by the plane wave passing through the optical system to be no longer an ideal sphere. The actual wavefront was tangent to the ideal wavefront at the exit pupil, and the difference between the two wavefronts was called wavefront aberration. The off-axial optical system wave aberration of the DMD imaging system could be expressed using Seidel polynomials [13] as follows:

$$W(H,\rho,\phi) = \sum_{j}\sum_{p}^{\infty}\sum_{n}^{\infty}\sum_{m}^{\infty}(W_{klm})_j H^k \rho^l \cos^m\phi,\ k = 2p + m,\ l = 2n + m, \tag{2}$$

In this equation, W represents the optical path difference between the actual wavefront and the ideal wavefront at the pupil position, W_{klm} the wave aberration coefficient, k, l, p, n, and m represent the power series coefficients of each expansion, j represents the order number of each surface in the optical system, H is the normalized field height, and ϕ is the angle between the field coordinates and the pupil coordinates.

When there is a decenter and tilt in the system, the contribution of the surface j to the aberration of any field coordinate vector $\vec{H}$ in the viewing field should be calculated with respect to the vector of aberration centers shifted by the vertex $\vec{\sigma}_j$ of the aberration, and the shifted vector, which is called the effective field of view $\vec{H}_{Aj}$.

$$\vec{H}_{Aj} = \left(\vec{H} - \vec{\sigma}_j\right), \tag{3}$$

From this, $\vec{H}_{Aj}$ is substituted into the wave aberration vector expression and yields:

$$\begin{aligned} W(H,\rho,\phi) = & \sum_j (W_{020})_j(\vec{\rho}\cdot\vec{\rho}) + \sum_j (W_{111})_j\left[\left(\vec{H}-\vec{\sigma}\right)\cdot\vec{\rho}\right] + \sum_j (W_{200})_j\left[\left(\vec{H}-\vec{\sigma}\right)\cdot\left(\vec{H}-\vec{\sigma}\right)\right] \\ & + \sum_j (W_{040})_j(\vec{\rho}\cdot\vec{\rho})^2 + \sum_j (W_{131})_j\left[\left(\vec{H}-\vec{\sigma}\right)\cdot\vec{\rho}\right](\vec{\rho}\cdot\vec{\rho}) + \sum_j (W_{222})_j\left[\left(\vec{H}-\vec{\sigma}\right)\cdot\vec{\rho}\right]^2 \\ & + \sum_j (W_{220})_j\left[\left(\vec{H}-\vec{\sigma}\right)\cdot\left(\vec{H}-\vec{\sigma}\right)\right](\vec{\rho}\cdot\vec{\rho}) + \sum_j (W_{311})_j\left[\left(\vec{H}-\vec{\sigma}\right)\cdot\left(\vec{H}-\vec{\sigma}\right)\right]\left[\left(\vec{H}-\vec{\sigma}\right)\cdot\vec{\rho}\right] \\ & + \Delta\sum_j (W_{020})_j(\vec{\rho}\cdot\vec{\rho}) + \Delta\sum_j (W_{040})_j(\vec{\rho}\cdot\vec{\rho})^2, \end{aligned} \tag{4}$$

The first-order characteristics include defocus (W_{020}), tilt (W_{111}) and translation (W_{200}), while the third-order characteristics include spherical aberration (W_{040}), coma (W_{131}), astigmatism (W_{222}), curvature (W_{220}), and distortion (W_{311}).

The generalized pupil function P can be expressed as follows:

$$P = p \cdot \exp(iW), \tag{5}$$

where p represents a transmittance function. The point spread function (PSF) of an optical system can be expressed as the generalized pupil function. Therefore, the relationship between PSF and wave aberration can be obtained as:

$$PSF = |FFT(P)|^2 = |FFT[p \cdot \exp(iW)]|^2, \tag{6}$$

where FFT represents Fourier Transformation. After encoding the image Y with PSF convolution and adding noise n, the low-resolution image can finally be obtained. The degradation process can be expressed by the following equation:

$$Y' = Y \otimes PSF + n. \tag{7}$$

Due to the alignment errors of the DMD equipment in the BCI system mainly including defocus caused by the front and rear position deviations, decenter caused by up, down, left, and right deviations, and tilt caused by mounting angle deviations, it was necessary to focus on the first order characteristics.

2.3. Optical System Design

To analyze the impact of system calibration errors on the super-resolution imaging optical system of DMD, we designed an imaging optical system. This system consists of a telescope group and two relay lens groups. The DMD is placed in the second intermediate image plane. The main parameters of the optical system are shown in Table 1, and the design diagram is shown in Figure 2a. The modulation transfer function (MTF) of the optical system is shown in Figure 2b. It can be seen that when there was no alignment error, the optical system had good imaging quality, and the MTF value was close to the diffraction limit.

Table 1. Parameters of the optical system.

Parameter	Value
Wavelength/nm	390–780
FOV(X/Y)/(○)	2.2°/2.2°
F/#	1
DMD array size/pixel	1024 × 1024
DMD pixel size/µm	7.6
Detector pixel size/µm	Detector pixel size/µm
3.45	3.45
Detector array size/pixel	256 × 256

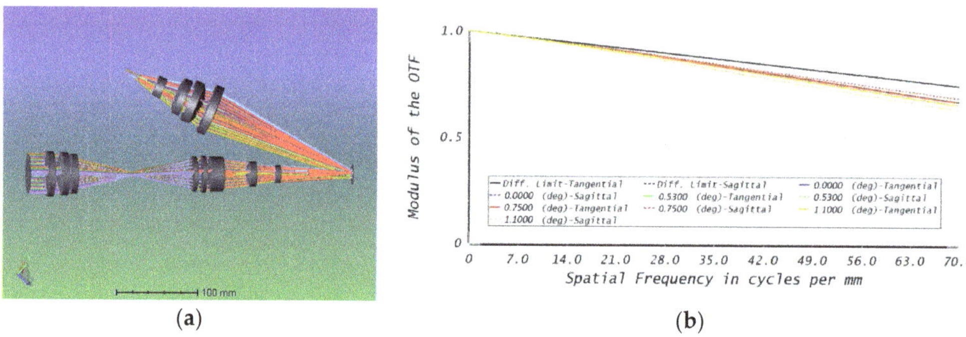

Figure 2. Schematic diagram of the DMD-based polarization optical system. (**a**) Structure of the optical system; (**b**) Modulation Transfer Function (MTF) of the optical system.

However, according to the analysis in the previous section, the optical system, when actually built, does not achieve an ideal imaging performance; therefore, we added alignment errors to the originally designed optical system, including the decenter, tilt, and defocus between the DMD and the detector. We obtained multiple PSFs with different error combinations, as shown in Figure 3. The position of the DMD in the actual system has undergone various deviations, resulting in a significant change in the PSF, which caused different degradation in the imaging quality.

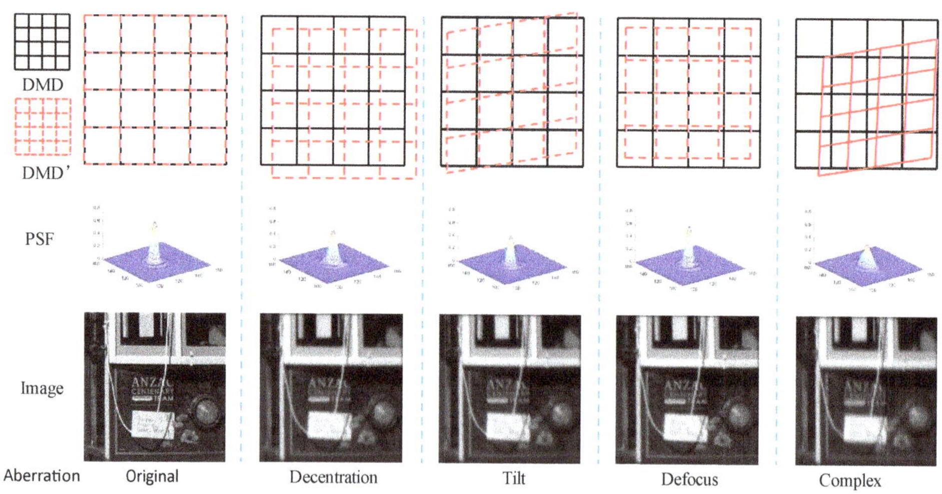

Figure 3. PSF and imaging results under different error conditions.

2.4. Super-Resolution Reconstruction Network

Different from the traditional compressed sensing image reconstruction algorithm, deep learning learns prior knowledge from the data, uses deep neural networks to establish a mapping relationship between the input and output, optimizes network parameters through large-scale data training, and inputs the sequence images collected by the detector into the compressed imaging system of the trained network, bypassing the complex computational process, resulting in a faster processing and the direct output of reconstructed high-resolution images [14].

Inspired by the Real-ESRGAN [15] network and the Joinput-CiNet network, we designed a compressed imaging super-resolution generational adversarial network based on error corrections. As shown in Figure 4, our network is mainly used to compensate for aberrations and optical alignment errors in optical imaging systems, achieving better reconstruction results. For this network, compressed encoding sampled low-resolution images with errors and high-resolution encoded images from which errors were input, and the reconstructed high-resolution images suppressed by error interference were output. Meanwhile, our network input multiple low-resolution images while maintaining the correlation among each block, effectively reducing the problem of the block effect in DMD compressed imaging and requiring no additional computational for the elimination of the block effect. We also used the U-Net as the discriminator for the network, which outputs the realness values for each pixel, and provides detailed per-pixel feedback to the generator.

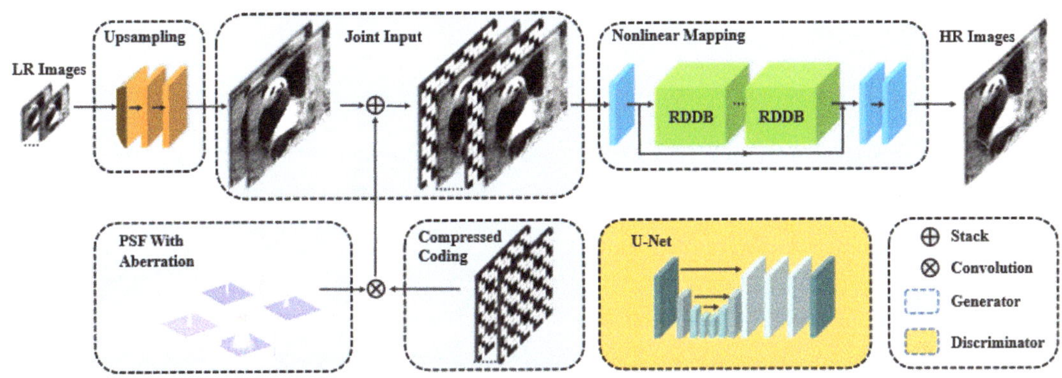

Figure 4. Flow chart of reconstructed network.

We improved the Real-ESRGAN network by using low-resolution compressed sampling data with alignment errors, DMD high-resolution encoded patterns, an optical system PSF with alignment errors as joint inputs to the network, and high-resolution images as the outputs. It has been verified that the combination of low-resolution images and compressed encoding, when input into the Joinput-CiNet network, can achieve better reconstruction results. Due to the four-fold difference between the low-resolution sampling and DMD size, it is difficult to train the network; therefore, we needed to align the two resolutions. We had to abandon the Joinput-CiNet network's PCA process to reduce the DMD encoding resolution and chose to enlarge low-resolution data to retain the full DMD encoding information and perform convolution with the encoding information and the PSF of the alignment mismatch to obtain the encoding matrix with errors. This approach improved the reconstruction results while simultaneously inputting the full encoding information and the alignment mismatch PSF.

2.5. Training Data

In order to train and evaluate the presented neural network, we used the DIV2K [8] and Flickr2K [9] datasets as high-resolution scene images. Data encoded sampling process is shown in Figure 5. Using the previously mentioned degradation process, we compressed the high-resolution image into multiple low-resolution images and then convolved them with the PSF of the optical system designed. Finally, we calculated the low-resolution compressed image with alignment errors and simulated the process of DMD compression imaging through encoding. The 40 sets of PSF were calculated, which were then convolved with the high-resolution image randomly. We added noise and reduced the resolution of the convolved images; the resulting degraded image is shown in Figure 6. The 3000 sets of high-resolution images were used as the training data, 300 sets of images were used as the validation data, and 100 sets were used as the testing data. In addition, random horizontal and vertical flips were also selected during training. To increase the speed of training, the training HR patch size was set to 256. Similar to Real-ESRGAN, our net was trained with a combination of L1 loss, perceptual loss, and GAN loss, with weights (1, 1, 0.1), respectively.

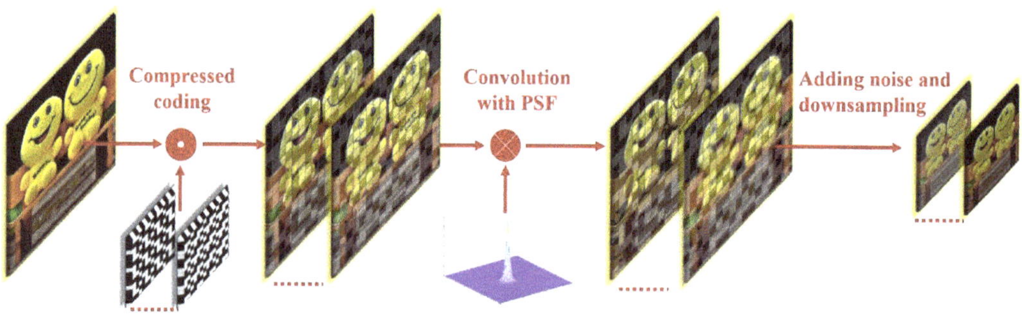

Figure 5. Data encoded sampling flow chart.

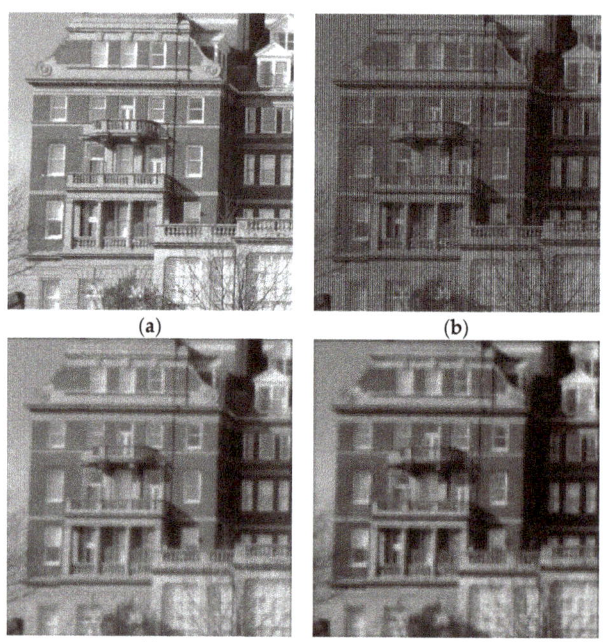

Figure 6. Effect of the degradation process of a high-resolution image. (**a**) Original image; (**b**) After DMD block encoding; (**c**) Image with alignment errors; (**d**) Image interpolated and amplified with noisy low-resolution detector.

3. Results

In order to apply this work to real scenarios, we conducted simulations and actual imaging experiments using our deep learning network to perform the super-resolution reconstruction of the collected images. The results from these experiments demonstrated the effectiveness of our approach.

3.1. Simulation

We divided the DMD into small blocks of size 4×4, with each block using the same 8-bit random Gaussian matrix as the sampling matrix. Then, we controlled the sampling rate by controlling the number of imaging times of the detector. Since the image magnification was $4\times$ and 16 images were acquired for complete sampling, the number of input low-resolution images corresponding to these three compression ratios was 1, 2, and 4. Super-resolution algorithms typically evaluate reconstruction results using PSNR and SSIM, and we calculated the values of the reconstructed images with different compression rates, as shown in Table 2. The reconstructed images with different compression rates of polarization are shown in Figure 7

Table 2. PSNR/SSIM values of reconstructed images with different compression rates of the test set.

CS Rate	1/16	1/8	1/4
PSNR/SSIM	23.3149/0.7649	26.9589/0.8846	28.0675/0.8723

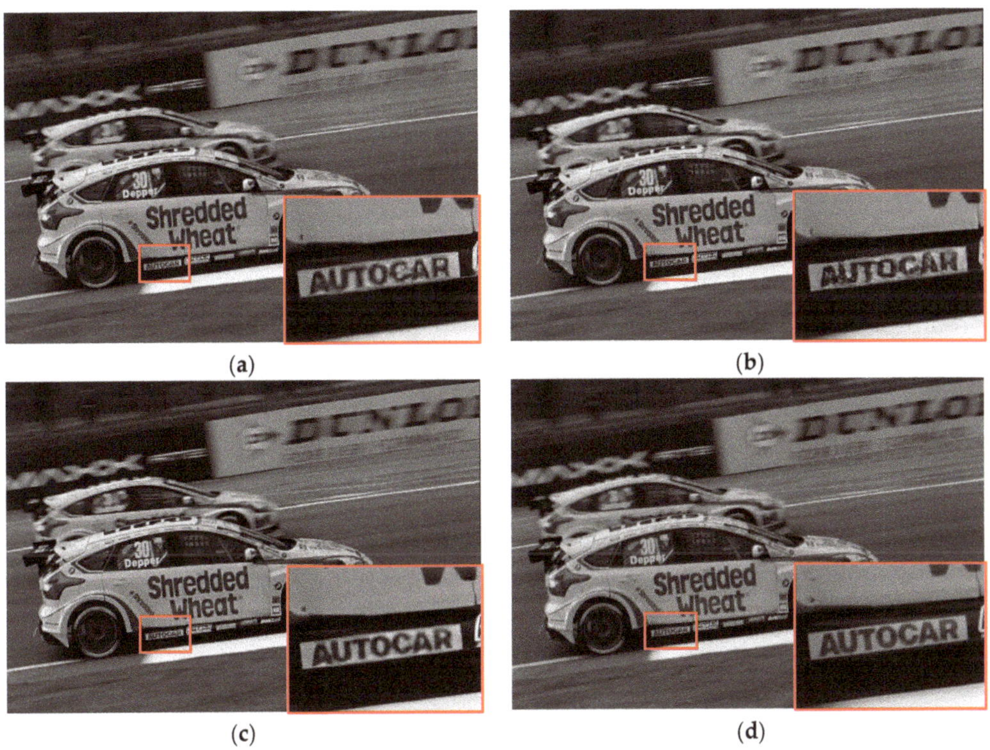

Figure 7. Reconstruction images with different sampling rates. (**a**) Sampling rates = 1/16; (**b**) Sampling rates = 1/8; (**c**) Sampling rates = 1/4; (**d**) Ground truth.

In order to evaluate the reconstruction effect of the algorithm, we compared our algorithm to the compressed sensing algorithm OMP and the deep learning networks Reconnet and Real-ESRGAN. The input of the OMP algorithm and Reconnet network on each block of data was obtained, directly outputting the reconstructed single block information and then combining each block of information to create a high-resolution image. Real ESRGAN is a single image super-resolution algorithm, with the input being a single low-resolution image that is directly collected without encoding, and the compression rate is equivalent to 0.0625. PSNR and SSIM are usually used to evaluate the reconstruction results. We evaluated the reconstruction effect of each algorithm under different alignment errors. The specific values of the reconstructed images under different alignment errors are shown in Table 3.

Table 3. The specific values of the reconstructed images under different alignment errors.

	Decenter	Tilt	Defocus	Complex
OMP	12.6821/0.0462	12.6631/0.0448	12.6633/0.0475	12.6673/0.0434
Reconnet	21.4278/0.5843	21.2601/0.5772	21.6159/0.5902	20.8833/0.5720
Real-ESRGAN	18.6232/0.6923	18.3760/0.6733	18.8864/0.6923	17.7600/0.6604
Ours	24.6103/0.8305	23.6994/0.8108	23.6765/0.8201	23.4484/0.7983

In Figure 8, the reconstruction images of different algorithms at a sampling rate of 0.0625 and magnification of four are shown. It can be seen that the OMP and Reconnet network could not be well applied to data reconstruction with alignment errors, and the inevitable presence of mosaic artifacts was caused; therefore, further correction is necessary. The Real-ESRGAN network performed well in a single-image super-resolution, but its original degradation model and alignment error model still had significant differences, resulting in the loss of reconstruction details. Our algorithm could reconstruct images with good imaging quality. More details of the reconstructed image in different error cases are shown in Figure 9.

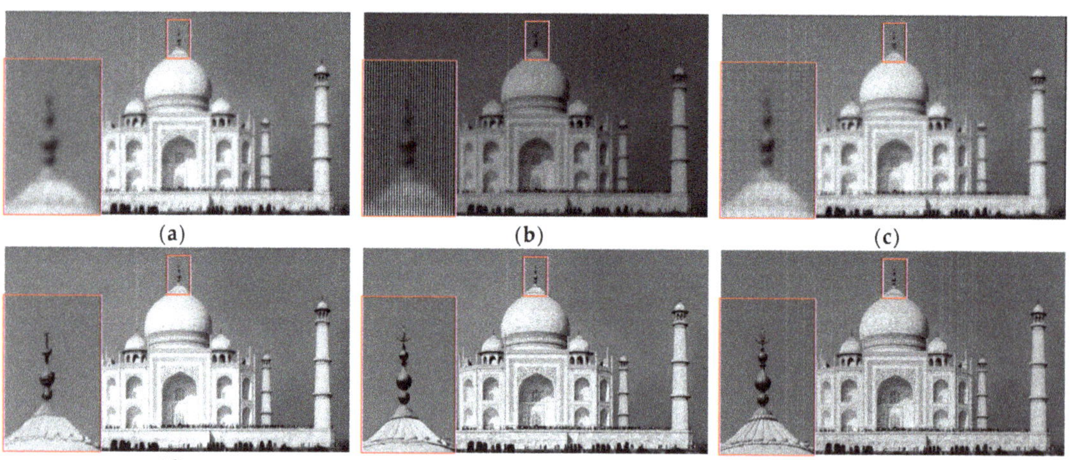

(a) (b) (c) (d) (e) (f)

Figure 8. Comparison of the results of different super-resolution methods. (**a**) Bicubic; (**b**) OMP; (**c**) Reconnet; (**d**) Real-ESRGAN; (**e**) Ours; (**f**) Ground truth.

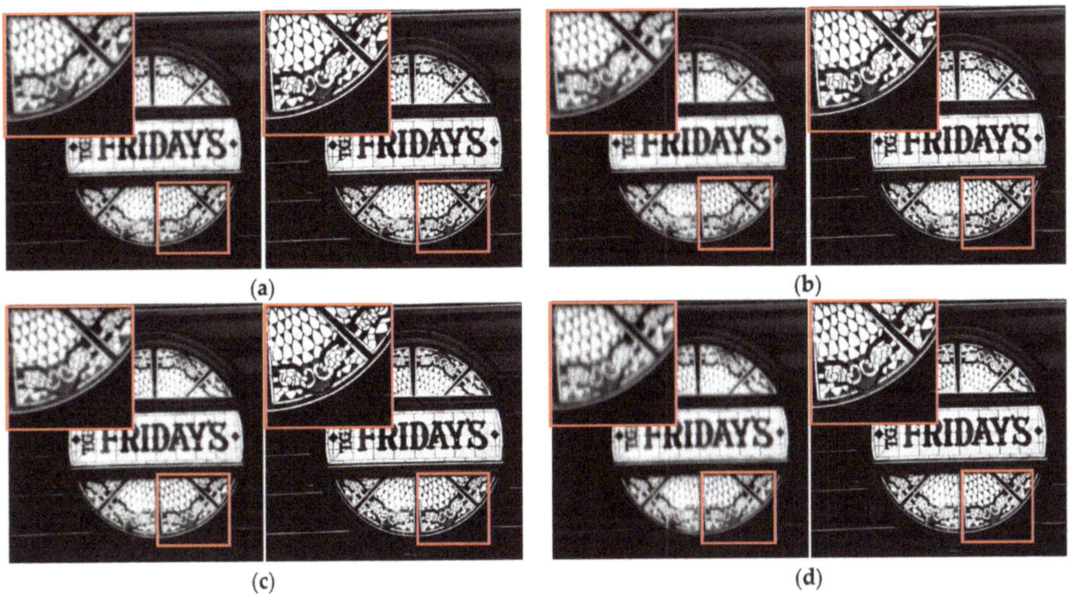

Figure 9. Low-resolution images with different errors and high-resolution reconstruction images. (**a**) Decenter (20 μm in both the X-axis and Y-axis directions); (**b**) Tilt (10′); (**c**) Defocus (20 μm); (**d**) Complex.

It can be seen that our algorithm exhibited excellent reconstruction effects at a low compression rate and dealt with different errors, both visually and numerically.

We used the Modulation Transfer Function (MTF) to analyze the reconstruction resolution. Our detector pixel size was 3.45 μm, and the corresponding system cutoff frequency was approximately 144 lp/mm. We tested black and white line pair images with different frequencies, added various alignment errors, and calculated the MTF of the reconstructed images, as shown in Figure 10. It can be seen that low-resolution images were difficult to visually distinguish at 25 lp/mm, while the images reconstructed by our algorithm still performed well at 50 lp/mm.

To evaluate the performance of our network, we tested the running time of OMP, ReconNet, Real ESRGAN, and ours in Figure 11. Reconstruction images with a size of (2040 × 1360) were used. The block size was (4 × 4). Our network included 16.6 million parameters for the generator and 4.37 million parameters for the discriminator. It can be seen that we performed well in the reconstruction effect, but there is still room for improvement in the running time. In future work, we hope to continue to optimize the reconstruction speed of the network.

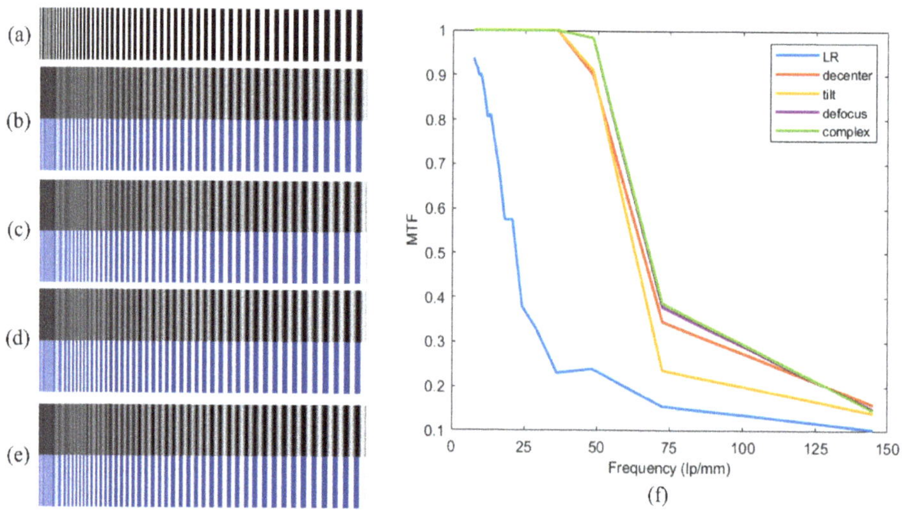

Figure 10. Low-resolution images and reconstruction results of black and white line pairs under different alignment error conditions. (**a**) Ground truth; (**b–e**) Low-resolution images with different errors and high-resolution reconstruction images; (**b**) Decenter; (**c**) Tilt; (**d**) Defocus; (**e**) Complex; (**f**) MTF images of reconstruction results with different alignment errors.

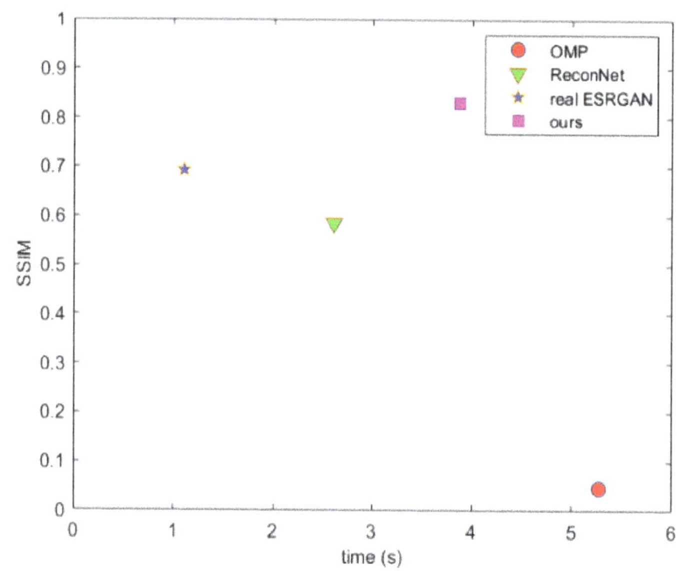

Figure 11. The runtime and SSIM of different reconstruction algorithms.

3.2. Experiment

To verify the application performance of our reconstruction network in a realistic scene, we selected core components such as TI Corporation's DMD, dual telecentric projection lenses, and LUCID Company's visible-light camera, and built a dual-arm reflective experimental setup, as shown in Figure 12a, the results of the experiment are shown in Figure 12b–f. Firstly, the scene was projected into the DMD by a convergent lens and a dual telecentric projection lens 1, with a resolution of 1920 × 1080. The size of each micro-mirror in DMD was 10.8 μm, and the lenses could be switched at a high speed between ±12° directions. Next, the reflected direction of different micromirrors on the DMD was controlled to encode and modulate the scene. Due to the diagonal flip of the micromirror, we rotated the DMD 45° and the angle between the two arms at 24°. Finally, the encoded scene image was collected by the camera using the dual telecentric projection lens 2. The camera was also rotated 45°. We selected a region of 1024 × 1024 pixels on the DMD, corresponding to the 256 × 256 regions of the detector, for image acquisition. The entire device was placed on a flat optical plate to take photos of the outdoor scene.

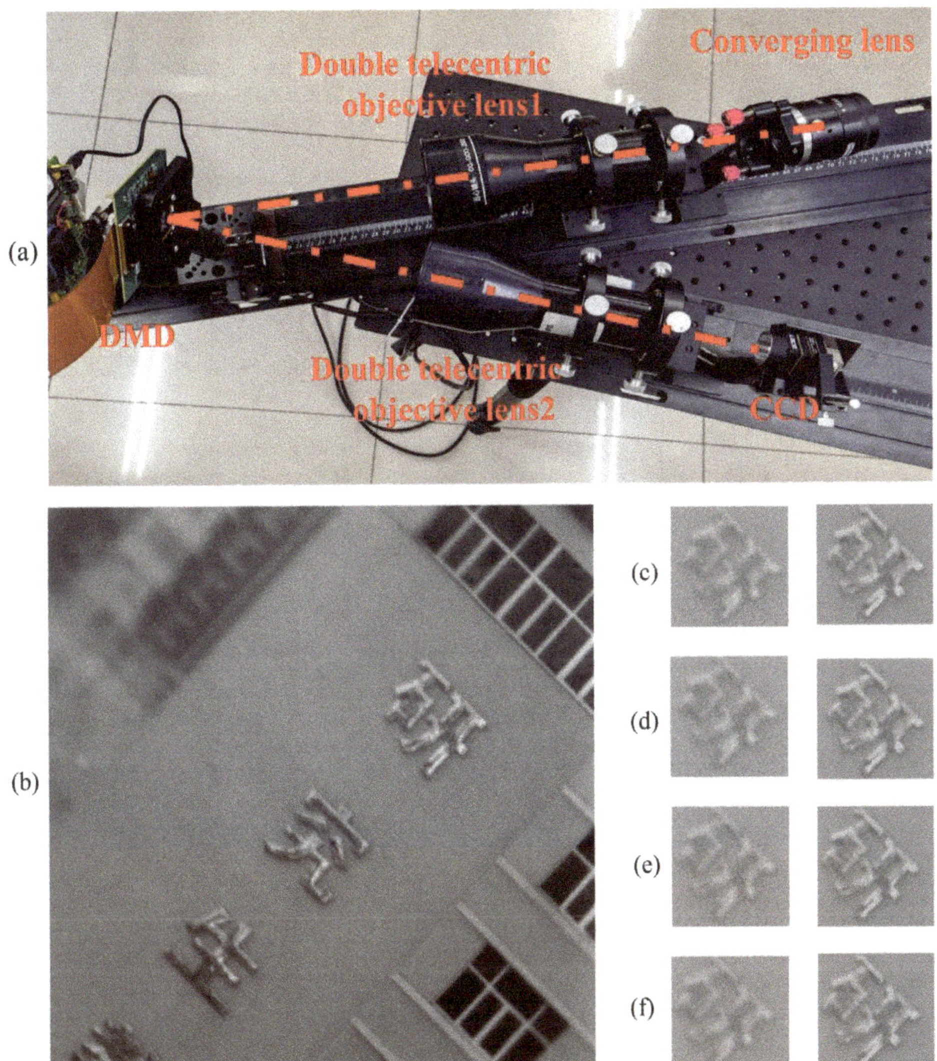

Figure 12. Experimental device and real imaging photography (**a**) Experimental device; (**b**) Reconstructed high-resolution image without alignment error; (**c**–**f**) Low-resolution images with different errors and high-resolution reconstruction images; (**c**) Decenter; (**d**) Tilt; (**e**) Defocus; (**f**) Complex.

Despite the large difference between the experimental data set and the actual installation error, we still improved the quality of the reconstructed image, indicating that our algorithm had certain robustness. This future improvement could be further improved by adding more real training data from actual photography.

4. Discussion

We built a deep learning network to achieve the super-resolution reconstruction of image compression sensing while compensating for the image degradation caused by alignment errors. During the reconstruction process, our network established a mapping relationship from low-resolution intensity images with alignment errors and compressed coding matrices to high-resolution intensity images. Compared with traditional methods, our deep compression and reconstruction algorithm not only had better reconstruction effects and faster reconstruction speeds but also the ability to correct the alignment error, making it more suitable for use in actual optical systems.

Currently, our method is limited by the resolution of DMD. In the visible wavelength range, our imaging method has not significantly improved its resolution. However, in the infrared detection dimension, by using appropriate training data sets, this resolution could be increased to approximately 2–4 times that of the original image. Our imaging method could use low-resolution infrared detectors to obtain high-resolution information, thereby reducing the cost of optical systems.

We designed a DMD compressed imaging optics system with Zemax optical design software and analyzed the effect of the alignment error on super-resolution reconstruction. The relationship between alignment error and the PSF can be explained with the wave aberration theory. We simulated the imaging degradation process caused by various alignment errors. When building the training data sets, we incorporated alignment errors into the optical system, generating a non-ideal PSF. When establishing the training dataset, we used our designed optical system to add alignment errors to generate nonideal PSF and used convolution to obtain images with errors, replacing the degradation model of common image algorithms. These methods are more suitable for real-world optics systems.

In order to better reconstruct the image, we took the compressed coding matrix as a priori knowledge and attempted to convolution the mismatch point spread function with the coding matrix to obtain the mismatch matrix. The mismatch matrix and low-resolution images were jointly input into the reconstruction network to improve the impact of optical system alignment errors on super-resolution imaging. The experiments have shown that our method significantly improved the reconstruction effect.

When the network trained from existing datasets was used to reconstruct images encoded and collected from real DMD optical systems, the reconstruction effect was already better than other traditional reconstruction algorithms, but the effect could still be further improved. Therefore, in future work, we look forward to collecting large amounts of data using DMD devices to meet specific demands, which will enable us to build a more accurate database for training.

5. Conclusions

In this article, we proposed a super-resolution compressed imaging method with an optical alignment error correction, which addressed the problem of image degradation caused by optical alignment errors in BCI. It took the encoded low-resolution image and high-resolution encoded matrix with optical alignment errors as joint inputs and reconstructed high-resolution images by GAN. Our method has been compared with existing methods, and it was found that the proposed method had a lower sampling rate, and the reconstructed images based on this method had more advantages in terms of evaluation indicators and visual effects. Additionally, our experimental setup demonstrated the potential of the imaging method and reconstruction network in future work, which will excite people's interest in applying deep learning and compressed sensing to practical imaging applications.

Author Contributions: Conceptualization, M.X. and C.W.; methodology, M.X. and C.W.; software, M.X.; validation, Q.F. and C.W.; data curation, H.S. and L.D.; writing—original draft preparation, M.X.; writing—review and editing, C.W.; project administration, Y.L.; funding acquisition, H.J. All authors have read and agreed to the published version of the manuscript.

Funding: This research was funded by [National Natural Science Foundation of China] 61805028, 61805027, 61890960, [Natural Science Foundation of Jilin Province] YDZJ202201ZYTS411, [Foundation strengthen domain technology fund] 2022-JCJQ-JJ-xx28, [Scientific and technological research projects of The Education Department of Jilin Province] JJKH20220742KJ, [Strategic Research Issues of Beijing Institute of Space Mechanics & Electricity], [Center of Space Exploration, Ministry of Education "Conceptual study on material exploration and in situ utilization of lunar underground lava tubes"] SKTC202101.

Institutional Review Board Statement: Not applicable.

Informed Consent Statement: Not applicable.

Data Availability Statement: Data underlying the results presented in this Letter are not publicly available at this time but may be obtained from the authors upon reasonable request.

Conflicts of Interest: The authors declare no conflict of interest.

References

1. Ke, J.; Lam, E.Y. Object reconstruction in block-based compressive imaging. *Opt. Express* **2012**, *20*, 22102. [CrossRef] [PubMed]
2. Zhang, X.; Li, C.; Meng, Q.; Liu, S.; Zhang, Y.; Wang, J. Infrared Image Super Resolution by Combining Compressive Sensing and Deep Learning. *Sensors* **2018**, *18*, 2587. [CrossRef] [PubMed]
3. Xu, M.; Wang, C.; Wang, K.; Shi, H.; Li, Y.; Jiang, H. Polarization Super-Resolution Imaging Method Based on Deep Compressed Sensing. *Sensors* **2022**, *22*, 9676. [CrossRef] [PubMed]
4. Wang, F.; Wang, C.; Chen, M.; Gong, W.; Zhang, Y.; Han, S.; Situ, G. Far-field super-resolution ghost imaging with a deep neural network constraint. *Light Sci. Appl.* **2022**, *11*, 1. [CrossRef] [PubMed]
5. Zhang, X.; Xie, J.; Li, C.; Xu, R.; Zhang, Y.; Liu, S.; Wang, J. MEMS-based super-resolution remote sensing system using compressive sensing. *Opt. Commun.* **2018**, *426*, 410–417. [CrossRef]
6. Wang, C.; Xing, S.; Xu, M.; Shi, H.; Wu, X.; Fu, Q.; Jiang, H. The Influence of Optical Alignment Error on Compression Coding Superresolution Imaging. *Sensors* **2022**, *22*, 2717. [CrossRef] [PubMed]
7. Dumas, J.P.; Lodhi, M.A.; Bajwa, W.U.; Pierce, M.C. Computational imaging with a highly parallel image-plane-coded architecture: Challenges and solutions. *Opt. Express* **2016**, *24*, 6145–6155. [CrossRef] [PubMed]
8. Ri, S.; Fujigaki, M.; Matui, T.; Morimoto, Y. Pixel-to-Pixel Correspondence Adjustment in DMD Camera by Moiré Methodology. *Exp. Mech.* **2006**, *46*, 67–75. [CrossRef]
9. Wu, Z.; Wang, X. DMD Mask Construction to Suppress Blocky Structural Artifacts for Medium Wave Infrared Focal Plane Array-Based Compressive Imaging. *Sensors* **2020**, *20*, 900. [CrossRef] [PubMed]
10. Kulkarni, K.; Lohit, S.; Turaga, P.; Kerviche, R.; Ashok, A. ReconNet: Non-Iterative Reconstruction of Images from Compressively Sensed Measurements. In Proceedings of the 2016 IEEE Conference on Computer Vision and Pattern Recognition (CVPR), Las Vegas, NV, USA, 26 June–1 July 2016.
11. Cui, C.; Ke, J. Spatial compressive imaging deep learning framework using joint input of multi-frame measurements and degraded maps. *Opt. Express* **2022**, *30*, 1235. [CrossRef] [PubMed]
12. Yang, J.; He, Q.; Liu, L.; Qu, Y.; Shao, R.; Song, B.; Zhao, Y. Anti-scattering light focusing by fast wavefront shaping based on multi-pixel encoded digital-micromirror device. *Light Sci. Appl.* **2021**, *10*, 149. [CrossRef] [PubMed]
13. Thompson, K. Description of the third-order optical aberrations of near-circular pupil optical systems without symmetry. *J. Opt. Soc. Am. A Opt. Image Sci. Vis.* **2005**, *22*, 1389–1401. [CrossRef] [PubMed]
14. Lohit, S.; Kulkarni, K.; Kerviche, R.; Turaga, P.; Ashok, A. Convolutional Neural Networks for Noniterative Reconstruction of Compressively Sensed Images. *IEEE Trans. Comput. Imaging* **2018**, *4*, 326–340. [CrossRef]
15. Wang, X.; Xie, L.; Dong, C.; Shan, Y. Real-ESRGAN: Training Real-World Blind Super-Resolution with Pure Synthetic Data. In Proceedings of the 2021 IEEE/CVF International Conference on Computer Vision Workshops (ICCVW), Montreal, QC, Canada, 11–17 October 2021; pp. 1905–1914.

photonics

Communication

Improved Target Laser Capture Technology for Hexagonal Honeycomb Scanning

Bing Jia, Fan Jin *, Qiongying Lv and Yubing Li

College of Mechanical and Electrical Engineering, Changchun University of Science and Technology, Changchun 130012, China
* Correspondence: jf1064yx@163.com

Abstract: In laser tracking systems, capturing moving targets is a prerequisite to guaranteeing a tracking system's performance. Previous studies have confirmed that the capture probability of hexagonal spiral scanning is higher than that for other scanning methods, but there is still room for improvement. This article proposes an improved hexagonal honeycomb scanning capture method based on hexagonal spiral scanning for a prior moving target model with a Gaussian distribution positioned within the scanning range of the visual threshold. Through experimental verification, it was found that, within the same scanning field of view, the capture probability can be increased by 3% compared to that in traditional hexagonal spiral scanning, making the capture probability greater than 98%. The improved hexagonal honeycomb structure scanning method proposed in this article provides a new solution for target acquisition problems in fields such as laser communication, laser docking, and airborne radar.

Keywords: moving target model capture; laser scanning; improved hexagonal honeycomb structure scanning

Citation: Jia, B.; Jin, F.; Lv, Q.; Li, Y. Improved Target Laser Capture Technology for Hexagonal Honeycomb Scanning. *Photonics* **2023**, *10*, 541. https://doi.org/10.3390/photonics10050541

Received: 17 March 2023
Revised: 20 April 2023
Accepted: 28 April 2023
Published: 6 May 2023

1. Introduction

Laser scanning is a topic that is extremely broad in scope. A scanning system can track objects by capturing reflected light or by fluorescing the image and acquiring the fluoresced light. The output system directs light to produce images for marking [1]. Laser scanning capture technology has been widely studied and applied in many engineering fields due to its ability to accurately and directly scan small targets.

In the field of biology, the main applications of laser scanning capture include subcellular protein localization [2], the imaging of biological surfaces [3], trapping light-absorbing microparticles in the air [4,5], bright-field light microscopy [6] and trapping nanoparticles on the surfaces of solutions [7]. In the medical field, research and development on technologies such as laser capture microdissection [8,9], the laser capture of individual cells in complex tissues [10], laser capture cell sampling [11], the analysis of gene expression profiles using laser capture microdissection [12], and the laser capture of blood cells [13] has demonstrated that laser capture technology has profound significance for medical progress and for the treatment of major diseases. In the engineering field, laser scanning acquisition is mainly used in laser radar, mechanical engineering, the space industry, and other fields, such as 3D nanofabrication [14], high angular resolution LiDAR [15], super-resolution laser probing [16], laser printing [17], and laser scanning [18,19]. However, in actual engineering applications, the problems of environmental complexity and spatial instability exist, posing new challenges for laser capture and tracking systems.

This article focuses on the field of laser communication and conducts research on the capture of moving targets. The measurement of dim or small moving targets in complex environments calls for high sensitivity, high precision and high angular deflection velocity in the visual axis for traditional single-station measuring equipment. The laser tracking measurement system has been proposed as a new type of measurement system with high measurement accuracy, a wide measurement range, and high pointing accuracy, and has thus become a global research hotspot in target tracking technology.

The premise of maneuvering tracking is to establish an accurate target model and then, based on a variety of filtering tracking algorithms, effectively estimate the target's position in the next step. Whether visual sensors or laser detection methods are used, target capture is the first step in the target tracking process.

Common scanning methods in laser tracking systems include rectangular branch scanning, rectangular spiral scanning, circular spiral scanning, hexagonal spiral scanning, etc. These scanning methods are used in laser communication, satellite docking, airborne radar, and other fields. The scanning mode and scanning parameters used in a photoelectric detection system are of great significance when it comes to improving the detection accuracy, detection efficiency, and target acquisition probability of the laser tracking system. In previous studies, photoelectric scanning detection models played an important role in monitoring low, slow, and small targets in the air, enabling the efficient capture of moving targets [20]. A rectangular spiral scanning method based on spiral scanning [21] has been proposed for space rendezvous lidar target acquisition technology that simplifies the system's structure, reduces system difficulty, and increases acquisition probability [22]. In previous studies on the use of radar scanning in space laser communication, it was concluded that uniform sinusoidal spiral scanning has greater advantages [23], but this scanning method has a complex control process for galvanometer laser scanning [24–26]. In addition, some studies have compared several common scanning methods and found through theoretical analysis and simulation experiments that hexagonal spiral scanning has advantages that are different from those of other scanning methods [27,28]. When the moving target's position is combined with a Gaussian distribution model, the capture probability using rectangular spiral scanning is 85.80%, and the required number of light feet is 81; the capture probability using rectangular branch scanning is 62.48%, and the required number of light feet is 81; and the acquisition probability using hexagonal spiral scanning is 87.79%, and the number of light feet required is 61. The acquisition speed of hexagonal spiral scanning is, therefore, the fastest.

The sampling frequency in laser tracking systems is also an important factor. When the sampling frequency of the target increases, the acquisition probability of hexagonal spiral scanning and rectangular spiral scanning increases steadily, and the acquisition probability of hexagonal spiral scanning is the highest out of all methods. Moreover, changes in the sampling frequency have a great impact on the rectangular branch scanning. Due to improvements in the sampling frequency, the branch scanning method starting from the edge can capture a moving target at the center of the field of view more quickly. If the laser sampling frequency is not high enough when the laser reaches the center of the field of view, the target will have moved outside of the field of view, that is, the acquisition probability will be extremely low. There is a peak before and after the sampling frequency reaches 30 kHz wherein the acquisition probabilities of hexagonal spiral scanning and rectangular spiral scanning show a certain decline. This indicates that when the sampling frequency increases, the spiral scanning spot moves faster to the edge of the field of view, resulting in a small reduction in the acquisition probability. The acquisition probability of rectangular branch scanning tends to be stable, indicating that it is not restricted by the sampling frequency but by the scanning method itself.

For a moving target with a high probability of appearing in the center of the field of view, the hexagonal spiral scanning mode and the target cannot be effectively matched in the time domain and space domain. Based on the above research, combined with the research background, this paper proposes an improved hexagonal scanning mode—called improved hexagonal honeycomb structure scanning—which establishes a prior motion model with a Gaussian distribution in the threshold scanning range. In the same scanning field of view, the acquisition probability is higher than that of traditional hexagonal spiral scanning.

2. Improved Scan Capture Model with a Hexagonal Honeycomb Structure

2.1. Establishment of the Improved Hexagonal Honeycomb Structure Scanning Model

At present, most hexagonal scanning methods available in the literature are hexagonal spiral scanning methods; the principle of this method is similar to spiral scanning and zigzag scanning [29–31]. They all scan from the center to the 2π direction. Based on the research of [20], it was concluded that when the position where the target enters the visual threshold range for the first time is modeled with a Gaussian distribution and different scanning methods are used to cover and determine the visual field, the hexagonal spiral scanning method has a higher acquisition probability than the circular and rectangular methods.

To further improve the target acquisition probability of laser tracking systems and make the detection spot appear in the center more often, this paper proposes a new hexagonal scanning method—the improved hexagon honeycomb structure scanning model. Different from traditional spiral scanning with an increasing hexagonal radius, the scanning of the hexagonal honeycomb structure is improved by first completing the hexagon of the smallest unit and then scanning from the inside to the outside. This method achieves detection in the center of the field of view more often and reduces the number of times target capture occurs outside the field of view. For moving targets with a Gaussian distribution, the scanning mode of the improved hexagonal honeycomb structure is the same as the spiral scanning mode, which starts scanning from the area with the highest probability of target occurrence. At the same time, this mode makes up for the defect of missing scanning of the outer circle of spiral scanning and can repeat the spiral scanning of the inner circle.

Assume that the pulse frequency matches the scanning frequency, that is, the laser sends out a pulse every time the galvanometer jumps. Then, define a visual scanning range threshold. The scanning mode is shown in Figure 1. In the figure, a circle is used to represent the light foot generated by a pulse, and an overlapping hexagon is used to represent the path of the laser spot's center. The darker the color, the more overlaps there are.

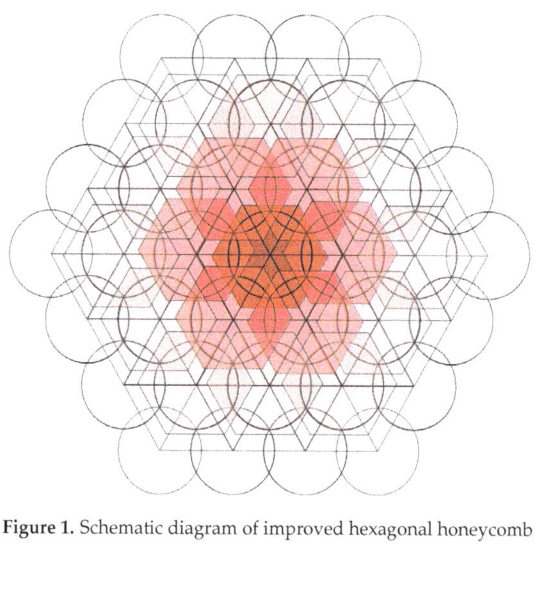

Figure 1. Schematic diagram of improved hexagonal honeycomb scanning structure.

The scanning path of the improved hexagonal honeycomb structure is briefly described below:

(1) First, form a light foot at the origin at the center of the scanning field, then scan along the regular hexagon edge of the center to form the first layer of light feet with six vertices as the centroid;
(2) Take the vertex of the central regular hexagon as the centroid and make six regular hexagons. The scanning track is the edge of the six regular hexagons, scanning out the second layer of light feet;
(3) The apex of the hexagon is formed after the second layer of scanning, and the bisection points of each side are taken as the hexagon centroid. Its edges are scanned in turn to generate the third layer of light feet, and so on.

According to the scanning mode shown in the figure, it can be seen that the optical foot density and coincidence rate in the center of the field of view are higher. Under a normal distribution of the target and determination of the field of view's size, compared with other scanning methods, the target is more likely to be captured using this scanning mode.

Therefore, the scanning path of the improved hexagonal honeycomb scanning structure is modeled. First, define the center position of the current flare. The vector groups A, B, C, D, E, and F of starting point O have included angles of 60° between their adjacent vectors, and the vectors' lengths are equal to the side length of the regular hexagon of the path. The model is shown in Figure 2.

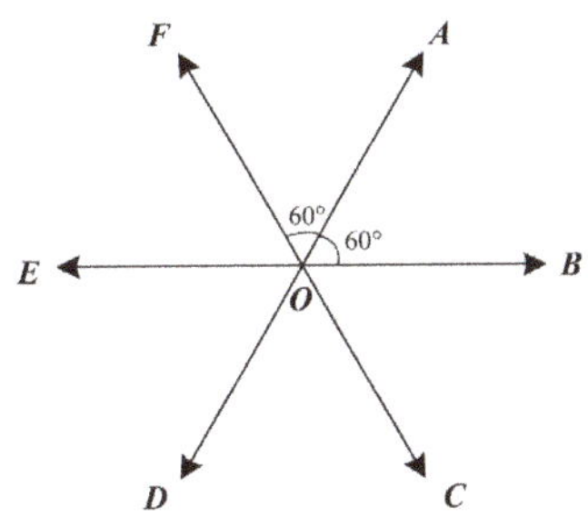

Figure 2. Six-direction scanning path vector.

The scanning path of the light spot is represented by a vector. If the included angle between the light foot and the horizontal right is 60° from the central position O of the light spot, and the length is r, it is represented by vector A. It can be seen that the scanning path vector satisfies the relationship of $A = -D$, $B = -E$, and $C = -F$. That is, if the trajectory follows the vector of $A + B + C + D + E + F$, and the light foot position returns to the starting point.

The scanning path vector of the first optical foot is shown in Figure 3.

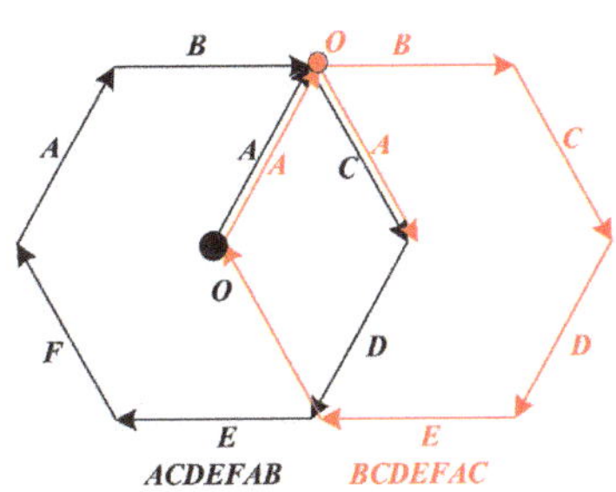

Figure 3. Vector representation example of the first layer optical foot scanning path.

Starting from the origin O of the scanning field of view, the optical foot conducts a scan around O to obtain the first optical foot. Its path description vector is $A + C + D + E + F + A + B$, abbreviated as ACDEFAB. From the above-mentioned path $A = -D$, $B = -E$, $C = -F$, the path description can be simplified to obtain the end position A of the optical foot relative to the origin after the path scanning as follows:

$$A + C + D + E + F + A + B = -D - F + D + E + F + A - E = A \quad (1)$$

This process is shown in the black path in Figure 3. Here, the accumulation of seven vectors is uniformly used to represent one regular hexagonal path scan, and the second regular hexagonal path scan is performed using the end position A of the first scan as the starting point. Then, the second path's description vector is $B + C + D + E + F + A + C$, abbreviated as *BCDEFAC*, as shown in the red path in Figure 3. Table 1 shows the scanning path of the first optical foot.

Table 1. Scanning path of first layer of optical feet.

Number	Staring Point	Path Vector
0	*O*	*ACDEFAB*
1	*A*	*BCDEFAC*
2	*B*	*CDEFABD*
3	*C*	*DEFABCE*
4	*D*	*EFABCDF*
5	*E*	*FABCDEA*
6	*F*	*ACDEFAB*

In the table, the optical foot scanning path No. 0 represents the hexagonal path with the scanning field origin as the centroid. The optical foot scanning path from No. 1 to No. 6 is the scanning path of the first layer of optical feet. The starting point is marked with vector coordinates relative to the origin. The starting point position of the current serial number is the last ending point position, which can be obtained by adding the last starting point vector and the path description vector. For example, the starting point of track No. 2 is the coordinate of the *B* vector, the path description vector is *CDEFABD*, and the starting point method for finding the No. 3 track is (Figure 4):

$$B + (C + D + E + F + A + B + D) = C \quad (2)$$

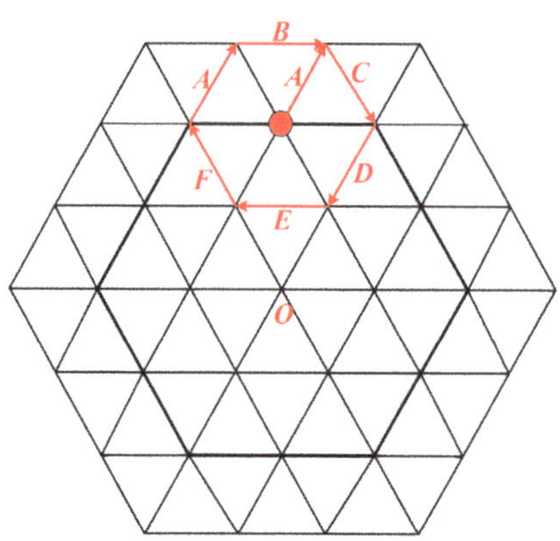

Figure 4. Schematic diagram of a regular hexagon in the first scan.

Table 2 shows the scanning track of the second optical foot.

Table 2. Scanning path of the second layer of optical feet.

Number	Starting Point	Path Vector
1	*AF*	*ACDEFAB*
2	*AA*	*BCDEFAC*
3	*AB*	*BCDEFAC*
4	*BB*	*BCDEFAC*
5	*BC*	*CDEFABD*
6	*CC*	*CDEFABD*
7	*CD*	*DEFABCE*
8	*DD*	*DEFABCE*
9	*DE*	*EFABCDF*
10	*EE*	*EFABCDF*
11	*EF*	*FABCDEA*
12	*FF*	*FABCDEA*

Due to the different methods of selecting the starting point of the second layer scanning path, the path vectors with serial numbers 1 and 2 have special properties. The other path description vectors are the same as the path description vectors with serial numbers 2 to 6 in the first layer, and they cycle once at the hexagon bisection point. This path description vector is extended to the scanning path of the third layer, and the corresponding path description vector of the first layer is cycled three times at the outermost regular hexagonal trisection point to obtain the three-layer light foot scanning path.

2.2. Capture Conditions

The controllable factors affecting the comprehensive probability of the laser tracking system capturing the target include the galvanometer scanning mode, the laser sampling frequency, and the laser divergence angle. The capture probability refers to the likelihood that the laser is scattered by the target and received by the receiving optical system within the laser's single frame scanning area. As long as the signal is received, it is determined that the target is captured. After the target enters the field of view, appropriate scanning system parameters are selected to complete the target acquisition, so that the maximum probability of the target falling into the visual range of the receiving system is achieved. Acquisition systems using other working modes, such as microwave radar, can scan the detection airspace in all directions, without considering the scanning speed. The laser acquisition and tracking system of the TOF measurement system selected in this paper is based on the matching of the time sequence and space to complete target acquisition. Therefore, the need to model the two kinds of independent random processes of the target and scanning systems based on a time sequence must be considered in the simulation and verification process.

The capture diagram of a moving target and the proposed detection model are shown in Figure 5, in which the measured target enters the detection area from the left side.

To simplify the complexity of the model, when a galvanometer-type photoelectric detection model is scanned for one frame, the time when the target enters the field of view is exactly the time when the detection model starts scanning; that is, the time when the random target enters the field of view and the time when the detection model scans for one frame are both T, which are set in the same detection area. When modeling a moving target, the following points should be obeyed:

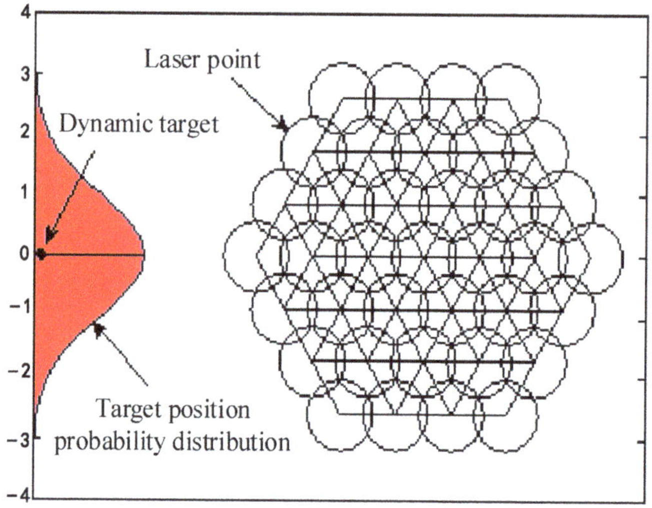

Figure 5. Schematic diagram of maneuvering target and detection model acquisition.

(1) The target motion model is an inertial target model;
(2) The scanning field of view position is the best estimation prior result;
(3) The target time location in the two-dimensional plane is Gaussian white noise, that is, it obeys a Gaussian distribution;
(4) The noise in the z direction is not considered.

2.3. Scanning Step Size Analysis

During galvanometer scanning, the scanning point's step size has a great impact on moving target acquisition. If the step size is too large, some scanning areas will be missed, reducing the acquisition probability, as shown in Figure 6a. Too small a step size will increase the scanning repetition rate and capture time, as shown in Figure 6b. To achieve fast and efficient scanning, the optimal scanning step size that completely covers the capture uncertainty area must be selected. The scanning beam can not only eliminate missed scanning areas but also reduce invalid overlapping scanning areas, as shown in the center circle of Figure 6c.

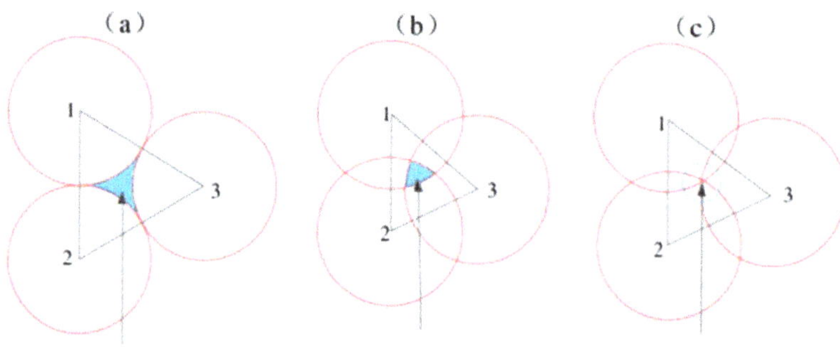

Figure 6. Schematic diagram of scanning results when different scanning steps are selected: (**a**) spot tangency, (**b**) spot intersection, (**c**) center intersection.

The numbers in Figure 6. represent the scanned light feet. When the points are tangent, a missed scan area in the blue area will appear in the middle, reducing the capture probability; When the points intersect, a repeating area with a blue part will appear in the middle, increasing the scanning time; When the centers intersect, the scanning efficiency is highest.

The projection of the laser light foot on the scanning area is shown in Figure 7. Assume that the laser beam's divergence angle is γ, the laser spot's diameter is d', and the step size of two adjacent spots is φ. At this time, the light spot offset distance is d'_x and d'_y. When φ is less than the maximum step size φ_{max}, the laser spots overlap. Since γ is of the order of mrad, the projection areas of two adjacent laser spots are regarded as circles with the same diameter. In rectangular spiral scanning, if the coverage area of the spot is rectangular, as shown in Figure 7a, the scanning field of view is S at this time:

$$S = d'_x \times d'_y \tag{3}$$

$$d'^2_x + d'^2_y = d'^2 \tag{4}$$

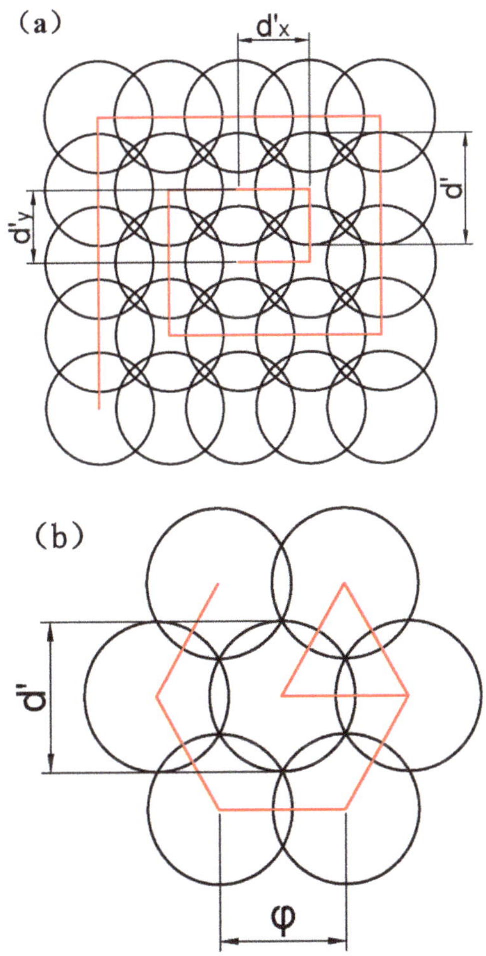

Figure 7. Schematic diagram of scanning compensation: (**a**) rectangular scanning area; (**b**) hexagonal scanning critical area.

Therefore, when and only when $d'_x = d'_y = \sqrt{2}/2d'$ has the largest area of S, according to the scanning relationship, the step size of the scanning point is $\varphi = \sqrt{2}/2d'$.

In hexagonal spiral scanning and hexagonal edge equidistant spiral scanning, the overlapping area of the light beam is small, as shown in Figure 7b. At this time, the step size of the scanning point is $\varphi = \sqrt{3}/2d'$ according to the scanning relationship.

3. Comparison and Analysis of Scanning Capture Models

To verify the above analysis, this paper adopts a TOF laser detection system and two-dimensional high-speed galvanometer scanning to obtain spatial information. On the one hand, the target acquisition probability should be considered in the scanning process, and on the other hand, the difficulty of the control method of a two-dimensional mechanical galvanometer should be considered. Therefore, the rectangular scanning and hexagonal scanning models should be analyzed and compared first, as follows (Figures 8–11):

(1) Simulation analysis of rectangular scanning

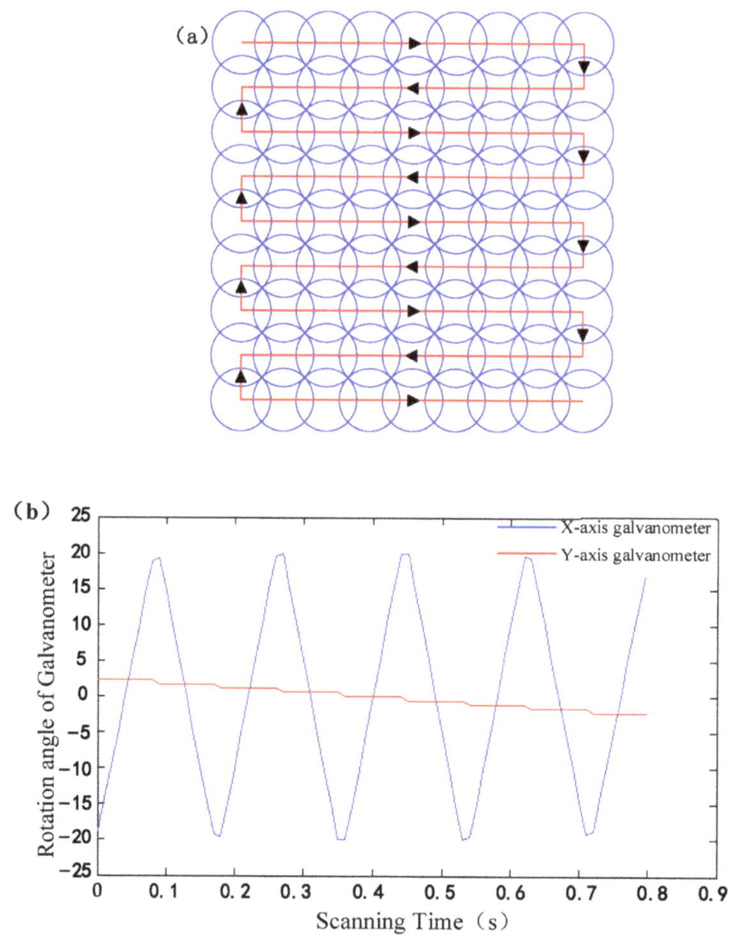

Figure 8. Rectangular scanning: (**a**) schematic diagram of rectangular scanning path; (**b**) schematic diagram of the galvanometer's deflection angle.

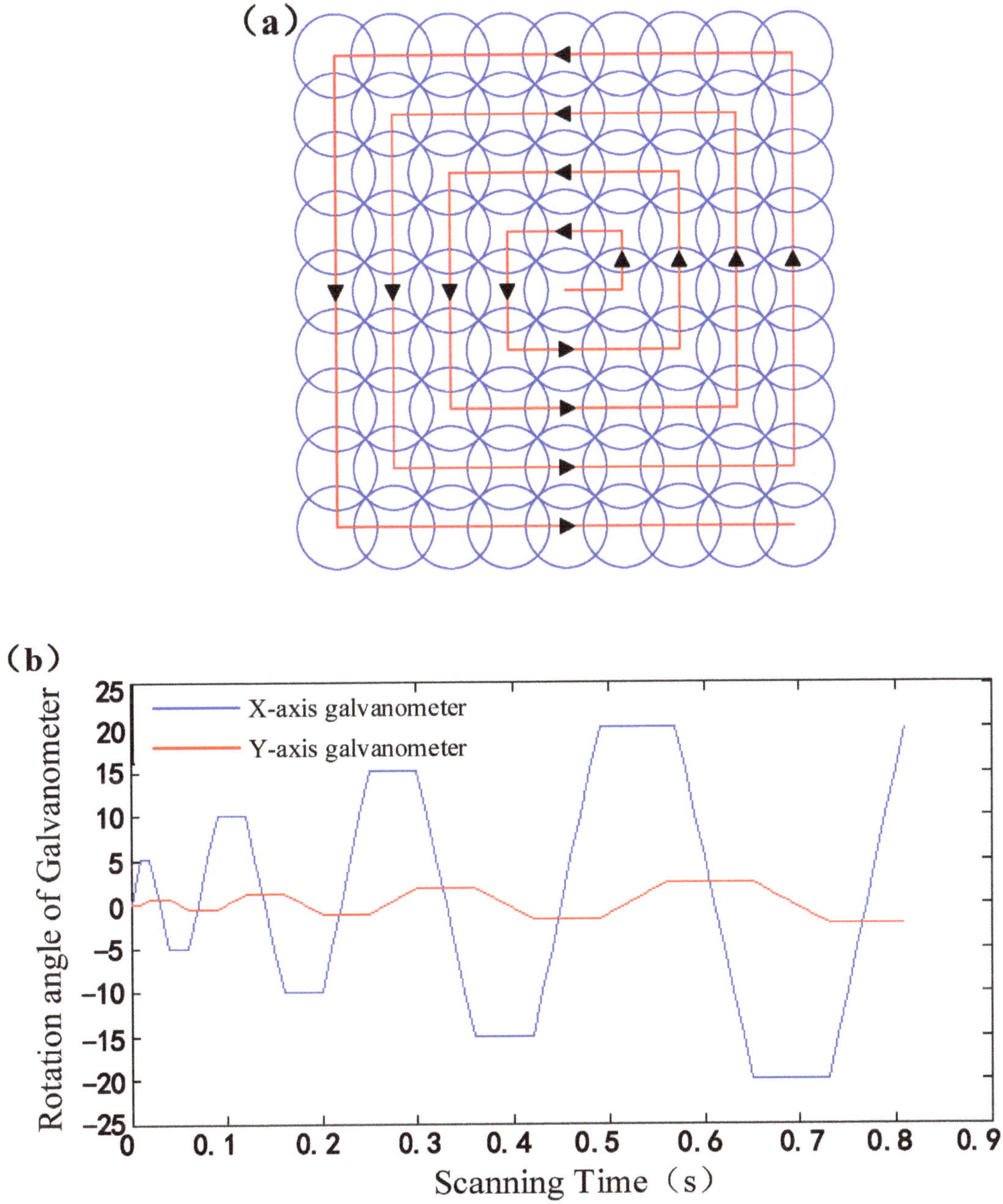

Figure 9. Rectangular spiral scanning: (**a**) schematic diagram of rectangular spiral scanning path; (**b**) schematic diagram of the galvanometer's deflection angle.

(2) Simulation Analysis of Hexagonal Scanning

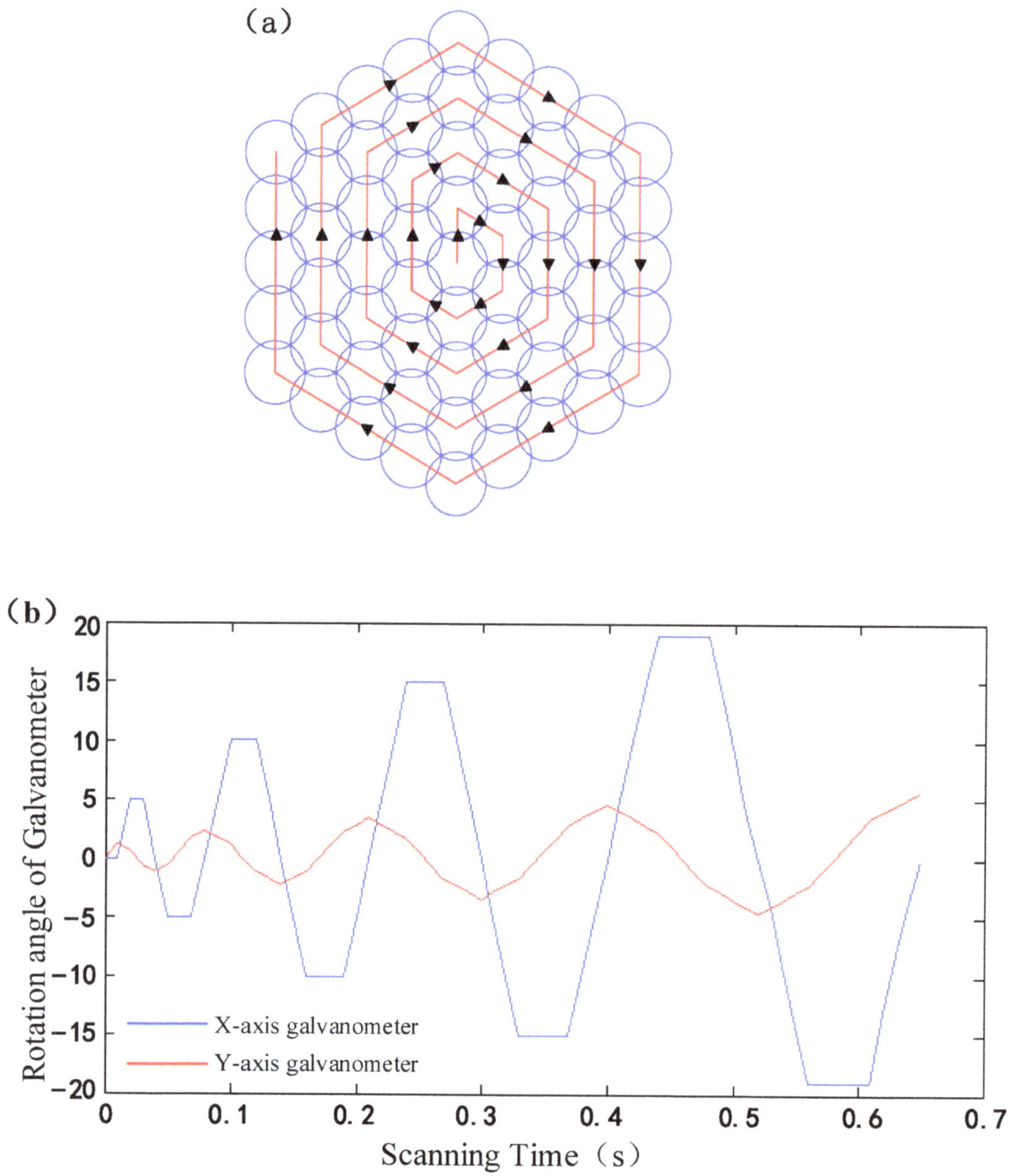

Figure 10. Hexagonal spiral scanning: (**a**) schematic diagram of hexagonal scanning path; (**b**) schematic diagram of the galvanometer's deflection angle.

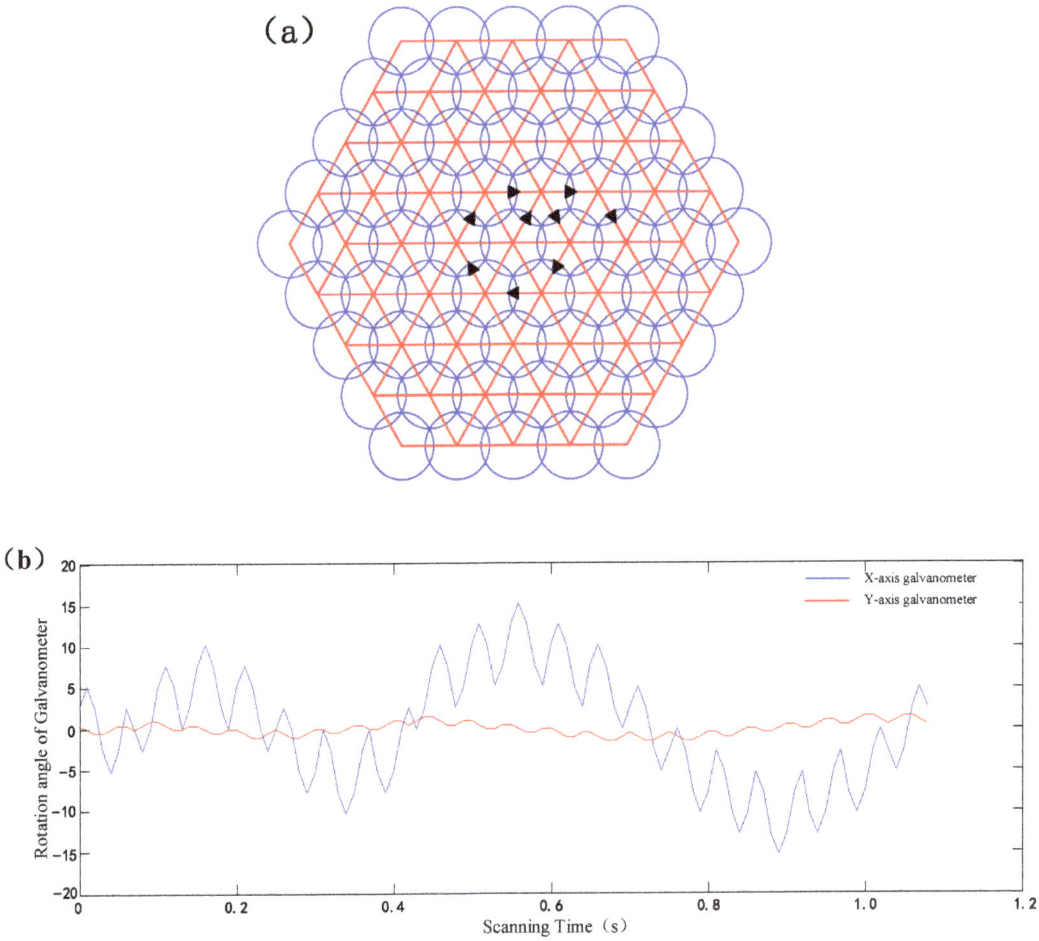

Figure 11. Improved hexagonal honeycomb structure scanning: (**a**) schematic diagram of scanning path of improved hexagonal honeycomb structure; (**b**) schematic diagram of the galvanometer's deflection angle.

The improved hexagonal honeycomb structure scans the inner circle more times within a visual threshold, so the scanning time is greater than traditional hexagonal spiral scanning. The simulation results show that the scanning amplitude of the honeycomb structure is smaller than that of traditional hexagonal spiral scanning, and the capture rate is also higher than that of the traditional method. The step size of the honeycomb structure does not change, which makes it easier to control the high-speed galvanometer system.

4. Experimental Simulation

To verify that the improved hexagonal honeycomb structure scanning method can improve the capture times of moving targets, this paper uses the control variable method to carry out a simulation and comparative analysis of four scanning methods.

Let the moving target speed of the CV model be 100 m/s, the laser output frequency be 10 kHz, the laser beam divergence angle be 10 mrad, the spot diameter be 25 mm, the step lengths of rectangular scanning and rectangular spiral scanning be 1.66 mm, and the step lengths of hexagonal spiral scanning and hexagonal honeycomb structure scanning be

3.15 mm. The distance R between the defined target point and the laser detection center point satisfies the formula:

$$|R| = \sqrt{(\Delta x)^2 + (\Delta y)^2} = \sqrt{|x(t_i) - x(t)|^2 + |y(t_i) - y(t)|^2} \tag{5}$$

where $y(t_i) = normrad(0,1,1,1)$ is a random number.

When the target is captured successfully, the spatial relationship between the photodetector and the target is shown in Figure 12, and the distance R between a single optical foot and the target point within the scanning threshold meets the following requirement:

$$|R| < |\lambda\delta| \tag{6}$$

where λ is the receiving field angle of the target detector, and δ is the plane distance between the target and the scanning visual threshold.

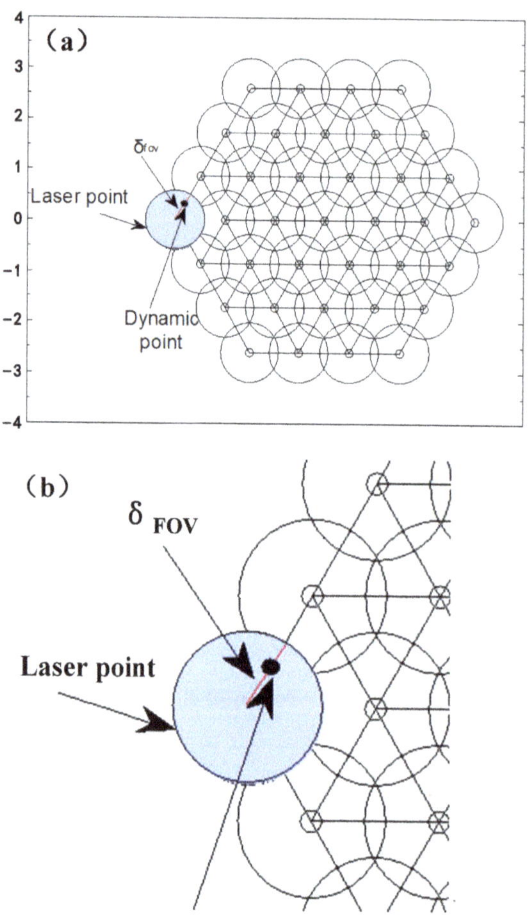

Figure 12. Schematic diagram of successful target acquisition. (**a**) Schematic diagram of spatial relationship between detector and target. (**b**) The relationship between a single light foot and the target point.

Matlab Simulink simulation software was used to establish simulation models for the four scanning methods; Monte Carlo simulations were used for capture analysis;

1000 single field scans were conducted for each scanning method; and the proportion of single field scanning capture times of each scanning method to the total scanning times was output.

The acquisition results obtained after the simulation are shown in Figure 13. From the simulation results in Figure 13, it can be concluded that when the target has been scanned 1000 times in a single field, the proportion of moving targets being captured by traditional hexagonal spiral scanning one to two times is large, 0.547 and 0.425, respectively. One moving target being captured by rectangular spiral scanning accounts for a large proportion, of 0.623. The proportions of two to three mobile targets being captured by rectangular branch scanning are 0.235 and 0.426. The proportion of two to four mobile targets captured by the improved hexagonal honeycomb structure scanning is larger, being 0.308, 0.285, and 0.298. The proportion of targets not captured after scanning a frame is smaller than that of other models. It can be seen from Figure 14 that when a target enters a fixed field of view, it is captured when it is detected more than or equal to one time. The maximum acquisition probability of the improved hexagonal honeycomb structure is 99%, followed by the hexagonal spiral scan at 96% and the rectangular spiral scan at 94%. The minimum acquisition probability is obtained by the rectangular branch scanning method, at 85%.

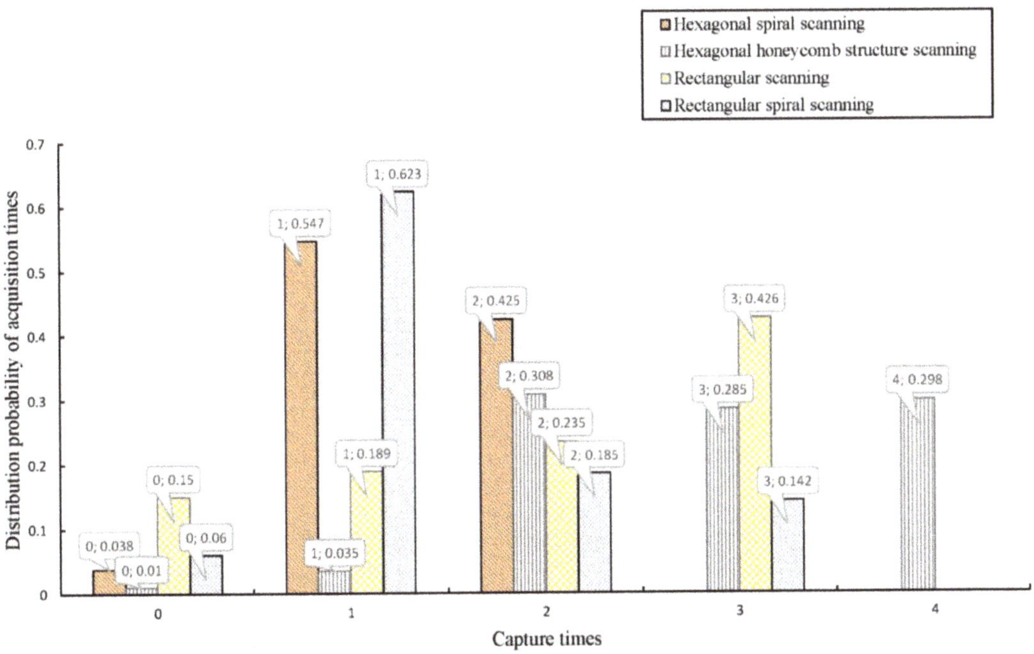

Figure 13. Capture probability simulation results of 1000 scans per field.

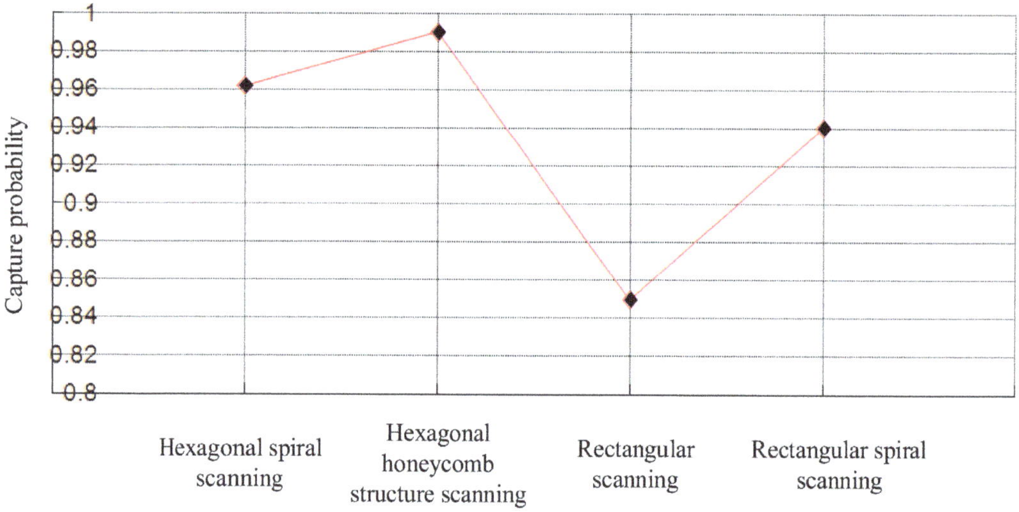

Figure 14. Analysis of capture probability simulation results.

When 1000 experiments were conducted, 988 were successfully captured, the probability of target capture in each experiment was $p = 0.9888$, and the probability of success was 98.88%. According to the binomial distribution, 988 obeyed the binomial distribution with parameters of 1000 and 0.9888, the sample average was $\bar{\mu} = np = 988.8$. Calculate the confidence interval with normal distribution: assuming the confidence level is 95%, the $z_{\alpha/2} = 1.96$, the standard deviation of the sample is $\sigma = \sqrt{np(1-p)} = 3.32$. The formula for the confidence interval is:

$$p - z_{\alpha/2} \cdot \sqrt{\frac{p(1-p)}{n}} \leq p \leq p + z_{\alpha/2} \cdot \sqrt{\frac{p(1-p)}{n}} \quad (7)$$

The confidence interval obtained through calculation is [0.9822, 0.9953].

5. Discussion

Based on the TOF laser measurement system and high-speed galvanometer scanning technology, this paper focused on the target acquisition problem for an a priori moving target model with a Gaussian distribution in the threshold scanning range. To improve the scanning times of the central area, an innovative and improved hexagonal honeycomb structure scanning model was proposed. Through simulation verification, within the visual threshold range of the same frame size, the acquisition probability of the target was 3% higher than that for the traditional hexagonal spiral scanning method, 5% higher than that for rectangular spiral scanning, and 14% higher than that for rectangular branch scanning. This scanning method provides a new solution for target acquisition in laser communication, laser docking, airborne radar, and other fields.

Author Contributions: Conceptualization, B.J. and Q.L.; methodology, F.J.; software, F.J.; validation, B.J., F.J. and Y.L.; formal analysis, Q.L.; data curation, Y.L.; writing—original draft preparation, F.J.; writing—review and editing, B.J. and Q.L.; visualization, F.J.; supervision, Y.L.; project administration, B.J.; funding acquisition, Q.L. All authors have read and agreed to the published version of the manuscript.

Funding: This research was funded by the "13th Five-Year Plan" Science and Technology Project of Jilin Provincial Education Department (Grant No. JJKH20200789KJ), Jilin Provincial Science and Technology Department Key R&D Project (Grant No. 20200401120GX) and Changchun University of Science and Technology (Grant No. 50923010501).

Institutional Review Board Statement: Not applicable.

Informed Consent Statement: Not applicable.

Data Availability Statement: Not applicable.

Conflicts of Interest: The authors declare no conflict of interest.

References

1. Marshall, G.F.; Stutz, G.E. *Handbook of Optical and Laser Scanning*, 2nd ed.; Marcel Dekker Inc.: New York, NY, USA, 2012.
2. Xie, S.; Du, Y.; Zhang, Y.; Wang, Z.M.; Zhang, D.L.; He, L.; Qiu, L.P.; Jiang, J.H.; Tan, W.H. Aptamer-based optical manipulation of protein subcellular localization in cells. *Nat. Commun.* **2020**, *11*, 1347. [CrossRef] [PubMed]
3. Abouakil, F.; Meng, H.; Burcklen, M.A.; Rigneault, H.; Galland, H.; LeGoff, L. An adaptive microscope for the imaging of biological surfaces. *Light Sci. Appl.* **2021**, *10*, 210. [CrossRef] [PubMed]
4. Porfirev, A.P.; Fomchenkov, S.A. Effect of laser radiation power on laser trapping of light-absorbing microparticles in air. *Procedia Eng.* **2017**, *201*, 48–52. [CrossRef]
5. Porfirev, A.P. Experimental investigation into the possibilities of using a spatial light modulator for laser-trapping light-absorbing micro-objects in air. In Proceedings of the XIII International Scientific and Technical Conference on Optical Technologies in Telecommunications, Ufa, Russian, 11–18 November 2015. [CrossRef]
6. Küppers, M.; Albrecht, D.; Kashkanova, A.D.; Lühr, J.; Sandoghdar, V. Confocal interferometric scattering microscopy reveals 3D nanoscopic structure and dynamics in live cells. *Nat. Commun.* **2023**, *14*, 1962. [CrossRef]
7. Lu, J.S.; Wang, H.Y.; Kudo, T.; Kudo, T.; Masuhara, H. Large Submillimeter Assembly of Microparticles with Necklace-like Patterns Formed by Laser Trapping at Solution Surface. *J. Phys. Chem. Lett.* **2020**, *11*, 6057–6062. [CrossRef]
8. Espina, V.; Heiby, M.; Pierobon, M.; Liotta, L.A. Laser capture microdissection technology. *Expert Rev. Mol. Diagn.* **2007**, *7*, 647–657. [CrossRef]
9. Simone, N.L.; Bonner, R.F.; Gillespie, J.W.; Emmert-Buck, M.R.; Liotta, L.A. Laser-capture microdissection: Opening the microscopic frontier to molecular analysis. *Trends Genet.* **1998**, *14*, 272–276. [CrossRef]
10. Suarez-Quian, C.A.; Goldstein, S.R.; Pohida, T.; Smith, P.D.; Peterson, J.I.; Wellner, E.; Ghany, M.; Bonner, R.F. Laser capture microdissection of single cells from complex tissues. *Biotechniques* **1999**, *26*, 328–335. [CrossRef]
11. Klee, E.W.; Erdogan, S.; Tillmans, L.; Kosari, F.; Sun, Z.F.; Wigle, D.A.; Yang, P.; Aubry, M.C.; Vasmatzis, G. Impact of sample acquisition and linear amplification on gene expression profiling of lung adenocarcinoma: Laser capture micro-dissection cell-sampling versus bulk tissue-sampling. *BMC Med. Genom.* **2009**, *2*, 13. [CrossRef]
12. Todd, R.; Kuo, M.W.L.W.P. Gene expression profiling using laser capture microdissection. *Expert Rev. Mol. Diagn.* **2002**, *2*, 497–507. [CrossRef]
13. Priezzhev, A.; Lee, K. Potentialities of laser trapping and manipulation of blood cells in hemorheologic research. *Clin. Hemorheol. Micro.* **2016**, *64*, 587–592. [CrossRef] [PubMed]
14. Ouyang, W.; Xu, X.; Lu, W.; Zhao, N.; Han, F.; Chen, S.C. Ultrafast 3D nanofabrication via digital holography. *Nat. Commun.* **2023**, *14*, 1716. [CrossRef]
15. Yang, D.; Liu, Y.; Chen, Q.; Chen, M.; Zhan, S.D.; Cheung, N.K.; Chan, H.Y.; Wang, Z.D.; Li, W.J. Development of the high angular resolution 360° LiDAR based on scanning MEMS mirror. *Sci. Rep.* **2023**, *13*, 1540. [CrossRef] [PubMed]
16. Ravikumar, V.K.; Chin, J.M.; Lua, W.; Linarto, N.; Ranganathan, G.; Trisno, J.; Pey, K.L.; Yang, J.K.W. Super-resolution laser probing of integrated circuits using algorithmic methods. *Nat. Commun.* **2022**, *13*, 5155. [CrossRef] [PubMed]
17. Abdelmoula, M.; Zarazaga, A.M.; Küçüktürk, G.; Maury, F.; Grossin, D.; Ferrato, M. Scanning Strategy Investigation for Direct Powder Bed Selective Laser Processing of Silicon Carbide Ceramic. *Appl. Sci.* **2022**, *12*, 788. [CrossRef]
18. Almukhtar, A.; Saeed, Z.O.; Abanda, H.; Tah, J.H.M. Reality Capture of Buildings Using 3D Laser Scanners. *CivilEng* **2021**, *2*, 214–235. [CrossRef]
19. Zhao, L.; Mbachu, J.; Wang, B.; Liu, Z.; Zhang, H. Installation Quality Inspection for High Formwork Using Terrestrial Laser Scanning Technology. *Symmetry* **2022**, *14*, 377. [CrossRef]
20. Liu, H.Z.; Ji, Y.F. Effect of aberration on performance of the bit error rate in an inter-satellite coherent optical communication receiving system. *Acta Opt. Sin.* **2012**, *32*, 38–43. [CrossRef]
21. Rana, M.S.; Pota, H.R.; Petersen, I.R. Spiral scanning with improved control for faster imaging of AFM. *IEEE Trans. Nanotechnol.* **2014**, *13*, 541–550. [CrossRef]
22. Kelley, K.P.; Ziatdinov, M.; Collins, L.; Susner, M.A.; Vasudevan, R.K.; Balke, N.; Kalinin, S.V.; Jesse, S. Fast scanning probe microscopy via machine learning: Non-rectangular scans with compressed sensing and gaussian process optimization. *Small* **2020**, *16*, 2002878. [CrossRef]
23. Rana, M.S.; Pota, H.R.; Petersen, I.R. Performance of sinusoidal scanning with MPC in AFM imaging. *IEEE/ASME Trans. Mechatron.* **2014**, *20*, 73–83. [CrossRef]
24. Pota, H.; Petersen, I.R. Reduction of phase error between sinusoidal motions and vibration of a tube scanner during spiral scanning using an AFM. *Int. J. Control Autom.* **2016**, *14*, 505–513. [CrossRef]

25. Habibullah, H.; Pota, H.R.; Petersen, I.R. A novel control approach for high-precision positioning of a piezoelectric tube scanner. *IEEE Trans. Autom. Sci. Eng.* **2016**, *14*, 325–336. [CrossRef]
26. Wang, T.; Yang, S.M.; Li, S.S.; Yuan, Y.; Hu, P.Y.; Liu, T.; Jia, S.H. Error analysis and compensation of galvanometer laser scanning measurement system. *Acta Opt. Sin.* **2020**, *40*, 2315001. (In Chinese)
27. Qi, S.; Zhang, Q.; Xin, X.J.; Tao, Y.; Tian, Q.H.; Tian, F.; Cao, G.X.; Shen, Y.F.; Chen, D.; Gao, Z.H.; et al. Research on Scanning Method in Satellite Laser Communication. In Proceedings of the International Conference on Optical Communications and Networks (ICOCN), Huangshan, China, 5–8 August 2019. [CrossRef]
28. Yuan, G.F.; Ma, X.Y.; Liu, S.; Yang, Q.L. Research on Lidar scanning mode. *High Power Laser Part. Beams* **2020**, *32*, 041001-1–041001-6. (In Chinese) [CrossRef]
29. D'Agostino, F.; Ferrara, F.; Gennarelli, C.; Guerriero, R.; McBride, S.; Migliozzi, M. Fast and accurate antenna pattern evaluation from near-field data acquired via planar spiral scanning. *IEEE Trans. Antennas Propag.* **2016**, *64*, 3450–3458. [CrossRef]
30. Bucci, O.M.; D'agostino, F.; Gennarelli, C.; Riccio, G.; Savarese, C. Near-field-far-field transformation with spherical spiral scanning. *IEEE Antennas Wirel. Propag. Lett.* **2003**, *2*, 263–266. [CrossRef]
31. Candra, R.; Madenda, S.; Sudiro, S.A.; Subali, M. The implementation of an efficient zigzag scan. *JTEC* **2017**, *9*, 95–98.

photonics

MDPI

Communication

Design of a Large Field of View and Low-Distortion Off-Axis Optical System Based on a Free-Form Surface

Bing Jia [1,2], Yubing Li [1,*], Qiongying Lv [1], Fan Jin [1] and Chunlin Tian [1]

1 College of Mechanical and Electrical Engineering, Changchun University of Science and Technology, Changchun 130022, China; 2015800071@cust.edu.cn (B.J.)
2 Chongqing Research Institute, Changchun University of Science and Technology, Chongqing 401120, China
* Correspondence: 13304456182@163.com; Tel.:+86-13304456182

Abstract: Free-form surfaces have good aberration correction capability and balance capability for nonrotationally symmetric imaging systems. In this study, we analyzed the quantitative relationship between *X–Y* polynomial combination and aberration for the efficient design of *X–Y* free-form optical systems. The purpose of this study was to provide an exhaustive design method for off-axis triple inverse optical systems with *X–Y* free-form surfaces. Finally, we designed a free-form off-axis optical system with a large field of view (FOV) and low distortion based on the off-axis triple inverse optical system without an intermediate image plane. The primary mirror (PM) of the system adopted an *X–Y* polynomial free-form surface to correct the aberration of different FOVs of the system and improve the image width and quality. The optical system had a focal length of 1000 mm, an F-value of 9.5, an FOV angle of $23^\circ \times 1^\circ$, a maximum distortion grid in the FOV less than or equal to -0.05%, and a full-field average wave aberration better than 0.055 λ (λ/18.2, λ = 632.8 nm). The analysis of the design results showed that the system had high-quality imaging and a compact structure. This design method can provide a technical reference for the study of such free-form off-axis systems.

Keywords: optical design; off-axis three mirrors; free-form surface; distortion; tolerance analysis

Citation: Jia, B.; Li, Y.; Lv, Q.; Jin, F.; Tian, C. Design of a Large Field of View and Low-Distortion Off-Axis Optical System Based on a Free-Form Surface. *Photonics* **2023**, *10*, 506. https://doi.org/10.3390/photonics10050506

Received: 17 March 2023
Revised: 21 April 2023
Accepted: 24 April 2023
Published: 27 April 2023

1. Introduction

With the rapid development of advanced precision optics, optical imaging systems with high resolution, large field of view (FOV) and wide spectral range have become the new development trend [1–3]. A three-mirror anastigmat system (TMAS) is a reflective optical system that evolved from a three-mirror coaxis system (TMCS) by biasing the FOV or aperture. The TMAS has received an increasing amount of attention in the domain of optical remote sensing because the system has the characteristics of long focal length, no chromatic aberration, good thermal performance, and comprehensive spectral band coverage. Thus, the TMAS has the advantages of a larger FOV and better imaging quality [4,5]. TMASs have been widely used in many international remote sensing optical systems, such as SPOT-6, Worldview-1, ALOS-3, PROB-V, and GOCI [6–9].

The TMAS avoids the central occlusion of the TMCS and improves the imaging quality. At the same time, the system loses its rotational symmetry, leading to a dramatic increase in advanced aberration of off-axis FOV. For large FOV optical systems using linear array push-scan mode imaging, aberrations in the push-scan direction can cause image shift, which leads to blurred imaging and reduces the imaging quality of the system [10,11]. However, it is challenging for off-axis reflective optical systems with conventional spherical or nonspherical shapes to balance large FOV, low distortion, and system modulation transfer function (MTF) because of the design degrees of freedom (DOF). Therefore, understanding how to increase the system FOV, reduce aberrations, and ensure a compact system structure with high-quality imaging has become an urgent problem in space optical systems.

Free-form surfaces, which rely on manufacturing and inspection technology developments, provide an excellent solution to meet the increasingly demanding performance specifications and large FOV requirements of optical systems [12–14]. The free-form surface is widely used in panoramic optical systems, head-mounted displays, and off-axis reflective optical systems [15–17]. Free-form optics are optical surfaces with nonrotationally symmetric characteristics that can improve optical performance while enabling more novel features, reducing the size of rotationally asymmetric optical systems, and correcting system aberrations [18–20]. To date, the main parametric representations for optical free-form design include *X–Y* polynomials, Zernike polynomials, Q-polynomials, radial basis functions, and spline functions [21]. *X–Y* polynomial surfaces and Zernike polynomial surfaces have the same free-form surface characterization capability, have interconversion relationships, and are more often used in machining [22]. In recent years, an increasing number of research institutions have launched corresponding studies on the spatial application of free-form surfaces. For example, the imaging spectrometer developed by the Institute of Optics at the University of Rochester takes advantage of the compactness of the free-form surface structure, reducing the size of the optical system by a factor of five compared with those using conventional spherical or aspheric designs, while expanding the spatial spread by a factor of two [23]. The free-form spatial off-axis reflective optical system developed by the Changchun Institute of Optics, Fine Mechanics, and Physics of the Chinese Academy of Sciences, has achieved 76° FOV imaging [24]. The free-form off-axis triple inverted infrared imaging system developed by Tsinghua University and Tianjin University takes advantage of the free-form design to greatly reduce the difficulty of mounting the system, which has a focal length of 138 mm and an FOV of 4° × 5° and works in the medium-wave and long-wave infrared wavelengths [16].

In this study, we proposed a large FOV and low-distortion TMAS based on an *X–Y* polynomial and described the method used to achieve this system. Compared to TMAS using conventional spherical or aspheric surface shapes, this system used the free-form surface to expand the design freedom of the system and to further improve the aberration balance capability of the system. Although the loss of rotational symmetry makes the measurement and processing of free-form surfaces challenging, the increasing sophistication of computer inspection technology and optical precision machining techniques have made it possible to fabricate components for imaging systems with larger free-form surfaces. This system has a certain reference value for designing a space optical system. It is suitable for a large FOV and low-distortion linear array push-broom imaging optical system. The remainder of this paper is organized as follows: In Section 2, the optical design, considerations, and initial configurations of the TMAS are described. In Section 3, the quantitative relationship between the *X–Y* polynomial, the Fringe Zernike polynomial, and the aberration is analyzed. In Section 4, a design example of a TMAS with a focal length of 1000 mm, an F-value of 9.5, a large FOV of 23° × 1°, and a distortion of less than 0.05% is given, and the imaging quality and tolerance of the system are analyzed. Section 5 discusses the results.

2. Optical System Design

The classic method of establishing the structure of the reflective optical system is based on the theory of near-axis optics, solving the structural parameters of the coaxial optical system using primary aberration theory, and then eliminating the central occlusion by FOV off-axis or aperture off-axis to establish the initial structure. The initial configuration of a TMAS evolved from a TMCS, a biased aperture, and FOV, and we adopted this configuration to eliminate the obstruction. The TMCS initial structure parameters include eight parameters: the radius of three surfaces, namely R_1, R_2, and R_3; the distance from the primary mirror (PM) to the secondary mirror (SM) is d_1; the distance from the SM to the tertiary mirror (TM) is d_2; and the quadratic surface coefficients of the three mirrors are K_1, K_2, and K_3. To simplify the process of solving the initial structural parameters, the PM and TM of the optical system overlapped when $d_1 = d_2$. The initial configuration of the TMCS that was used in this study is shown in Figure 1. The calculation method for the initial configuration of the TMCS was based on the third-order aberration theory, which has been introduced in many literature references [25–27] and is not discussed in detail here.

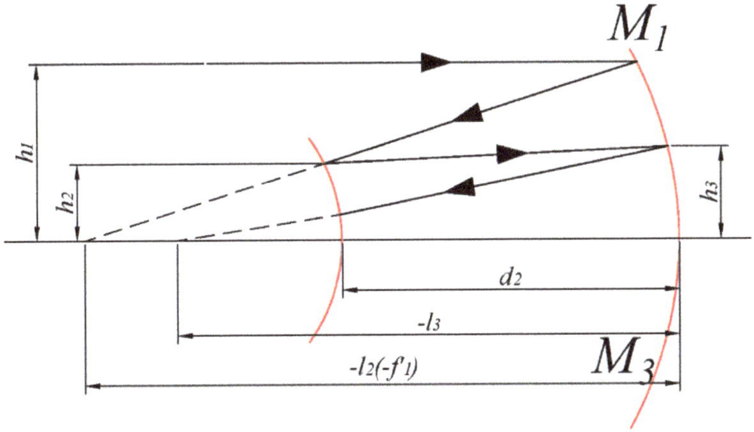

Figure 1. Initial TMCS configuration.

Two common forms of TMAS are applied in the optical remote sensing domain: relayed TMAS, shown in Figure 2a, and nonrelayed TMAS, shown in Figure 2b.

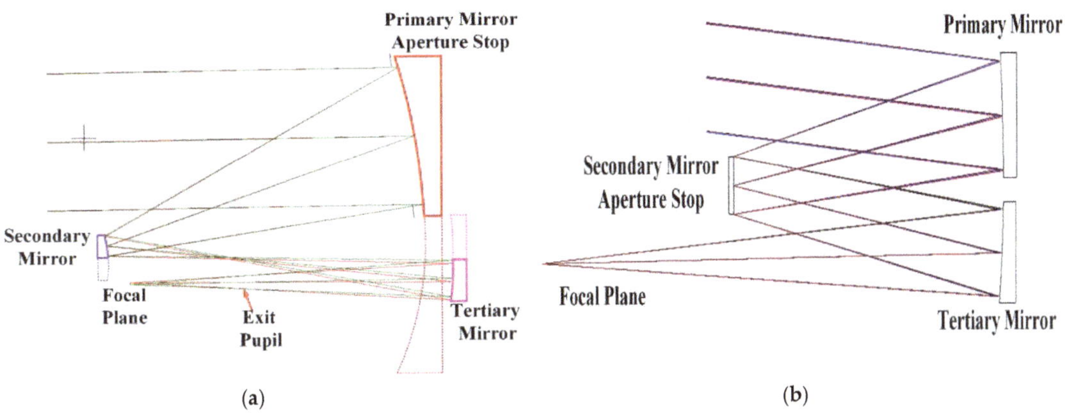

Figure 2. (**a**) Relayed TMAS; and (**b**) nonrelayed TMAS.

In a relayed TMAS [28], shown in Figure 2a, the aperture stop is located at the PM or the entrance pupil of the system. Because the system is the off-axis part of the TMCS, it inherits the advantage of the highly compact ratio of the TMCS. It has an intermediate imaging plane between the SM and the TM, and a real exit pupil in the back focal length (BFL). These places are suitable for a field stop and a rear stop, which can ensure that the system has good performance in the suppression of stray radiation. Because of the real exit pupil, the system is also suitable for multispectral imaging with a common aperture. However, because the relayed TMAS inherited the configurable characteristics of the TMCS, it has been difficult to achieve a large FOV.

In a nonrelayed TMAS [29], shown in Figure 2b, the aperture stop is located at the SM, and a biased FOV is adopted to remove the obstruction of the SM. However, because the biased FOV moves the real FOV away from the field center, more nonsymmetric aberrations are generated when the biased angle increases. The PM and the TM are symmetrically relative to the optical axis, similar to the projection of the subpupil of each FOV. As a result, forms of the PM and TM are usually rectangular with high aspect ratios (ARs). This configuration is beneficial to lateral aberration correcting and larger FOV imaging. As the FOV increases, however, the aspect ratio of the PM and TM will gradually increase.

In contrast to a relayed TMAS, a nonrelayed TMAS can easily implement large FOV imaging, and it is more suitable for optical remote sensing with a drift-scan pattern. At the same time, a rectangular mirror manufactured with metal can achieve a low surface shape error with a much higher lightweight level, which can make up for the deficiency in size and weight of a nonrelayed TMAS. Therefore, in this study, we adopted a nonrelayed TMAS for the initial configuration to further optimize the large FOV and the low-distortion TMAS.

3. Quantitative Analysis of the Free-Form Surface

A free-form surface is a nonrotationally symmetrical optical surface type. Compared with conventional spherical and aspherical surfaces, a free-form surface can offer more DOFs for optical designers and enhance the corrective abilities of an off-axis aberration. This type of optical imaging system mainly uses Zernike polynomials and *X–Y* polynomials [30]. An *X–Y* polynomial was first used for low-order free-from surfaces and most commonly is used to describe free-form surfaces. An *X–Y* polynomial adds a more refined shape description to a surface sag as well as more effective DOFs for higher-order aberration correction. An *X–Y* polynomial is also a classic nonorthogonal polynomial that is suitable for three-dimensional modeling and numerically controlled optical manufacturing. Because it can achieve high precision in mirror shape, it is widely used in metal-based mirrors, such as the JSS-56 imager [31]. The expression of the *X–Y* polynomial is as follows:

$$Z = \frac{cr^2}{1+\sqrt{1-(1+k)c^2r^2}} + \sum_{i=1}^{N} A_i E_i(x,y) \quad (1)$$

where Z is the sag of the surface parallel to the z-axis, c is the vertex curvature, r is the radial coordinate, k is the conic constant, N is the polynomial sequence, A_i is the polynomial coefficient, and $E_i\ (x, y)$ is the polynomial-like difference term expression.

A Zernike polynomial is a complete set of functions that is orthogonal over a circle with a unit radius. This characteristic makes it much more suitable for describing and fitting an aberration. A fringe Zernike polynomial has a uniform equation and uniform characteristics, except that the terms of the Fringe Zernike are remapped according to an interferometric data analysis program. Thus, the Fringe Zernike is widely used in the design, detection, and alignment of optical systems. The expression of the Zernike polynomial is as follows:

$$Z = \frac{cr^2}{1+\sqrt{1-(1+k)c^2r^2}} + \sum_{j=1}^{N} C_{(j+1)} ZP_j \quad (2)$$

where $C_{(j+1)}$ is the polynomial coefficient and ZP_j is the polynomial-like difference term expression.

As seen in Equation (1), the *X–Y* polynomial is usually expressed in Cartesian coordinates. The *X–Y* polynomial is consistent with the computer numerical control machine expression and is highly practical for the machining of metal reflectors [32]. Because the free-form surface corresponds to a specific geometric aberration, it has a more powerful off-axis field aberration correction capability. The design of an optical imaging system involves the process of an aberration balance to produce the minimum residual wavefront error (WFE). Optimization of optical systems using *X–Y* polynomials is to obtain better free-from surface solutions by solving for the coefficients of each order, which in turn optimizes the system aberration to improve the imaging quality. As expressed by Equation (2), the Zernike polynomial is usually expressed in polar coordinates. The Zernike polynomial corresponds with the geometric aberration, which can describe the aberration of the system more easily. Thus, this translates into the problem of solving the relationship between the X–Y polynomial and the Fringe Zernike polynomial.

To better illustrate the relationship between the X–Y polynomial and the aberration, we converted each aberration term in the Fringe Zernike polynomial to a Cartesian coordinate system, and the conversion equation is shown in Equation (3).

$$\begin{aligned} X &= R \times cos\theta \\ Y &= R \times sin\theta' \end{aligned} \tag{3}$$

where R is the radial distance and θ is the azimuth angle.

Different X–Y polynomial combinations are added to the single parabolic mirror surface to characterize the aberrations in the transformed Fringe Zernike polynomials. This method is used to verify the aberration correction ability of the X–Y polynomial. We analyzed only third-order aberrations, and the aberration results are shown in Table 1.

Table 1. Coordinate transformation.

No.	Aberration	Radial Coordinate	Cartesian Coordinate
1	Piston	1	1
2	Tilt (X)	$R\cos\theta$	X
3	Tilt (Y)	$R\sin\theta$	Y
4	Defocus	$2R^2 - 1$	$2X^2 + 2Y^2 - 1$
5	Astig (0°)	$R^2\cos(2\theta)$	$X^2 - Y^2$
6	Astig (45°)	$R^2\sin(2\theta)$	$2XY$
7	Coma (X)	$(3R^3 - 2R)\cos\theta$	$3X^3 + 3XY^2 - 2X$
8	Coma (Y)	$(3R^3 - 2R)\sin\theta$	$3X^2Y + 3Y^3 - 2Y$
9	Spherical	$6R^4 - 6R^2 + 1$	$6X^4 + 12X^2Y^2 + 6Y^4 - 6X^2 - 6Y^2 + 1$
10	Trefoil (X)	$R^3\cos(3\theta)$	$X^3 - 3XY^2$
11	Trefoil (Y)	$R^3\sin(3\theta)$	$3X^2Y - Y^3$

The analysis and results of the relationship between the combination of X–Y polynomials and each of the aberrations using the specified approach revealed the following: (1) The sign of the monomial only affected the value of the aberration, without affecting the type of aberration, and the variation in the coefficient of A_{ij} led to an absolute value change in the aberration. (2) In the polynomial combinations used to represent the different aberrations, the sign of various monomials affected the types of aberration. Considering items 4 and 5 in Table 2 as an example, when A_{02} had a positive value, the system was defocused. The wavefront aberration map for this scenario is shown in Figure 3a. When A_{02} had a negative value, the system had an astigmatism. The wavefront aberration map for this scenario is shown in Figure 3b. (3) The combination of X–Y polynomials, when characterizing a particular aberration, also produced other aberration terms, but the magnitude of the other aberration terms usually was small.

Table 2. Relationship between *X–Y* polynomial and Fringe Zernike polynomial.

No.	Aberrations	Coefficient	Value	*X–Y* Polynomial Combinations	Coefficient	Value
1	Piston	0	0	0	0	0
2	Tilt (*X*)	0	0	A_{10}	0	0
3	Tilt (*Y*)	0	0	A_{01}	0	0
4	Defocus	1×10^{-3}	−6.3209	$2A_{20} + 2A_{02}$	1×10^{-3}	Piston: −3.1605 Defocus: −3.129
5	Astig (0°)	1×10^{-3}	−3.1531	$A_{20} - A_{02}$	1×10^{-3}	−3.1534
6	Astig (45°)	1×10^{-3}	−3.1530	$2A_{11}$	1×10^{-3}	−3.1528
7	Coma (*X*)	1×10^{-3}	Tilt(*X*): −6.322 Coma(*X*): −3.105	$3A_{30} + 3A_{12} - 2A_{10}$	1×10^{-3}	Tilt(*X*): −6.322 Coma(*X*): −3.105
8	Coma (*Y*)	1×10^{-3}	Tilt(*Y*): −6.322 Coma(*Y*): −3.105	$3A_{21} + 3A_{03} - 2A_{01}$	1×10^{-3}	Tilt(*X*): −6.322 Coma(*X*): −3.105
9	Spherical	1×10^{-3}	Piston: 3.1604 Spherical: −3.146	$6A_{40} + 12A_{22} + 6A_{04} - 6A_{20} - A_{02}$	1×10^{-3}	Piston: 3.1446 Spherical: −3.146
10	Trefoil (*X*)	1×10^{-3}	−3.1547	$A_{30} - 3A_{12}$	1×10^{-3}	−3.1548
11	Trefoil (*Y*)	1×10^{-3}	−3.1547	$3A_{21} - A_{03}$	1×10^{-3}	−3.1548

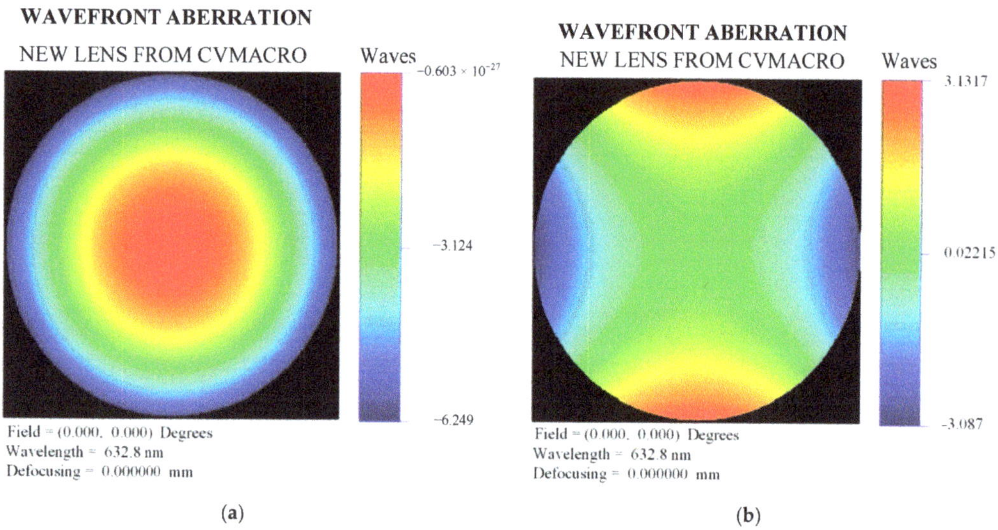

Figure 3. The aberration when A_{20} is positive: (**a**) A_{02} with positive value, aberration out of focus; (**b**) A_{02} with negative value, aberration-like scattering.

Based on this analysis, the generated aberration of the *X–Y* polynomial combinations was consistent with the aberration of the Fringe Zernike polynomial, and these two types had the same value, which proved the quantitative relationship between the *X–Y* free-form surface and the aberration. This was an instructive method for aberration balancing and optical designing based on an *X–Y* free-form surface. The 3D diagram of the free-form surface is shown in Figure 4. As could be seen from the figure, the free-form surface had a gentle variation, which was suitable for Computer-Generated Hologram (CGH) for mounting and inspection, with less processing difficulty and error, and easy to achieve high inspection accuracy. Therefore, we used the free-form surface described by the *X–Y* polynomial in the designed off-axis triple inverse optical system to increase the effective FOV of the system and improve the correction capability of off-axis aberration, especially the correction of off-axis asymmetric aberration.

Figure 4. Three-dimensional view of the free-form surface.

4. System Design and Analysis

4.1. Design Results

In this study, we selected the optical simulation software Code V to optimize the aberrations of the TMAS without an intermediate image plane by using the *X*–*Y* polynomial surface shape analyzed earlier. In the optimization process, astigmatism and coma were the most important initial aberrations in the parallax field imaging system. Therefore, we selected A_{20}, A_{02}, A_{21}, A_{03}, and A_{01} to correct these two aberrations. Throughout the optimization process, the aberration values of the system were controlled. Next, we selected A_{40}, A_{22}, and A_{40} for spherical aberration correction. Finally, we used other higher-order terms to minimize the RMS wave aberrations in the full FOV. After optimizing the design according to the optical system parameters, we obtained a TMAS with large FOV and low distortion. The optical system structure diagram is shown in Figure 5, the optical path diagram after adding the fold mirror is shown in Figure 6, and the system parameters are given in Table 3. The PM of the system used an *X*–*Y* polynomial free-form surface to correct for aberrations in the different FOVs of the system. The system's SM was parabolic, which reduced the difficulty of machining and setting up the system. The TM of the system was an ellipsoidal surface with a high secondary term. To further compress the volume, the focal plane could be folded into the lower part of the TM by adding a fold mirror (400 mm × 52 mm) to the rear intercept of the system. The final optical system had an overall length of $f'/2.6$, a height of $f'/2.9$, and a width of $f'/2.25$, creating a compact system. The configuration parameters are given in Table 4, and the *X*–*Y* polynomial of the PM is given in Table 5.

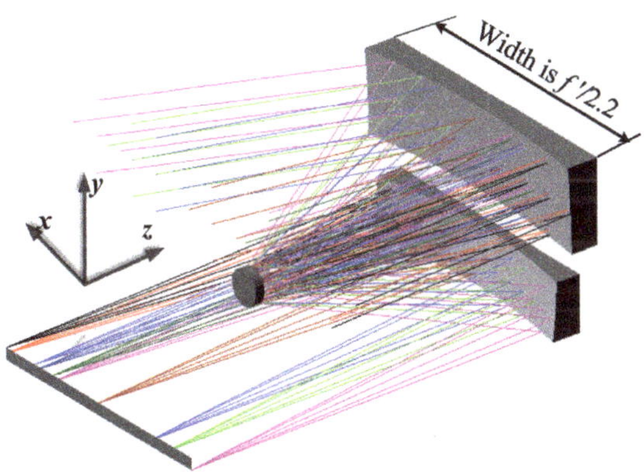

Figure 5. Optical system configuration.

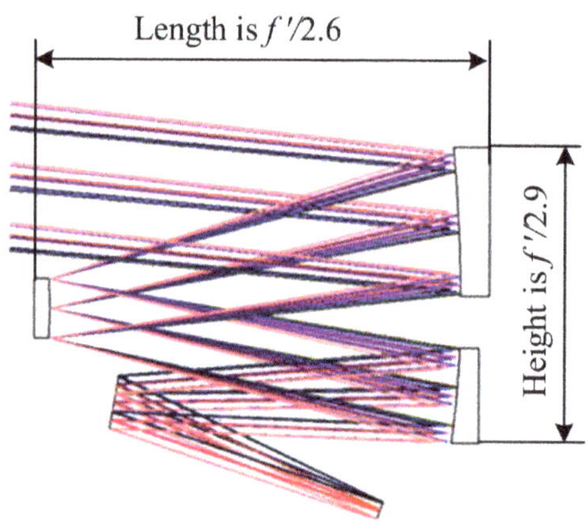

Figure 6. System optical path diagram after adding fold mirror.

Table 3. Parameters of optical system.

No.	Parameter	Value
1	Focal length (mm)	1000
2	F#	9.5
3	FOV (°)	23 × 1
4	Spectrum (nm)	450–850
5	Nyquist frequency (lp/mm)	71.4

Table 4. Configuration parameters of reflective mirror.

Mirror	Surface Type	Radius (mm)	Distance (mm)	Conic	Size (mm)
PM	*X–Y* polynomial	−1538.978	−384.8	−1.52	444 × 125
SM	Paraboloid	−592.75	384.8	−1	Φ58
TM	Even asphere	−935.58	−648.81	−0.178	392 × 84

Table 5. Polynomial parameters of free-form primary mirror.

No.	Item	A_{ij}	No.	Item	A_{ij}	No.	Item	A_{ij}
1	X^1Y^0	0	10	X^4Y^0	2.09×10^{-11}	19	X^1Y^4	0
2	X^0Y^1	-3.97×10^{-4}	11	X^3Y^1	0	20	X^0Y^5	-1.63×10^{-13}
3	X^2Y^0	-1.41×10^{-7}	12	X^2Y^2	3.8×10^{-11}	21	X^6Y^0	-2.98×10^{-17}
4	X^1Y^1	0	13	X^1Y^3	0	22	X^5Y^1	0
5	X^0Y^2	-1.37×10^{-7}	14	X^0Y^4	4.47×10^{-11}	23	X^4Y^2	-7.32×10^{-18}
6	X^3Y^0	0	15	X^5Y^0	0	24	X^3Y^3	0
7	X^2Y^1	7×10^{-10}	16	X^4Y^1	-2.01×10^{-14}	25	X^2Y^4	6.83×10^{-18}
8	X^1Y^2	0	17	X^3Y^2	0	26	X^1Y^5	0
9	X^0Y^3	-1.39×10^{-9}	18	X^2Y^3	-2.63×10^{-14}	27	X^0Y^6	3.4×10^{-16}

The structural form of the system shows that the TMAC was symmetrical with respect to the y-axis direction of the meridional plane. To ensure that the imaging quality of the system was symmetrical with respect to the meridional plane, we could not use an expression with odd symmetry for the x-axis in the design process for optimization. Therefore, we used 13 of the first 28 terms of the X–Y polynomial as variables to optimize the imaging quality of the system. The X–Y polynomial expressed as follows:

$$Z = \frac{cr^2}{1 + \sqrt{1 - (1+k)c^2r^2}} + A_{01}x^0y^1 + A_{20}x^2y^0 + \cdots + A_{06}x^0y^6 \quad (4)$$

4.2. Tolerance Analysis

The FOV of the TMAC designed in this paper was −11.5° to + 11.5° in the arc vector direction and −5° to −6° in the meridian direction. We selected 10 FOVs within the effective FOV and evaluated the image quality of the optical system using the operating wavelength of the optical system λ = 632.8 nm as the reference wavelength, as shown in Table 6. From the design results, it could be seen that the maximum value of system aberration occurs in the (−11.5°, −6°) FOV, and the ideal ray tracing position (Parax X, Parax Y) of this FOV was calculated to be (−203.597599, −105.18107), and the actual ray tracing position (Real X, Real Y) was (−203.4958002, −105.086116). According to the distortion formula where distortion = (real value − ideal value)/ideal value × 100%, we could get the system in this FOV arc to where the vector direction distortion was −0.05%, and the meridian direction distortion was 0.01%, so the system maximum distortion was −0.05%.

Table 6. Field of view (FOV).

No.	x/(°)	y/(°)	No.	x/(°)	y/(°)
1	0	−5	6	8.05	−5.5
2	5.75	−5	7	11.5	−5.5
3	8.05	−5	8	−5.75	−6
4	11.5	−5	9	−8.05	−6
5	5.75	−5.5	10	−11.5	−6

The MTF is the main parameter used to evaluate the imaging quality of an optical system. As the MTF value increased, the spatial frequency sharpness increased, and the pixel resolution increased. The MTF curve is shown in Figure 7, and the MTF of the system was better than 0.403 at the Nyquist frequency (71.4 lp/mm). The system used a free-form surface to compensate for the asymmetric aberration, which at the same time led to poor uniformity of imaging quality between FOVs. Therefore, the image quality had to be evaluated for the whole FOV. The wave aberration of each FOV is shown in Figure 8, and the maximum wave aberration of the system was 0.055 λ (λ/18.2, λ = 632.8 nm), and the system imaging is good. The aberration grid is shown in Figure 9, which is the visual response of the image aberration. As can be seen from the figure, after optimized design, there was basically no deviation between the ideal imaging position and the real image point position. Thus, the aberration was well corrected for the large FOV imaging optical system. Because of the influence of the system phase difference, the light passed through the optical system and presented a diffuse spot on the image surface, and the size of the diffuse spot determined the perfection of the system. As shown in Figure 10, the spot column diagram of each FOV of the system was within the Airy spot, and the imaging quality of the system was completely proportional to the meridional plane. A comprehensive analysis of the data showed that the system had good imaging quality in the effective FOV, and the system had high balance correction capability and low distortion in phase difference. The designed off-axis triple-reverse optical system met the parameters specified in Table 3.

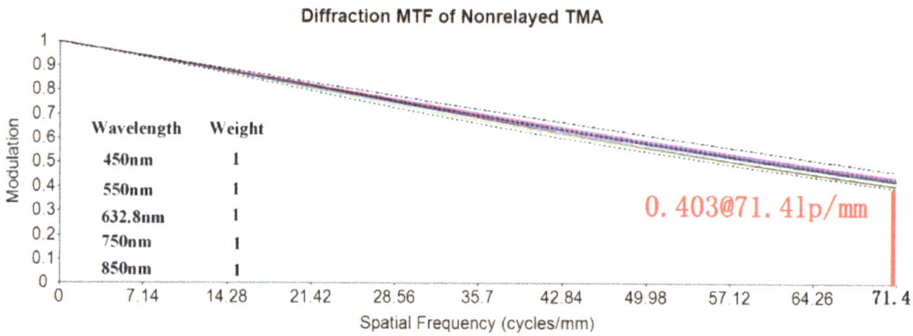

Figure 7. The modulation transfer function (MTF) curves: the dashed line indicates the direction of the arc vector; the solid line indicates the meridian direction.

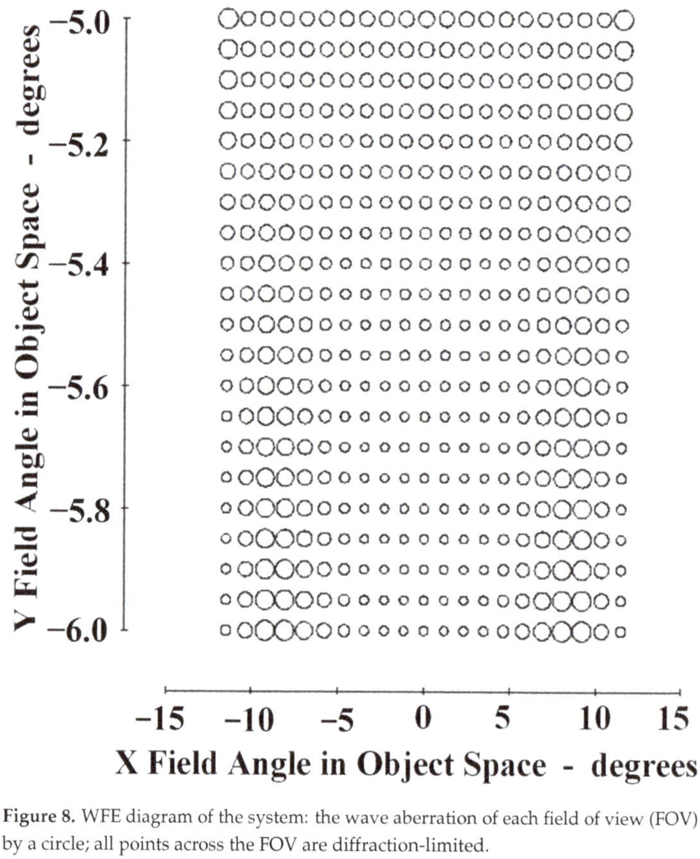

Figure 8. WFE diagram of the system: the wave aberration of each field of view (FOV) is represented by a circle; all points across the FOV are diffraction-limited.

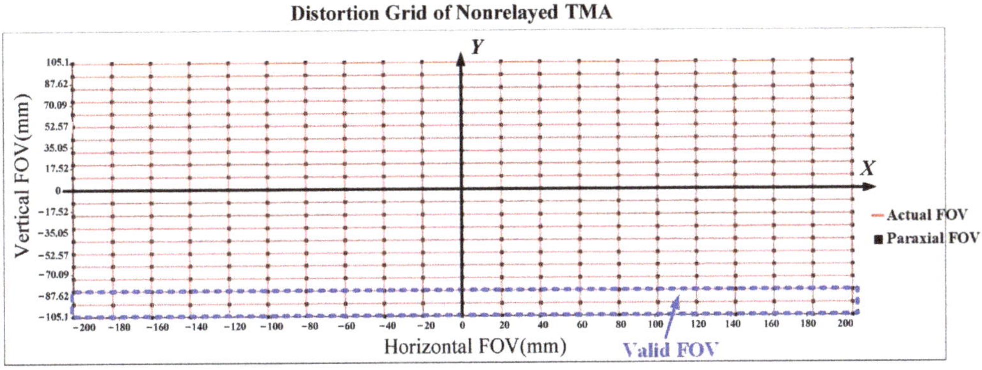

Figure 9. Grid distortion.

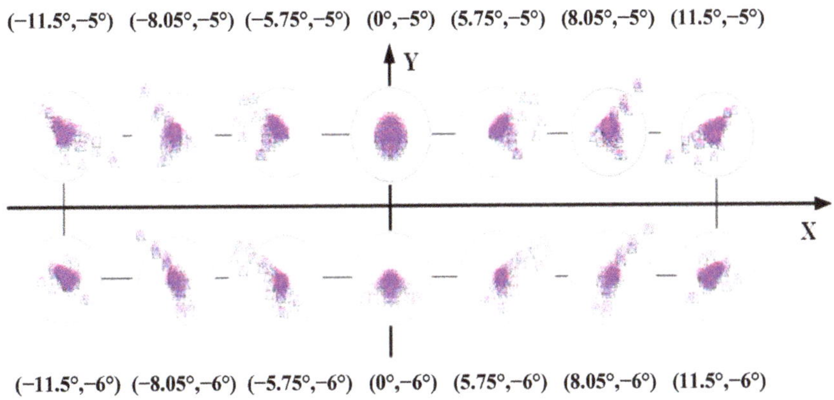

Figure 10. The spot column diagram of each field of view (FOV) of the system.

4.3. System Tolerance Analysis

We analyzed the manufacturing and assembly tolerances of the TMA system using the Monte Carlo method. We used the complex color MTF in Code V software as the performance index and used the FOV in Table 6 as the feature reference point. The image plane position was used as the compensation parameter. The machining tolerances included curvature radius tolerance, secondary surface coefficient tolerance, high secondary aspheric surface tolerance, and face shape tolerance. In the system-mounting process, we used the PM as the reference, and the SM and TM had six DOFs—that is, six position tolerances, including translation and rotation tolerances along the x-, y-, and z-axes. The results of tolerance allocation of the system are shown in Table 7.

After bringing the above tolerances into the optical system, we obtained the MTF curves of different FOVs, as shown in Figure 11. Thus, the results showed that 80% of the MTF was better than 0.37 at Nyquist frequency, and the tolerances were reasonably allocated to meet the design index requirements.

Table 7. Tolerance allocation result.

Type	Item	Primary Mirror	Secondary Mirror	Tertiary Mirror
	Displacement x/mm	-	0.05	0.1
	Displacement y/mm	-	0.05	0.1
Assembling	Displacement z/mm	-	0.2	0.3
	Tilt α/(″)	-	20	20
	Tilt β/(″)	-	20	30
	Tilt γ/(″)	-	40	60
	ΔR/mm	0.8	0.3	0.3
Manufacturing	ΔK	0.001	0.002	0.001
	Surface error RMS (λ = 632.8 nm)	λ/50	λ/50	λ/50

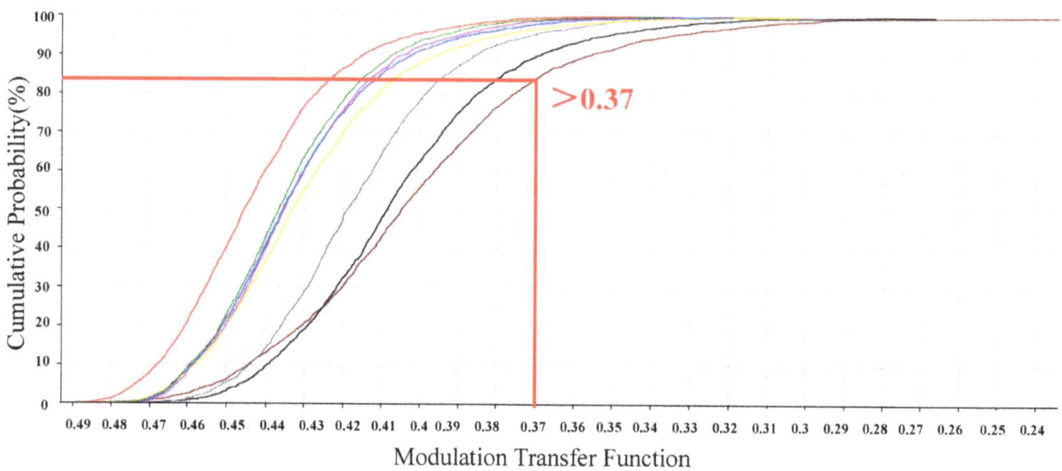

Figure 11. Curves of tolerance probability.

5. Results and Discussion

In this study, we analyzed and compared the TMAS with an intermediate phase plane and the TMAS without an intermediate image plane. We found that the TMAS without an intermediate image plane was suitable for large FOV imaging, which provided a theoretical basis for the selection of a large FOV imaging optical system design. We verified the relationship between the *X–Y* polynomial, Fringe-Zernike polynomial, and the aberration based on the quantitative relationship between the *X–Y* polynomial and aberration. The results provided a guideline for the balance of aberration and the application of the *X–Y* polynomial free-form surface shapes in optical system design. A three-dimensional diagram of the designed *X–Y* polynomial free-form surface shape was also created for the purpose of visual observation. Finally, based on the off-axis triple-reversal optical system without an intermediate image plane and the analyzed free-form surface shape, we designed a large FOV low-distortion off-axis triple-reversal optical system with a focal length of 1000 mm, an F-value of 9.5, and an FOV angle of $23^{\circ} \times 1^{\circ}$. The PM adopted an *X–Y* polynomial free-form surface to increase the freedom of the system while also correcting the aberration of each FOV. The SM adopted a paraboloidal surface to reduce the difficulty of processing and mounting the system, and the TM was ellipsoidal with a high secondary term. After optimizing the design, the imaging quality of the system was close to the diffraction limit, the maximum distortion grid in the FOV was less than or equal to −0.05%, and the average wave aberration in the whole FOV was better than 0.055 λ. Meanwhile, after adding the plane reflector to fold the rear optical path, the total optical length of the system reached $f'/2.6$, the height reached $f'/2.9$, and the width reached $f'/2.25$, creating a compact system structure. The application of the free-form surface expanded the FOV and corrected the system aberrations. After the image quality evaluation and tolerance analysis, we determined that the system achieved good imaging quality. This design method has good application prospects in the design of high-resolution, large FOV, with a long focal length and a light and small push-scan imaging optical load.

Author Contributions: Conceptualization, B.J. and Q.L.; methodology, Y.L.; software, Y.L.; validation, B.J., Y.L. and F.J.; formal analysis, Q.L.; investigation, C.T.; resources, C.T.; data curation, F.J.; writing—original draft preparation, Y.L.; writing—review and editing, B.J., Q.L. and C.T.; visualization, Y.L.; supervision, F.J.; project administration, B.J.; funding acquisition, Q.L. All authors have read and agreed to the published version of the manuscript.

Funding: This research was funded by the "13th Five-Year Plan" Science and Technology Project of Jilin Provincial Education Department (Grant No. JJKH20200789KJ), Jilin Provincial Science and Technology Department Key R&D Project (Grant No. 20200401120GX), and Changchun University of Science and Technology Project (Grant No. 50923010501).

Institutional Review Board Statement: Not applicable.

Informed Consent Statement: Not applicable.

Data Availability Statement: Not applicable.

Acknowledgments: The authors would like to thank the laboratory and university for their support.

Conflicts of Interest: The authors declare no conflict of interest.

References

1. Gorevoy, A.V.; Machikhin, A.S.; Khokhlov, D.D.; Batshev, V.I. Optimization of prism-based stereoscopic imaging systems at the optical design stage with respect to required 3D measurement accuracy. *Opt. Express* **2020**, *28*, 24418–24430. [CrossRef] [PubMed]
2. Kim, D.W.; Suratwala, T.I.; Schröder, S. Special Section Guest Editorial: Ultraprecision Optics Fabrication and Characterization. *Opt. Eng.* **2019**, *58*, 092601. [CrossRef]
3. Yang, T.; Zhu, J.; Wu, X.F.; Jin, G.F. Direct design of freeform surfaces and freeform imaging systems with a point-by-point three-dimensional construction-iteration method. *Opt. Express* **2015**, *23*, 10233–10246. [CrossRef]
4. Shen, Z.H. Design of Large Field of View Off-axis Three-mirror Reflective Optical System. *Electro. Opt. Technol. Appl.* **2020**, *35*, 24–27, 64. [CrossRef]
5. Yan, J.H.; Hu, Z.J.; Zhu, D.Y. Design of Compact Off-axis Three-mirror Afocal System Based on Freeform Surface. *Acta Photonica Sin.* **2022**, *51*, 297–307. [CrossRef]
6. Figoski, J.W. Alignment and test results of the QuickBird telescope using the Ball Optical System Test Facility. *Proc. SPIE* **1999**, *3785*, 99–108. [CrossRef]
7. Nakajima, M.; Kojima, Y.; Moriyama, T. Mission overview and instrument concept of the Global Imager (GLI). In Proceedings of the SPIE's 1994 International Symposium on Optics, Imaging, and Instrumentation, San Diego, CA, USA, 14 September 1994. [CrossRef]
8. Mazzoli, A.; Holbrouck, P.; Houbrechts, Y.; Maresi, L.; Stockman, Y.; Taccola, M.; Versluys, J. Baffles design of the PROBA-V wide FOV TMA. In Proceedings of the International Conference on Space Optics—ICSO 2010, Rhodes Island, Greece, 20 November 2017. [CrossRef]
9. Ohgi, N.; Iwasaki, A.; Kawashima, T.; Inada, H. Japanese hyper-multi spectral mission. In Proceedings of the 2010 IEEE International Geoscience and Remote Sensing Symposium, Honolulu, HI, USA, 25–30 July 2010; pp. 3756–3759. [CrossRef]
10. Hugot, E.; Wang, X.; Valls-Gabaud, D.; Lemaître, G.; Agócs, T.; Wang, J.Y. A freeform-based, fast, wide-field, and distortion-free camera for ultralow surface brightness surveys. *Proc. SPIE* **2014**, *9143*, 9143X. [CrossRef]
11. Liu, Y.X.; Zhang, D.D.; Niu, X.H. Design of Multi-spectrum Push-broom optical system. *Opt. Tech.* **2022**, *48*, 139–143. [CrossRef]
12. Miao, L.; Zhu, L.; Fang, C.; Yan, N.; Yang, X.; Zhang, X. Modeling and Analysis of System Error for Highly Curved Freeform Surface Measurement by Noncontact Dual-Axis Rotary Scanning. *Sensors* **2021**, *21*, 554. [CrossRef] [PubMed]
13. Fang, F.Z.; Zhang, X.D.; Weckenmann, A.; Zhang, G.X.; Evans, C. Manufacturing and measurement of freeform optics. *CIRP Ann. Manuf. Technol.* **2013**, *62*, 823–846. [CrossRef]
14. Chen, S.Y.; Xue, S.; Zhai, D.D.; Tie, G.P. Measurement of Freeform Optical Surfaces: Trade-Off between Accuracy and Dynamic Range. *Laser Photonics Rev.* **2020**, *14*, 1900365. [CrossRef]

15. Ma, T.; Yu, J.C.; Liang, P.; Wang, C.H. Design of a freeform varifocal panoramic optical system with specified annular center of field of view. *Opt. Express* **2011**, *19*, 3843–3853. [CrossRef]
16. Zhu, J.; Hou, W.; Zhang, X.D.; Jin, G.F. Design of a low F-number freeform off-axis three-mirror system with rectangular field-of-view. *J. Opt.* **2015**, *17*, 015605. [CrossRef]
17. Pan, J.W.; Che-Wen, C.; Huang, K.D.; Wu, C.Y. Demonstration of a broad band spectral head-mounted display with freeform mirrors. *Opt. Express* **2014**, *22*, 12785–12798. [CrossRef]
18. Schober, C.; Beisswanger, R.; Gronle, A.; Pruss, C.; Osten, W. Tilted Wave Fizeau Interferometer for flexible and robust asphere and freeform testing. *Light Adv. Manuf.* **2022**, *3*, 48. [CrossRef]
19. Schiesser, E.M.; Bauer, A.; Rolland, J.P. Effect of freeform surfaces on the volume and performance of unobscured three mirror imagers in comparison with off-axis rotationally symmetric polynomials. *Opt. Express* **2019**, *27*, 21750–21765. [CrossRef] [PubMed]
20. Zhang, B.Q.; Jin, G.F.; Zhu, J. Towards automatic freeform optics design: Coarse and fine search of the three-mirror solution space. *Light Sci. Appl.* **2021**, *10*, 65. [CrossRef]
21. Ye, G.F.; Chen, L.; Li, X.H.; Yuan, Q.; Gao, Z.S. Review of optical freeform surface representation technique and its application. *Opt. Eng.* **2017**, *56*, 110901. [CrossRef]
22. Sun, Y.H.; Sun, Y.Q.; Chen, X.Y.; Wang, F.; Yan, X.; Zhang, X.N.; Cheng, T.L. Design of a free-form off-axis three-mirror optical system with a low f-number based on the same substrate. *Appl. Opt.* **2022**, *61*, 7033–7040. [CrossRef] [PubMed]
23. Reimers, J.; Bauer, A.; Thompson, K.; Rolland, J. Freeform spectrometer enabling increased compactness. *Light Sci Appl.* **2017**, *6*, e17026. [CrossRef] [PubMed]
24. Zhang, X.; Zheng, L.G.; He, X.; Wang, L.G.; Zhang, F.; Yu, S.Q.; Shi, G.W.; Zhang, B.Z.; Liu, Q.; Wang, T.T. Design and fabrication of imaging optical systems with freeform surfaces. *Proc. SPIE* **2012**, *8486*, 46–55. [CrossRef]
25. Meng, Q.Y.; Wang, W.; Ma, H.T.; Dong, J.H. Easy-aligned off-axis three-mirror system with wide field of view using freeform surface based on integration of primary and tertiay mirror. *Appl. Opt.* **2014**, *53*, 3028–3034. [CrossRef] [PubMed]
26. Chang, J.; Weng, Z.C.; Cong, X.J. Design of compact high-resolution optical system for space and stray-light analysis. *Proc. SPIE* **2022**, *4927*, 71–76. [CrossRef]
27. Ni, D.W.; Li, X.Y.; Ren, Z.G.; Wang, Y.M.; Wang, L.; Tian, C.L. Design of off-axis three-mirror freeform optical system with wide field of view. *Proc. SPIE* **2019**, *10837*, 108370Z. [CrossRef]
28. Cook, L.G. Three mirror anastigmatic optical system. U.S. Patent 4265510, 5 May 1981.
29. Rodgers, J.M. Unobscured mirror designs. In Proceedings of the International Optical Design Conference 2002, Tucson, AZ, USA, 3–5 June 2002. [CrossRef]
30. Yang, T.; Ye, Y.Z.; Cheng, D.W. Freeform Imaging Optical System Design: Theories, Development, and Applications. *Acta Opt. Sin.* **2021**, *41*, 0108001. [CrossRef]
31. Kirschstein, S.; Koch, A.; Schöneich, J.; Döngi, F. Metal mirror TMA, telescopes of Jena spaceborne scanners: Design and analysis. *Proc. SPIE* **2005**, *5962*, 59621M. [CrossRef]
32. Atchison, D.A.; Scott, D.H.; Cox, M.J. Mathematical treatment of ocular aberrations: A user's guide. In *Vision Science and Its Applications*; Lakshminarayanan, V., Ed.; Optica Publishing Group: Santa Fe, NM, USA, 2000; pp. 110–130. [CrossRef]

photonics

MDPI

Communication

Automatic Inhomogeneous Background Correction for Spatial Target Detection Image Based on Partition Processing

Chun Jiang [1,2], Tao Chen [1], Changzheng Lu [1,2], Zhiyong Wu [1], Changhua Liu [1], Meng Shao [1] and Jingtai Cao [1,3,*]

1 Changchun Institute of Optics, Fine Mechanics and Physics, Chinese Academy of Sciences, Changchun 130033, China
2 University of Chinese Academy of Sciences, Beijing 100049, China
3 College of Communication Engineering, Jilin University, Changchun 130012, China
* Correspondence: caojt@ciomp.ac.cn

Abstract: High-resolution imaging with wide field of view (FoV) ground-based telescopes is often affected by skylight background and noise due to the detector, resulting in an inhomogeneous background. In this paper, we propose an improved method for spatial image non-uniformity correction based on partition processing. First, an evaluation metric is introduced to evaluate the partition size and automatically iterate a suitable partition value for different scenarios based on the different operating conditions of the telescope. Then, we iteratively calculate the mean and variance in each partitioned region to extract the background of each partitioned region. Finally, after applying bilinear interpolation to the background extracted from each region, the inhomogeneous background is obtained and removed from the original image. The experiments on the simulated and real images show that the proposed method can effectively remove the inhomogeneous background of spatial images and meet the requirements of the real-time processing of high-resolution images under long exposure conditions.

Keywords: partition processing; inhomogeneous background correction; image pre-processing; optical telescope image

Citation: Jiang, C.; Chen, T.; Lu, C.; Wu, Z.; Liu, C.; Shao, M.; Cao, J. Automatic Inhomogeneous Background Correction for Spatial Target Detection Image Based on Partition Processing. *Photonics* **2023**, *10*, 433. https://doi.org/10.3390/photonics10040433

Received: 27 February 2023
Revised: 22 March 2023
Accepted: 3 April 2023
Published: 12 April 2023

1. Introduction

Following the growing number of space debris, space target detection becomes a complex and challenging subject. Ground-based photoelectric telescopes are the most commonly used equipment in space target detection; several organizations have carried out observations with ground-based photoelectric telescopes and published their results [1–5]. Meanwhile, the removal of the inhomogeneous background from their captured images is an important prerequisite for improving space target detection capabilities. Image degradation not only affects image quality and the image signal-to-noise ratio (SNR), but it also has a serious impact on the subsequent image segmentation and target detection. Therefore, the inhomogeneous background correction of spatial images becomes a necessary pre-processing step. In addition, for applications where small telescopes search for space targets, remote unattended automatic observation with multi-station deployment can be achieved if the equipment has the characteristics of a lower cost, automatic observation, and easy deployment [6].

Methods of geometric modification, image smoothing and filtering, image morphological operations, and image arithmetic operations are used to process spatial images taken by ground-based telescopes. Each of these operations enhances or highlights the region of interest for later viewing, processing, etc. However, image processing operations can cause some damage to the original information of the image, which is known as the process of image degradation.

Although ground-based telescope systems can use adaptive optics or phase modulation to improve image quality, these systems are usually expensive and complex [7–9].

Even more effective image enhancement techniques are applied, such as deconvolution and image processing. It still cannot be achieved in a way that satisfies the real-time processing requirements [10–12]. A lot of research has been published in the field of image processing on the issue of inhomogeneous background correction. A simple method is the flat-field correction under uniform illumination [13,14]. Although it can solve the problem in a fixed scene, the correction must be performed again when the observation environment changes. For ground-based telescopes, obtaining a flat-field image under uniform illumination is more problematic. Another way to solve the inhomogeneous background is image processing. Depending on the number of images processed, the methods can be classified as multi-frame image methods and single-frame image methods. The widely used multi-frame image method extracts the inhomogeneous background from multiple images of the same scene. Yuan et al. [15] proposed an improved radial gradient correction method with a better radial uniformity, achieving more accurate details in the corrected images. Litvinov et al. [16] proposed a method to simultaneously estimate the camera radiometric response, camera gain, and vignetting using multiple consecutive images. However, a limitation of the multi-frame processing methods is adapting to the drastic changes in the scene, such as observing a fast-moving target in low earth orbit (LEO).

Single-frame inhomogeneous background correction methods are more flexible because they do not require the accumulation of multi-frame information. The single-frame inhomogeneous background correction studies mainly concentrate on the field of computed tomography, which does not require high real-time processing [17,18]. In the field of spatial target detection, Zhang et al. [19] used maximum likelihood estimation for correction. Partitioning the image into segments is currently the most commonly used method for the real-time correction of an inhomogeneous background in a single-frame image. However, the segmentation threshold is mostly a fixed empirical value, which cannot be adjusted correspondingly over the frames.

This paper presents a single-frame method for automatically removing the inhomogeneous background from the images taken with telescopes. It is more suitable for images captured by small telescopes of an integrated design and meets the real-time processing requirements within 2.5 s at 4 Kpx × 4 Kpx resolution. By introducing a split-size evaluation index, the segmentation value can be adjusted over different scenes. Experiments with simulated images and captured images in this paper show that the method has a better quality and performance compared to the processing method with fixed segmentation thresholds.

2. Equipment and Methods

We first describe the data capture device information in Section 2.1. Next, the workflow of the method is described. The complete workflow of the method is illustrated in Figure 1. The method can be divided into 3 steps, which are the automatic partition, the background-estimating, and the background-removing step. The automatic partition step determines the split value and partitions the image into several regions. The background-estimating step extracts the background within each region. The background-removing step calculates the full image background and then removes it from the original image. The steps will be explained in detail in Section 2.2, Section 2.3 and Section 2.4, respectively.

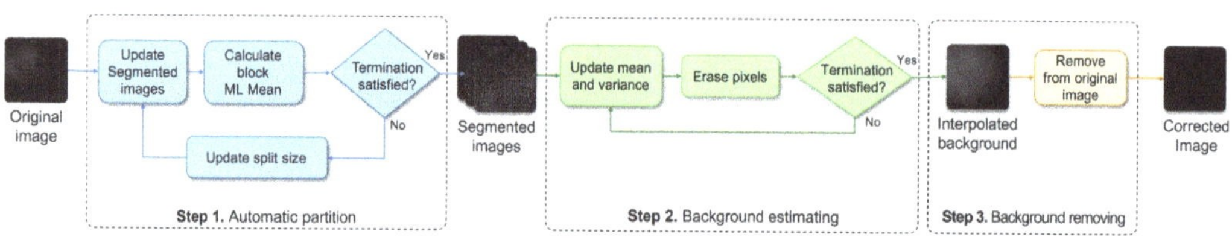

Figure 1. Flowchart of the method.

2.1. Experimental Equipment

The image acquisition equipment for the experiment is shown in Figure 2. The system consists of an optical system, a camera, a focusing mechanism, a motion control system, and an integrated control box. We chose a scientific CMOS detector with a resolution of 4096 px × 4096 px and a pixel size of 9 × 9 μm. The detector was cooled to −40 °C to improve imaging and increase the telescope's observing power. The telescope's optical system is designed as a transmissive imaging system, which reduces the size and weight of the optical system. The optical system has an aperture of 150 mm and a field of view of 6.5° × 6.5°. For the integrated design of the telescope, the software system, including the image processing and motion control modules, runs on NVIDIA Jetson AGX Xavier. At the same time, the real-time processing of the telescope is challenged by the high resolution of the images and the low computing power of the processing devices.

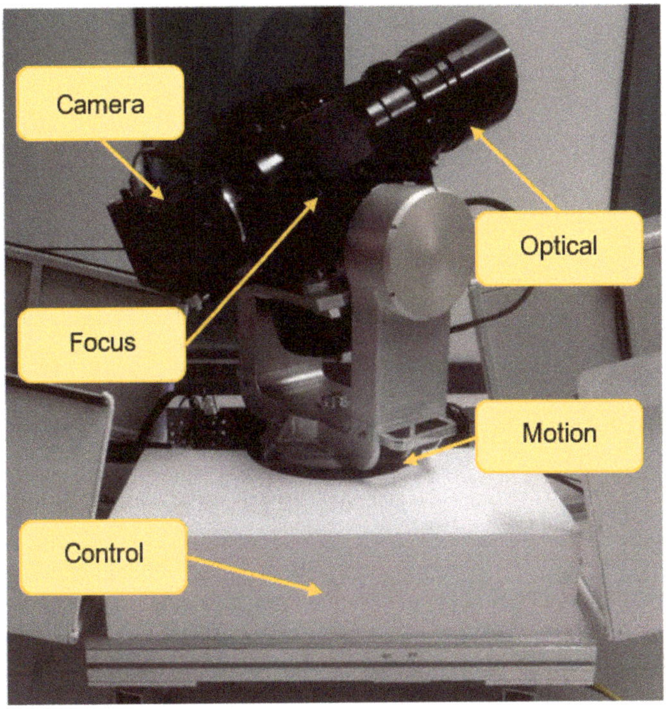

Figure 2. The structural overview of the image acquisition equipment.

The method in this paper is designed to meet the real-time processing requirements of these devices, and we use images taken with each of these devices in a city with a strong skylight background and in a countryside with a weak skylight background. To facilitate the presentation of the results of our method, we store the captured images and then process them offline. At the same time, the images were not compressed in any way to preserve their original information.

2.2. Determination of Region Split Size

The main causes of image degradation are the atmospheric scattering noise and the CMOS detector noise [20]. Mohamed et al. [21] and Sonnenschein et al. [22] pointed out that the noise generated by atmospheric scattering is mainly the Poisson noise. Pain et al. [23] and Nguyen et al. [24] illustrated the sources of CMOS noise, where the dark current noise follows a Poisson distribution, and the readout noise follows a Gaussian distribution. An inhomogeneous image is a non-exposure distorted image. In other words, a large number of pixels are non-saturated, and only the space target or star is the pixel under the maximum grey level. An imaging system with optical system interference can be expressed as:

$$\mathrm{I}' = I + V \times \mathcal{S} + N \tag{1}$$

where I and I' are the pure image and the image with the inhomogeneous background. V is the change caused by the optical system, $\mathcal{S}$ is the inhomogeneous background caused by stray light or the detector, and N is the noise of the detector. Assuming that the changes caused by the optical system are slow and uniform, stray light does not cause abnormal exposures in some images. The corrected image can be obtained by the following equation:

$$I = \mathrm{I}' - \mathfrak{B} \tag{2}$$

where $\mathfrak{B}$ is an inhomogeneous background. Our goal is to determine a method $\mathfrak{F}$ to extract information about the inhomogeneous background from the image.

$$\mathfrak{B} = \mathfrak{F}(\mathrm{I}') \tag{3}$$

Although the distribution of the abovementioned noise is inhomogeneous, the variation in the inhomogeneous background within a certain small image area is very flat. Therefore, we split the image into several small area blocks R_{block}. A larger block size can reduce the processing time but it cannot accurately estimate the background of the original image. A smaller split block size can estimate the background of the original image more accurately but requires a correspondingly longer processing time.

Our method determines the split size in step 1. First, a small initial split size is set to split the image I'. This initial split size is set differently for different scenes and different image resolutions. In this article, the initial split size is set to 10 px. Then, the mean value of pixels within each segmentation block is calculated to obtain the evaluation image $S(m,n)$ of size $m \times n$, which is used to evaluate the segmentation quality. Since the processing is low-pass filtering, the smoothness between each region is a characterization of the changing trend of the high-frequency part of the image. Therefore, an effective smoothness-evaluating operator must perform high-pass filtering on the image. One way to high-pass filter an image is to determine its second-order derivative. For a two-dimensional image, the Laplace operator $\nabla^2 S(m,n)$ can be defined as follows:

$$\nabla^2 S(m,n) = \frac{\partial^2 S(m,n)}{\partial m^2} + \frac{\partial^2 S(m,n)}{\partial n^2} \tag{4}$$

We note that the second-order derivatives of the Laplace operator in the m and n directions may have opposite signs and tend to cancel each other out. This phenomenon occurs frequently on telescope-taken images, meaning the Laplace operator may be unstable. We overcome this problem by modifying the Laplace operator ${\nabla^2}_M S(m,n)$ as:

$${\nabla^2}_M S(m,n) = \left|\frac{\partial^2 S}{\partial m^2}\right| + \left|\frac{\partial^2 S}{\partial n^2}\right| \tag{5}$$

The modified Laplace operator is always larger or equal to the pre-correction Laplace operator. The discretization of a Laplace operator is usually approximated as a 3×3 operator. Therefore, the discrete approximation of the modified Laplace (ML) operator can be obtained as:

$$ML(m,n) = |2S(m,n) - S(m-1,n) - S(m+1,n)| + |2S(m,n) - S(m,n-1) - S(m,n+1)| \tag{6}$$

$ML(m,n)$ evaluates the undulation between small regions of the discriminated image [25]. Further, the mean ML value between all segmented regions in the evaluation image $S(m,n)$ is regarded as the evaluation value E_1 of the original image at the current segmentation size.

$$E_1 = \frac{1}{m \times n} \sum_{m,n} [|2S(m,n) - S(m-1,n) - S(m+1,n)| + |2S(m,n) - S(m,n-1) - S(m,n+1)|] \quad (7)$$

We iteratively scale up the size of the segmented regions with the preset step size then update the evaluation value until the kth evaluation value E_k satisfies the following termination condition.

$$\frac{E_k - E_{k-1}}{E_{k-1}} < 0.01 \quad (8)$$

If this condition is satisfied, the rate of change in the evaluation value at this time is small. We consider that the split size is suitable for the current frame.

2.3. Region Background Extraction

In Section 2.2, the mean value of the region is simply taken as the background of the block, but the calculation of the block background should exclude the bright stars and faint points to obtain an accurate background. For star maps taken by telescopes, most of the image elements belong to the background grayscale, and the range is concentrated. At the same time, stars and faint points are presented with high and low grayscale values, respectively. To extract the regional background of the image accurately, the dark and bright spots should be removed when processing the inhomogeneity correction. The removal is performed as follows. First, by calculating the mean value m^1_{block} and the standard deviation σ^1_{block} in a block.

$$m^1_{block} = \frac{\sum_{(x,y)\in R_{block}} f_{block}(x,y)}{N_{block}} \quad (9)$$

$$\sigma^1_{block} = \sqrt{\frac{\sum_{(x,y)\in R_{block}} \left(f_{block}(x,y) - m^1_{block}\right)^2}{N_{block}}} \quad (10)$$

where $f_{block}(x,y)$ means the gray value at (x,y) in the region block R_{block}, and N_{block} represents the number of pixels in the R_{block}. Then, we calculate the segmentation thresholds $T^1_{block}L$ and $T^1_{block}H$ for the removal:

$$T^1_{block}L = m^1_{block} - 3 \times \sigma^1_{block} \quad (11)$$

$$T^1_{block}H = m^1_{block} + 3 \times \sigma^1_{block} \quad (12)$$

Next, the images are culled according to the following rules:

$$g^1_{block}(x,y) = \begin{cases} 0, f_{block}(x,y) \le T^1_{block}L \\ f_{block}(x,y), T^1_{block}L < f_{block}(x,y) < T^1_{block}H \\ 0, f_{block}(x,y) \ge T^1_{block}H \end{cases} \quad (13)$$

$g^1_{block}(x,y)$ is the image after the first-time removal of bright and dark pixels. Continue analyzing $g^1_{block}(x,y)$ to calculate the mean m^2_{block} and standard deviation σ^2_{block} of the remaining points after the removal.

$$m^2_{block} = \frac{\sum_{(x,y)\in R_{block} \neq 0} g^1_{block}(x,y)}{N_{block} - N_0} \quad (14)$$

$$\sigma^2_{block} = \sqrt{\frac{\sum_{(x,y)\in R_{block} \neq 0} \left(g^1_{block}(x,y) - m^2_{block}\right)^2}{N_{block} - N_0}} \quad (15)$$

where N_0 denotes the number of pixels where $g^1_{block}(x, y) = 0$. Our method extracts the block background by removing pixels in each iteration. We update the pixel values iteratively until the image standard deviation satisfies the condition as follows:

$$\frac{\sigma^{n-1}_{block} - \sigma^{n}_{block}}{\sigma^{n-1}_{block}} < 0.01 \tag{16}$$

σ^{n-1}_{block} and σ^{n}_{block} denote the standard deviation of the image after the $n-1$st and nth removal. When the condition in Equation (16) is satisfied, the standard deviation of the image after the nth removal is close to the previous one, and the dark and bright spots in the image are considered to be removed cleanly. The mean value of m^n_{block} at this point is the gray value of the background in the current region.

2.4. Full Image Background Estimation

There are stepped differences in gray values between the blocks of the background image acquired in Section 2.3. However, the gray value undulation of the background in the actual image is usually continuous. To recover the background of the image accurately, an interpolation filter is required. We choose bilinear interpolation as the method for full image recovery. Compared with nearest neighbor interpolation, bilinear interpolation can provide smoother results. It also performs significantly better than bicubic spline interpolation and meets the real-time requirement.

Similar to the up-sampling, our method takes the center pixel of each region as a sample point, then the estimated background can be obtained by interpolating the gray value over the pixels between the sample points. After, a corrected image can be obtained by differencing the original image from the estimated background.

3. Results and Discussion

In this section, we conduct experiments on the method to verify the effectiveness and practicality of the proposed method. We first introduce the evaluation indicators in Section 3.1. In Section 3.2, we evaluate the effectiveness using the simulated images and then analyze the experimental results. Then, we test our method on the image data captured by the experimental equipment in Section 2.1 and list the results of our method in Section 3.3. The experimental results of both simulated and captured images demonstrate the effectiveness of our method.

3.1. Evaluation Indicators

It should be noted that ground truth images are present in the simulated images but not in the captured images. Therefore, the evaluation indicators are applied to the simulated images. Structural similarity (SSIM) and mean square error (MSE) are used to evaluate the correction quality of the simulated inhomogeneous background. The MSE is defined as follows:

$$\mathrm{MSE} = \frac{1}{m \times n} \sum_{i=0}^{m-1} \sum_{j=0}^{n-1} (I(i,j) - K(i,j))^2 \tag{17}$$

where m and n are the resolutions of the image, respectively; both of them are set to 4096. I and K are the real images and the corrected image, respectively. The SSIM is defined as follows:

$$\mathrm{SSIM} = \frac{\left(2\mu_x\mu_y + c_1\right)\left(\left(2\sigma_{xy} + c_2\right)\right)}{\left(\mu_x^2 + \mu_y^2 + c_1\right)\left(\sigma_x^2 + \sigma_y^2 + c_2\right)} \tag{18}$$

where μ_x and μ_y represent the average of the input x and the output y, σ_x, and σ_y represent the standard deviation of x and y, and σ_{xy} is the covariance between x and y. c_1 and c_2 are the constant to avoid the error when the denominator is zero.

A smaller MSE indicates that the error between the ground truth image and the corrected image is smaller, which means the background noise is better removed. SSIM closer to 1 declares that the ground truth image and the corrected image are more similar, which denotes that the information of the original image is better preserved during the correction.

3.2. Experiment on Simulated Images

We generated a set of simulated images to evaluate the effectiveness of the proposed method. The simulated inhomogeneous backgrounds are generated according to the model [26]. We generated a background-free image and then composite the background onto it to obtain the test image. In this way, the ground truth image, the inhomogeneous background, and the simulated image are obtained. We apply the method in Section 2 to the simulated images to obtain the estimated background and corrected images; the results are shown in Figure 3.

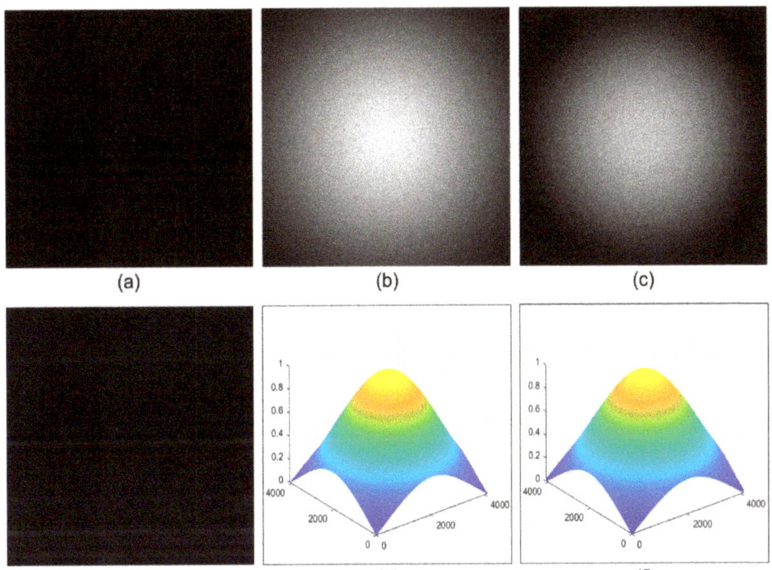

Figure 3. An example of correcting a simulated image with our method. (**a**) The ground truth image; (**b**) the simulated inhomogeneous background; (**c**) the simulated image by compositing the ground truth and the simulated background; (**d**) the clear image after the correction; (**e**) the grey-level distribution of (**b**); and (**f**) the estimated background by our method.

Figure 4 compares the grey-level distribution in Figure 3a,d on the same row (scan line 350). Figure 4a shows the grey level distribution of the ground truth image on the scan line; Figure 4b shows the corrected image on the grey level distribution. We can see that the average grey level distribution of the corrected image remains the same as the ground truth image, and as for the superimposed noise, the method removes it well.

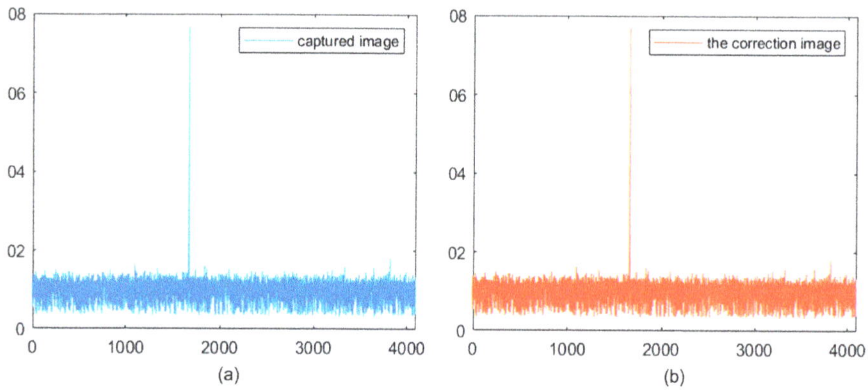

Figure 4. Comparison of the grey-level distribution on the same row in the ground truth image and corrected image. (**a**) ground truth image grey level distribution; (**b**) corrected image grey level distribution.

The effectiveness of the proposed method can be evaluated quantitatively on simulated images. We evaluate the effectiveness of the proposed method by comparing it with the method for a fixed empirical split region size. The results in Table 1 show that the proposed method can find a reasonable split region size compared to the method based on empirical values. Additionally, it has a better correction result on the simulated degraded image, thus obtaining a noise-suppressed image.

Table 1. Performance of different partition areas in the test set.

Split Area Size	MSE	SSIM
10 px × 10 px	33.62	0.9437
20 px × 20 px	23.02	0.9813
30 px × 30 px	8.65	0.9827
40 px × 40 px	9.55	0.9805
Ours	7.46	0.9837

3.3. *Experiment on Captured Images*

We have demonstrated the effectiveness of the method in Section 3.2. Different from the simulated images, it is difficult to obtain ground truth images of the captured images to evaluate our method quantitatively. However, the method processing effect can be identified using the improvement of the image quality before and after the method processing.

In Figure 5, the first column shows the captured images, and the second column lists the corrected images by our method. The third and fourth column represent the background estimated by the method and its gray-level distribution, respectively. The grey scale of the image in the figure has been stretched to present the detail. It is worth noting that the observations in rows 1–2 of the figure are set in cities, where the images contain background stray light scattered by the atmosphere due to the high level of city lighting. The observations in rows 3–4 are set in the countryside, where the stray light scattered by the atmosphere is weaker and the background is mainly caused by detector readout noise and diffraction from bright stars. The comparative results show that our method can correct the inhomogeneous background well in both strong and weak sky backgrounds, demonstrating the universality of the method of this paper.

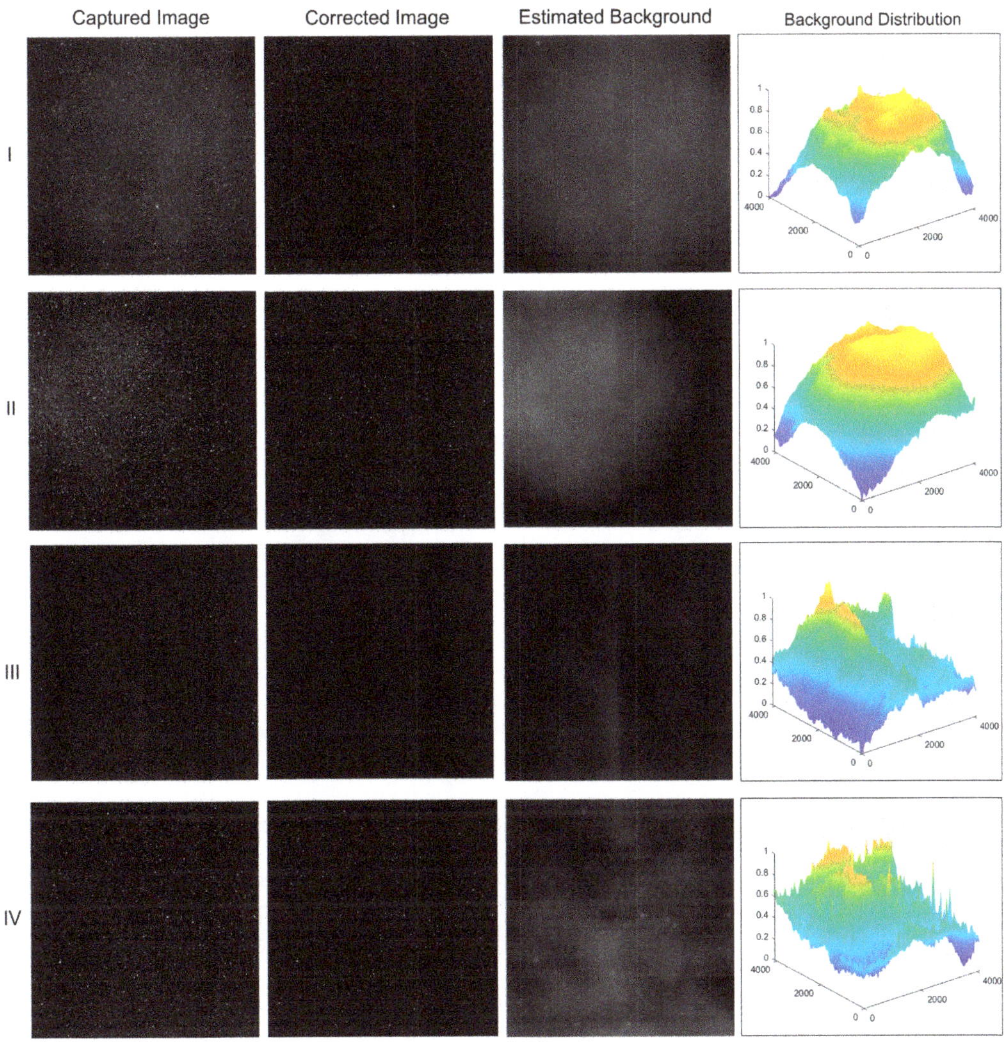

Figure 5. Correction results on images captured with the image acquisition equipment mentioned in Section 2.1. I–IV show different images.

We have also processed the case when a small amount of cloud cover is obscured. As shown in Figure 6, the method also works well, since the variations in the clouds are usually smooth over small areas.

Figure 7a,b displays the captured and corrected images, respectively, Figure 7c shows the background corrected by the method, and Figure 7d is the normalized background distribution. Due to the low elevation angle, more stray light is refracted and scattered by the atmosphere near the horizon, causing large undulations in the vertical direction of the shot. This variation is still relatively smooth in small areas. It can therefore be well removed by the method proposed in this paper.

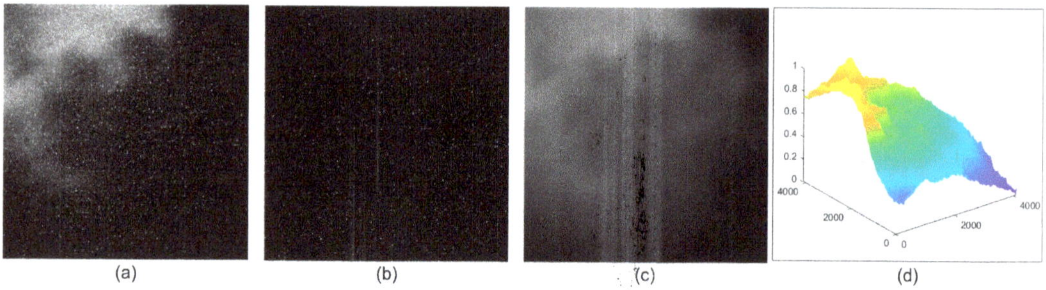

Figure 6. In the case of a fewer number of clouds, our model can still obtain reliable result. (**a**) The captured image with the image acquisition equipment mentioned in Section 2.1; (**b**) the corrected image that the clouds are removed; (**c**) the estimated background by our method; and (**d**) the grey-level distribution of (**c**).

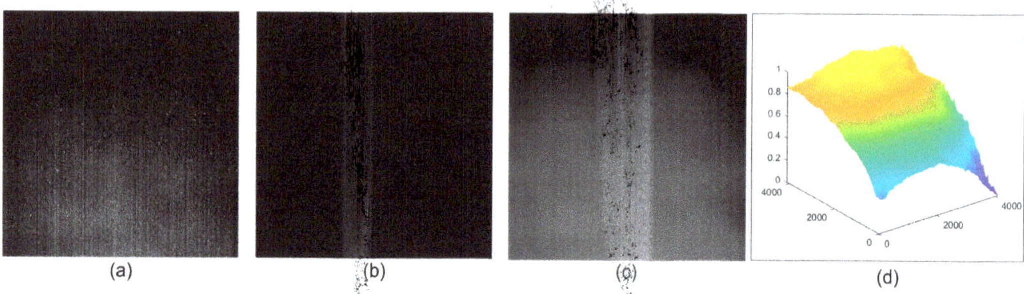

Figure 7. The results of removing inhomogeneous backgrounds using our method under a low elevation angle. (**a**) The captured image; (**b**) the corrected image; (**c**) the estimated background; (**d**) the grey-level distribution of (**c**).

We also compare the performance of the method in this paper with the single-image correction method of Zhang et al. [19]. It is worth noting that our method is run in MATLAB R2018a with a single thread. Therefore, only one core was used when testing our method. As shown in Table 2, in the sample average running time of a set of 40-star images, the average running time of our method is less than Zhang et al.'s method, under the same resolution. Meanwhile, for long-exposure images, the processing time requirement (2.5 s) can also be met under 4 Kpx × 4 Kpx resolution.

Table 2. Comparison of average execution time.

Image-Size	CPU	Average Running Time	Method
2 kpx × 2 kpx	8Core@2.10 GHz	9.29 s	Zhang
2 kpx × 2 kpx	6Core@3.20 GHz	1.25 s	Ours
4 kpx × 4 kpx	6Core@3.20 GHz	2.36 s	Ours

4. Conclusions

This paper presents a method for the inhomogeneous background correction of single-frame images from small telescopes. The method removes the skylight background and camera noise from the images. The method is a model-free method that can adapt to complex environments and extract background information from a single-frame image. The method has the advantages of a high performance, high correction accuracy, and the ability to deal with the heterogeneity of complex sky backgrounds, showing a strong practicality.

The experiments on the simulated and captured images prove the effectiveness of the method. The method can significantly improve image quality under the conditions of complex noise and complicated skylight backgrounds. It applies to astronomical observations and remote sensing images and is of great value for these applications. It can be applied to agricultural observations, ocean observations, and ground-based observations. This robust method can also be extended for medical detection and analysis and high dynamic range imaging. In future research, we will speed up our method with an optimized C++ implementation or GPU-accelerated computation.

Author Contributions: Conceptualization, T.C., Z.W. and J.C.; methodology, J.C., C.J., Z.W. and C.L. (Changhua Liu); software, C.J.; validation, M.S. and C.L. (Changhua Liu); formal analysis, C.J. and C.L. (Changzheng Lu); data curation, C.J. and C.L. (Changzheng Lu); writing—original draft preparation, C.J.; writing—review and editing, C.J. and C.L. (Changzheng Lu); visualization, C.J. and M.S.; project administration, T.C.; funding acquisition, J.C. All authors have read and agreed to the published version of the manuscript.

Funding: This work was supported by the National Natural Science Foundation of China, Grant Nos. 62001448 and U21A20451.

Institutional Review Board Statement: Not applicable.

Informed Consent Statement: Not applicable.

Data Availability Statement: Not applicable.

Acknowledgments: The authors are thankful to Risong Zhang for assistance with the language and figures, and to every member of the Changchun Changguang Orion Optoelectronic Technology Co., Ltd.; all their contributions were necessary for the completion of this paper.

Conflicts of Interest: The authors declare no conflict of interest.

References

1. Gao, Y.; Zhao, J.-Y. Space object detecting ability improvement method based on optimal principle. *Optoelectron. Lett.* **2019**, *15*, 459–462. [CrossRef]
2. Steindorfer, M.A.; Kirchner, G.; Koidl, F.; Wang, P.; Jilete, B.; Flohrer, T. Daylight space debris laser ranging. *Nat. Commun.* **2020**, *11*, 3735. [CrossRef] [PubMed]
3. Beck, P.; Robson, L.; Gallaway, M.; Jones, H.R.; Campbell, D. Efficient follow-up of exoplanet transits using small telescopes. *Publ. Astron. Soc. Pac.* **2019**, *131*, 084402. [CrossRef]
4. Yibin, R.; Yu, Z.; Xuejian, N. Application of large shipborne theodolite in space target measurement. In Proceedings of the 17th International Conference on Optical Communications and Networks (ICOCN2018), Zhuhai, China, 16–19 November 2018; SPIE: Bellingham, WA, USA, 2019; p. 110482J.
5. Sun, R.-Y.; Zhan, J.-W.; Zhao, C.-Y.; Zhang, X.-X. Algorithms and applications for detecting faint space debris in GEO. *Acta Astronaut.* **2015**, *110*, 9–17. [CrossRef]
6. Luo, H.; Zheng, J.-H.; Wang, W.; Cao, J.-J.; Zhu, J.; Chen, G.-P.; Zhang, Y.-S.; Liu, C.-S.; Mao, Y.-D. FocusGEO II. A telescope with imaging mode based on image overlay for debris at Geosynchronous Earth Orbit. *Adv. Space Res.* **2022**, *69*, 2618–2628. [CrossRef]
7. Paolo, C.; Guido, A.; Matteo, A.; Carmelo, A.; Andrea, B.; Andrea, B.; Maria, B.; Andrea, B.; Marco, B.; Lorenzo, B.; et al. MAORY: The adaptive optics module for the Extremely Large Telescope (ELT). In *Adaptive Optics Systems VII*; SPIE: Bellingham, WA, USA, 2020; p. 114480Y.
8. Ryan, D.; Mark, C.; Yutaka, H. On the feasibility of using a laser guide star adaptive optics system in the daytime. *J. Astron. Telesc. Instrum. Syst.* **2019**, *5*, 019002. [CrossRef]
9. Dong, L.; Wang, B. Research on the new detection method of suppressing the skylight background based on the shearing interference and the phase modulation. *Opt. Express* **2020**, *28*, 12518–12528. [CrossRef] [PubMed]
10. Sanders, T.; Hedges, R.; Schulz, T.; Abijaoude, M.; Peters, J.; Steinbock, M.; Arreola, A.; Holmes, T. Real Time Deconvolution of Adaptive Optics Ground Based Telescope Imagery. *J. Astronaut. Sci.* **2022**, *69*, 175–191. [CrossRef]
11. Torben, E.A.; Mette, O.-P.; Anita, E. Image-based wavefront sensing for astronomy using neural networks. *J. Astron. Telesc. Instrum. Syst.* **2020**, *6*, 034002. [CrossRef]
12. Hickman, S.; Weddell, S.; Clare, R. Image Correction with Curvature and Geometric Wavefront Sensors in Simulation and On-sky. In Proceedings of the 2019 International Conference on Image and Vision Computing New Zealand (IVCNZ), Dunedin, New Zealand, 2–4 December 2019; pp. 1–6.
13. Yu, W. Practical anti-vignetting methods for digital cameras. *IEEE Trans. Consum. Electron.* **2004**, *50*, 975–983.

14. Tek, F.B.; Dempster, A.G.; Kale, I. Computer vision for microscopy diagnosis of malaria. *Malar. J.* **2009**, *8*, 153. [CrossRef] [PubMed]
15. Yuan, X.; Wang, X.; Dou, A.; Ding, X. Vignetting Correction of Post-Earthquake UAV Images. In Proceedings of the IGARSS 2018—2018 IEEE International Geoscience and Remote Sensing Symposium, Valencia, Spain, 22–27 July 2018; pp. 5704–5707.
16. Litvinov, A.; Schechner, Y.Y. Addressing radiometric nonidealities: A unified framework. In Proceedings of the 2005 IEEE Computer Society Conference on Computer Vision and Pattern Recognition (CVPR'05), San Diego, CA, USA, 20–25 June 2005; pp. 52–59.
17. Kaiqiong, S.; Yaqin, L.; Shan, Z.; Jun, W. Hybrid active contour model for inhomogeneous image segmentation with background estimation. *J. Electron. Imaging* **2018**, *27*, 023018. [CrossRef]
18. Vopalensky, M.; Czech Academy of Sciences (CAS); Kumpova, I.; Vavrik, D. Suppression of residual gradients in the flat-field corrected images. *E-J. Nondestruct. Test.* **2019**, *25*. [CrossRef] [PubMed]
19. Zhang, D.; Yang, Q.Y.; Chen, T. Vignetting correction for a single star-sky observation image. *Appl. Opt.* **2019**, *58*, 4337–4344. [CrossRef] [PubMed]
20. Zhang, Y.; Wang, J.; Su, J.; Cheng, X.; Zhang, Z.J.T.A.J. Astrometric observations of a near-Earth object using the image fusion technique. *Astron. J.* **2021**, *162*, 250. [CrossRef]
21. Mohamed, E.H.; Michael, C.R.; Durdu, O.G. Reconstruction of images degraded by aerosol scattering and measurement noise. *Opt. Eng.* **2015**, *54*, 033101. [CrossRef]
22. Sonnenschein, C.M.; Horrigan, F.A. Signal-to-Noise Relationships for Coaxial Systems that Heterodyne Backscatter from the Atmosphere. *Appl. Opt.* **1971**, *10*, 1600–1604. [CrossRef] [PubMed]
23. Pain, B.; Cunningham, T.; Hancock, B.; Wrigley, C.; Sun, C. Excess noise and dark current mechanisms in CMOS imagers. In Proceedings of the IEEE Workshop on CCD's and Advanced Image Sensors, Karuizawa, Nagano, Japan, 9–11 June 2005.
24. Nguyen, T.-K.; Kim, C.-H.; Ihm, G.-J.; Yang, M.-S.; Lee, S. CMOS low-noise amplifier design optimization techniques. *IEEE Trans. Microw. Theory Tech.* **2004**, *52*, 1433–1442. [CrossRef]
25. Nayar, S.K.; Nakagawa, Y. Shape from focus. *IEEE Trans. Pattern Anal. Mach. Intell.* **1994**, *16*, 824–831. [CrossRef]
26. Kang, S.B.; Weiss, R. Can we calibrate a camera using an image of a flat, textureless Lambertian surface? In Proceedings of the Computer Vision—ECCV 2000: 6th European Conference on Computer Vision, Dublin, Ireland, 26 June–1 July 2000; pp. 640–653.

photonics

MDPI

Article

Multiobjective Optimization of SiC Mirror Based on Dual-Parameter Coupling

Quanliang Dong [1,2], Qianglong Wang [1,2], Chong Wang [1], Yunjie Luan [1,2], Xiaoxun Wang [1] and Xiaoming Wang [1,*]

1 Changchun Institute of Optics, Fine Mechanics and Physics (CIOMP), Chinese Academy of Sciences, Changchun 130033, China
2 University of Chinese Academy of Sciences, Beijing 100049, China
* Correspondence: 13604311978@163.com; Tel.: +86-13604311978

Abstract: For photoelectric theodolite, the mirror is the core optical component, so it is of great significance to design and optimize a mirror with excellent overall performance. In order to comprehensively consider the contradictory objectives of mass, natural frequency, and RMS under gravity, a multiobjective optimization method based on the performance analysis of two-parameter coupling was proposed. On the basis of the performance law, a suitable solution for balancing multiple objective functions is obtained by introducing manual intervention. The results show that compared with the traditional empirical design of mirrors, the first-order natural frequency, mass, and RMS performance of the optimized mirror are improved by 18.64%, 0.1%, and 15.58%, respectively. The frequency/Mass ratio and 1/(RMS*Mass) ratio are increased by 18.72% and 18.59%, respectively. Its comprehensive performance has been improved. This method is effective and provides a reference for the design of photoelectric theodolite and other mirrors.

Keywords: SiC mirror; parameters coupling; structural optimization

Citation: Dong, Q.; Wang, Q.; Wang, C.; Luan, Y.; Wang, X.; Wang, X. Multiobjective Optimization of SiC Mirror Based on Dual-Parameter Coupling. *Photonics* **2023**, *10*, 171. https://doi.org/10.3390/photonics10020171

Received: 19 November 2022
Revised: 19 January 2023
Accepted: 1 February 2023
Published: 7 February 2023

1. Introduction

Photoelectric theodolite is a typical photoelectric tracking measurement device, which can accurately measure the external ballistic parameters of the airborne target in real time and record its flight attitude at the same time. It has the advantages of strong anti-interference ability and high measurement accuracy. As an important part of the optical system, the surface precision of the primary mirror plays a key role in imaging quality. With the improvement of observation resolution and light-gathering ability, the aperture of theodolite becomes larger and larger, increasing the aperture of the mirror will make its shape accuracy more susceptible to the influence of gravity and ambient temperature. In order to ensure the accuracy and system stability of the photoelectric theodolite, it is necessary to design the primary mirror structure to meet the index requirements in order to reduce the deformation of the primary mirror during the working process [1]. This improves the surface shape accuracy and image quality of the primary mirror. The general requirement for the face shape accuracy of the primary mirror is that the RMS value of the machined mirror is less than $\lambda/40$ (λ = 632.8 nm). Therefore, many efforts are focused on the research of high-performance, low-cost, lightweight mirrors [2–4]. One difficulty in developing optical devices is to minimize the quality of mirrors without compromising their stiffness. To solve this problem, some methods should be adopted during the development of mirror design, such as the choice of materials and structures [5,6]. In terms of materials, Zerodur and ULE are traditional optical materials, while silicon carbide (SiC) has been considered the most promising optical material due to its excellent properties [7–9]. Silicon carbide is a semiconductor compound with high covalent bond strength. Due to its unique thermal and mechanical properties, such as good wear resistance, good thermal stability, small coefficient of thermal expansion, and high hardness characteristics, it has a wide range of technical applications. Silicon carbide has been widely used in optical devices, nanotechnology, and nuclear material science for decades [10,11]. For example, since 1990, foreign telescopes such as Hubble, SST, and KST have used Be and ULE materials [12–16]. Since 2010, better-performing SiC material was chosen as the material for the primary mirror for the Herschel telescope [17–19]. In terms of structure, the main structures of the reflector are the rib, flanging, etc. The coupling between these structural shapes plays a crucial role in the influence of mirror performance. Therefore, it is an important research work of engineering significance to explore the mechanism of the influence of the coupling of structural parameters on performance and to seek the optimization method of the mirror from the essential law.

Many studies have analyzed and optimized lightweight mirrors. Kihm et al. [20] proposed a design method for a 1 m lightweight mirror in a space optical system. He divided the dimensional design into two separate problems, i.e., mirror design, and flexible design, and applied a multiobjective genetic algorithm to optimize the reflector. Butova et al. [21] used the method of parameter optimization to optimize the mirror, which is more than 1 m. A light mirror with a small optical surface deflection is obtained. Inspired by the topology optimization method considering casting constraints, Liu et al. [22] proposed an optimization model for mirror back configuration design, through which the distribution and height of ribs on the mirror back could be optimized simultaneously. Shao et al. [23] obtained key parameters by analyzing the sensitivity of primary mirror size parameters on performance. The parameter optimization is realized by the photomechanical analysis under multiple loads. Qu et al. [24] proposed a method that combines topological optimization with multiobjective function and parameter optimization. The new configuration design has an obvious advantage. Chen et al. [25] proposed an integrated photomechanical design method and optimized the parameters of the mirror and the bipod curved supports. All of the above research on mirror analysis and optimization have laid a foundation for subsequent improvement. Through previous studies, it is found that the traditional parameter-optimized method is to change a certain variable of the mirror through the computer, and the other parameter values are unchanged. Then, the influence of each parameter on mirror performance is analyzed one by one. However, only a simple analysis model can be obtained without considering the influence of different structural parameters on the overall performance of the mirror. With the change of parameters, the performance of one kind is improved while the performance of another kind is degraded. The optimization results cannot be directly applied to the practice. Moreover, only a certain structure can be obtained, according to the set objective function, which can only optimize a certain performance of the mirror. This optimization structure cannot guarantee simultaneous improvement of multiple performances. There are also some primary mirrors that are lightened using the topology optimization method. The method has the limitation that it cannot simultaneously optimize the global design of the key structural dimensions of the primary mirror.

On the basis of previous studies, this paper intends to solve the following research problems. The primary mirror of Φ672 mm in a photoelectric theodolite was taken as the research object, and the mirror configuration was designed by the empirical formula. Through the method of parametric modeling, the main structural parameters (such as rib, outer wall, center hole, etc.) were analyzed by parameter coupling, so as to summarize the law of the influence of parameter coupling on the performance. Compared with the sensitivity of a single structural variable, this method can obtain a more intuitive numerical representation and variation trend of the parameters on the performance. This model design is more common in other structural performance analyses. In addition, the problem to be solved is how to carry out a multiobjective size optimization based on the results of the parameter coupling analysis to achieve a comprehensive improvement of mirror performance. The expected performance improvement of the mirror was set by human intervention, and the feasible region was planned to meet the expected performance analysis results. In the different performance results obtained by multiple parameter variables, the intersection of the feasible region is selected. By selecting a series of parameter values in the intersection set, the data of different structure sizes can be obtained correspondingly and the performance of the mirror obtained by any group of data is better than that of the initial mirror. A multiobjective optimization of stiffness, mass, and surface shape was achieved. This method is based on the analysis content to guide the structural optimization and can be applied to the structural analysis and improvement under different working conditions.

2. The Initial Structure

2.1. Design of the Initial Structure of the Mirror

In this paper, according to the structure of a certain type of photoelectric theodolite, the semienclosed mirror was designed as the initial structure according to the actual working conditions and empirical formula of the mirror. The inside of the mirror is a traditional triangular lightweight hole, which is equipped with three support holes and a center hole. The initial mirror structure is shown in Figure 1.

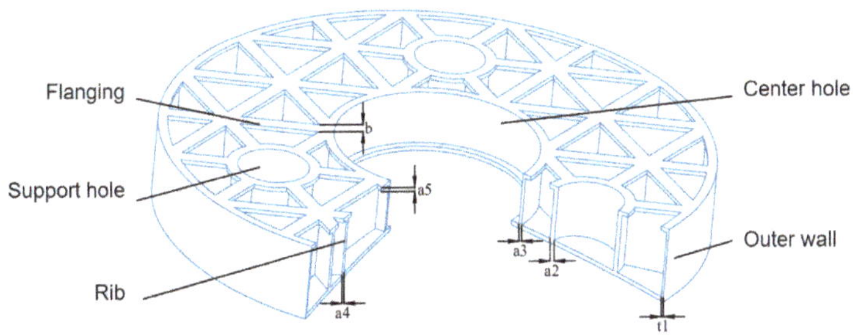

Figure 1. Semienclosed initial mirror model.

The main structural parameters of the mirror include the thickness of the mirror (A), the thickness of the outer wall (t1), the radius of the support hole (r2), the thickness of the support hole (a2), the position of the center of the supporting hole (l2), the radius of the center hole (r3), the thickness of the center hole (a3), the thickness of the rib (a4), the thickness of the flanging (a5), the width of the flanging (b), and the height of the mirror (H). The structural parameters of the initial mirror are shown in Table 1. After the initial mirror structure is determined, COMSOL software is used for parametric modeling. Thus, the time of repeated modeling during the process of analysis is reduced when the parameter values of different structures are changed, which is convenient for analysis and comparison.

Table 1. Structural parameter values of the initial mirror (mm).

Parameter	Variable	Value
The thickness of the mirror	A	5.5
The thickness of the outer wall	t1	4
The radius of the support hole	r2	40
The thickness of the support hole	a2	9
The position of the center of the supporting hole	l2	220
The radius of the center hole	r3	125
The thickness of the center hole	a3	4
The thickness of the rib	a4	3.5
The thickness of the flanging	a5	5
The width of the flanging	b	7.5
The height of the mirror	H	93.8

2.2. Material of the Mirror

The material properties of the mirror are also very important to its performance. The existing commonly used mirror materials are SiC, ultralow expansion glass ULE, microcrystalline materials, Be, etc. The inherent properties of materials, such as density, Young's modulus, Poisson's ratio, and thermal expansion coefficient, directly affect the inherent properties of mirrors in static and dynamic environments. So the primary consideration at the beginning of structural design is material selection [26]. According to the actual working condition requirements of the photoelectric theodolite, the distribution and transportation are mostly based on vehicles, ships, or aircraft. It is necessary to measure the adaptability of the mechanical properties of different materials to working conditions. In addition, theodolite is mostly located in a harsh outdoor environment, which is obviously affected by temperature and wind load factors. Therefore, the material thermal physical properties are also an important index affecting the working accuracy of the mirror. Gravity is one of the main reasons to influence the precision of surface shape, so it is necessary to select materials with a high stiffness and a high lightweight degree as much as possible. The properties of common materials are shown in Table 2 [27].

Table 2. Properties of common materials.

Material	Density (g/cm^3)	Young's Modulus (G Pa)	Specific Stiffness (E/ρ)	Thermal Conductivity ($m \cdot K$)	Coefficient of Linear Expansion ($10^{-6}/K$)
SiC	3.20	400	125.00	155	2.40
Be	1.85	287	155.14	216	11.40
Zerodur	2.53	91	35.97	1.64	0.05
ULE	2.21	67	30.31	1.31	0.03

SiC material has the advantages of moderate density, nontoxicity, small coefficient of linear expansion, and high thermal conductivity. It is a widely used ceramic material [28]. Currently, there are more than 200 crystalline modifications of silicon carbide. Different crystalline modifications have different properties. By stacking several identical structures in different orders, different crystalline modifications can be obtained. For example, α-SiC is the most common crystalline form of silicon carbide and it has a hexagonal crystal structure similar to wurtzite, as shown in Figure 2a. β-SiC has a diamond-like sphalerite crystal structure, as shown in Figure 2b. Another common silicon carbide, 4H-SiC, which has a hexagonal crystal structure, is shown in Figure 2c. However, as a compound with a strong covalent bond (up to 88% covalent bond composition), the difference between grain boundary energy and the surface energy of SiC is very small, so it is difficult to form a grain boundary in the sintering process. Usually, the sintering of silicon carbide needs to be achieved by means of admixtures, pressure, or siliconizing. Reaction-bonded silicon carbide (RB-SC) generally uses α-SiC and carbon as raw materials. In the process of sintering, carbon reacts with liquid silicon or silicon vapor to form a secondary β-SiC in the primary position, which is combined with initial α-SiC particles to form dense SiC [29]. With the development of modern optical technology, the photoelectric theodolite field needs more and more urgent optical systems with a large aperture, that are ultralightweight, and have a complex shape. Compared with the other materials in Table 2, SiC has the characteristics of high specific stiffness (it is only lower than Be), low thermal expansion rate, high thermal conductivity, good chemical stability, good optical processing, and so on. So it is a lightweight mirror material with excellent comprehensive performance [30]. These properties have made SiC the preferred material for spatial and theodolite mirrors in recent years. According to the actual working conditions of photoelectric theodolite, and the current methods of preparing materials commonly used in my organization, SiC material was selected for mirror analysis.

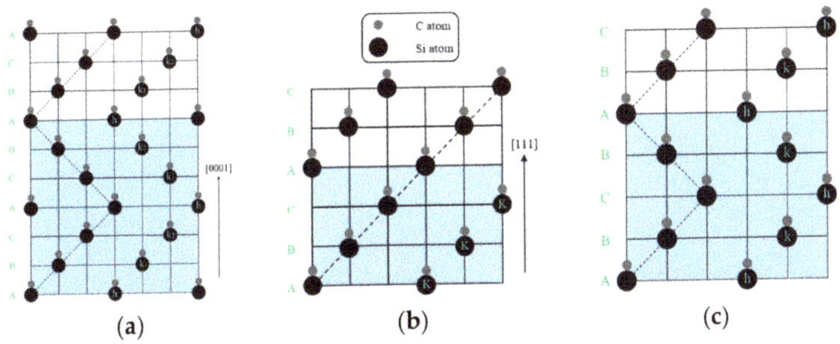

Figure 2. Three main crystalline forms of silicon carbide. (**a**) (α)6H-SiC; (**b**) (β)3C-SiC; (**c**) 4H-SiC.

2.3. *Performance Analysis of the Initial Structure*

The natural frequency of the mirror under free vibration is an important performance index. In the free state, the natural frequency of the mirror and the position of the vibration are analyzed, in order to avoid resonance in the design of the structure. The initial mirror has meshed. After the finite element mesh is obtained, its free modes are analyzed. The vibration equation of its discrete form is shown as follows:

$$M\ddot{u}+B\dot{u}+Ku=0, \tag{1}$$

where M is the overall quality matrix. K is the overall stiffness matrix. B is the damping matrix. The node displacement vector is u. The velocity vector is $\dot{u}$. The acceleration vector is $\ddot{u}$. Usually, the system can be viewed as undamped free simple harmonic vibration. Therefore, the form of the solution of Equation (1) is shown as follows:

$$u = A\sin(w_i+\varphi), \tag{2}$$

where A is the system amplitude vector and w_i is the free vibration frequency. Substituting Equation (2) into Equation (1), the mode shape equation is

$$(K - w_i^2 M)A = 0, \tag{3}$$

Equation (3) has the existence of nonzero solutions. Then, the determinant is zero. That is the following equation:

$$\left|K - w_i^2 M\right| = 0. \tag{4}$$

Equation (4) is the system characteristic equation or frequency equation. The eigenvalue w_i is the natural frequency.

So, without any constraints, according to the above formula, by calculating the first 12-order characteristic frequencies of the initial mirror, after removing the first six-order rigid body displacements, the seventh-order characteristic frequency is the first-order natural frequency. Its natural frequency value is 1169.7 Hz. Figure 3 shows the free modes of orders 1–6. The first two free modes are approximately the same and they have similar natural frequencies. They bend inward from the edges of the mirror and the deformation is maximum at the four edges of the circle. The third mode shows that the center support hole has the largest deformation, and its surrounding structure protrudes outward. Its natural frequency value is about two times larger than the first natural frequency. The fourth free mode shows a total of six obvious inward or outward local deformations. Modes five and six are local modes. The natural frequency and mode shape of the mirror can be obtained by modal analysis, which can reflect the stiffness of the mirror and avoid resonance of the structure.

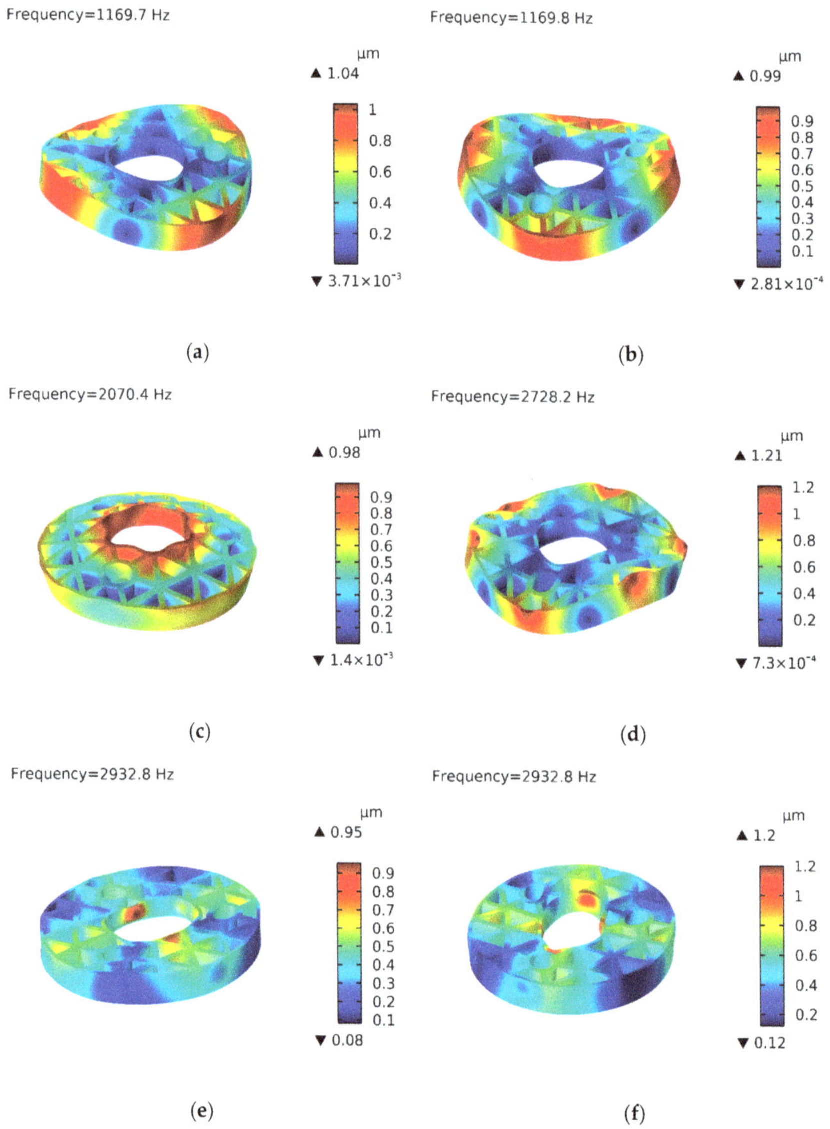

Figure 3. Free mode analysis of the 1st–6th order of the initial mirror. (**a**) The first free mode shape; (**b**) The second free mode shape; (**c**) The third free mode shape; (**d**) The fourth free mode shape; (**e**) The fifth free mode shape; (**f**) The sixth free mode shape.

According to the above, the free modes of the initial mirror are calculated without any constraints. Next, the accuracy of surface shape and deformation under the condition of dead weight will be analyzed. As can be seen from the mirror model in Figure 1, it has three circular support holes. Simple cylinders were then added at the positions of each support hole to simulate the support structure, using them as fixed constraints. In other words, boundary conditions are set at the support hole, and its displacement is constrained to be constant at zero. The condition is that the optical axis of the SiC mirror points vertically to the zenith so that it is subjected to a 1 g gravity load under the action of three-point fixed constraints. The boundary conditions, constraints, and deformation results of the mirror under the action of dead weight are shown in Figure 4. When the mirror is in static equilibrium under its gravity, it satisfies the following equation:

$$Ku = F. \quad (5)$$

where K is the overall stiffness matrix, u is the displacement vector, and F is the external force vector.

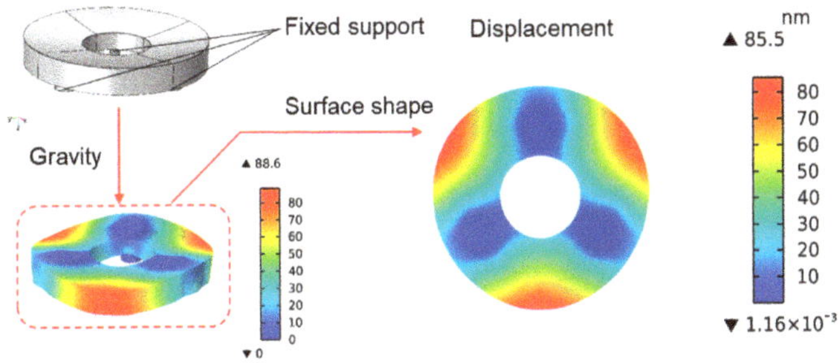

Figure 4. Displacement of an initial SiC mirror based on three−point support (RMS = 39.56 nm).

When photoelectric theodolite is working, the face shape accuracy is related to the imaging quality. The root means square (RMS) of the surface shape is usually used to evaluate the accuracy of the surface shape. It is defined as the root mean square of the mirror normal displacement after removing the structural rigid body displacement [31,32]. RMS indicates the root mean square value of the deviation between each node of the deformed surface shape relative to the fitted surface shape:

$$RMS = \sqrt{\sum_{i=1}^{n} \frac{1}{n}(x_i - \overline{x})^2}. \quad (6)$$

where x_i is the distance from the ith node to the fitted surface after deformation, and $\overline{x}$ is the average distance of all nodes to the fitted surface. By calculating the mirror deformation, the quality of the mirror's face shape can be evaluated.

The deformation of its dead weight and RMS (root mean square value of face shape) are analyzed as the reference value of the performance under the initial structure. Figure 4 shows the deformation of the initial mirror based on gravity. The maximum value is 88.6 nm, and the RMS value is 39.56 nm.

3. Parametric Coupling Analysis

There are many factors affecting the structural stiffness and surface shape accuracy of the mirror, such as the form of the lightweight hole, the degree of back closure, the thickness of the rib, the size of the support hole, etc. By analyzing the main structure size of the mirror, the stiffness of the mirror is obtained by comparing its first natural frequency (The seventh-order characteristic frequency). The comparison between stiffness and mass can analyze the influence of structural parameters on mirror performance and guide subsequent research. The RMS value under the action of gravity was analyzed to evaluate the imaging quality. The comparison between the RMS value and the mass can reflect the relationship between shape accuracy and light weight. Many factors affect the structure and performance of the mirror, such as the thickness of the flanging (a5), the width of the flanging (b), the thickness the of support hole (a2), the thickness of the center hole (a3), the thickness of the rib (a4), and the thickness of the outer wall (t1). In order, the two-parameter couplings with high correlation are grouped into a group. They were divided into three groups for parameter coupling analysis.

This paper presents the results of the influence of each parameter in a certain range on the mirror performance in the form of two-dimensional graphs. In each set of parameters, four different types of two-dimensional graphs are defined to show the results.

The first type of figure expresses the first-order natural frequencies at different sizes for a certain set of parameter combinations. It is *frequency*, which can show the stiffness of the mirror. When the value of the coordinate in the figure is larger, it means that the stiffness under this parameter is better. The second type of figure expresses the ratio between the natural frequency and mass of the mirror. It is *frequency/mass*. When the natural frequency is higher or the mass is smaller, the ratio is larger. The corresponding coordinates reflect the high "cost performance", which not only improves the stiffness, it also reduces the quality. However, a higher value does not directly indicate better performance. Other factors should be considered. The third type of figure expresses the surface shape accuracy of the mirror based on the three-point fixed support constraint and the application of gravity. It is *RMS*. When the value is smaller, it means that the surface shape accuracy under the size is better. Usually, the deformation caused by gravity is one of the main reasons affecting the surface shape accuracy. The gravity depends on the quality of the mirror itself, and the light weight is a key component in the optimization design of the mirror. The purpose of light weight is not only to reduce the mass, it is also to ensure stiffness and shape accuracy. Therefore, the fourth type of graph is needed to express the relationship between the RMS value and the quality of the mirror. It is 1/(*RMS*Mass*). Since both RMS value and quality are seeking the minimum and appropriate value, this value needs to comprehensively consider the issue of "cost performance". A larger value indicates better performance.

Through the above method, four kinds of results in the form of graphs can be obtained for every two related variables. They can clearly show the influence of parametric coupling on the mirror performance and the distribution law.

3.1. Coupling Analyzing the Properties of Flanging's Thickness and Width

The flanging is located on the back of the mirror. The size of the flanging parameters determines whether the mirror is open, semiclosed, or fully closed. The thickness of flanging (a5) and the width of flanging (b) were combined. Then the combination was parametrically scanned and analyzed. The value of a5 ranges from 2 to 11 mm. The value of b ranges from 4 to 15 mm. After calculating the natural frequency and RMS values by the finite element method, a series of two-dimensional graphs of mirror performance results are obtained.

As can be seen from Figure 5, the influence of the two parameters on the performance of the mirror is different. The analysis results are shown as follows.

First of all, according to the results in Figure 5a,b, under the mutual influence of the two parameters, the ratio of the first-order natural frequency to frequency/mass changes obviously. When the trend of stratification approaches the level, the results show that the width of the flanging has a high sensitivity to its corresponding performance, and the size parameter has an obvious influence on the structural performance. Increasing the value of this parameter will increase the natural frequency.

Then, since the size of the mirror along the optical axis is significantly smaller than its radial size, the mirror is more prone to deformation along the optical axis. In order to improve the bending stiffness of the mirror, the design of the mirror generally increases the flanging width value, so that it is a closed back or semiclosed structure. Under the action of gravity, Figure 5c,d show that the RMS value becomes smaller, and the surface shape accuracy is better when the increase of the flanging's thickness and width are increased. However, if the value is too large, there will be problems such as excessive mass and manufacturing difficulties, which hinder the optimization of the mirror.

In addition, the corresponding regions of the optimal values of the four kinds of performance have intersections. It shows that the parameters in the public area can simultaneously improve the mirror structure performance.

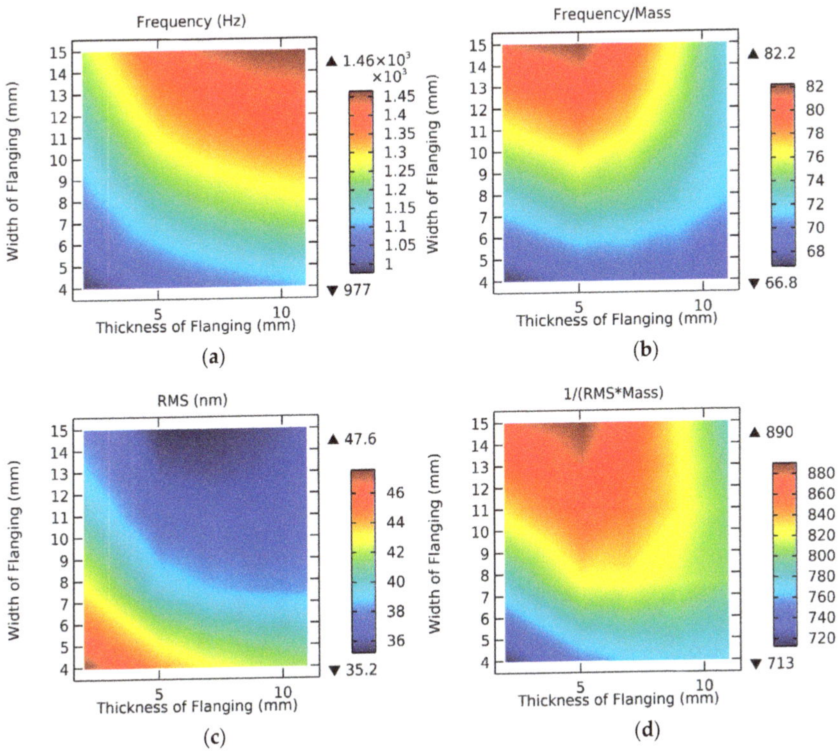

Figure 5. Performance analysis of flanging's thickness and width under different parameters. (**a**) Frequency; (**b**) Frequency/Mass; (**c**) RMS; (**d**) 1/(RMS*Mass).

There is one more point. When viewing Figure 5a,b longitudinally and Figure 5c,d horizontally, the coupling law of the two parameters can be clearly found. When the value of one parameter is different, the influence of another parameter's value on the performance is different. The value of performance does not change at the same rate. That is, when one parameter takes different values, the gradient of the influence of the other coupling parameter on the performance is different.

3.2. Coupling Analyzing the Properties of the Thickness of the Support Hole and Center Hole

The support modes commonly used in mirrors include three-point support, multipoint support, center support, side support, etc. In this mirror model, the thickness of the three-point support hole (a2) and the thickness of the center hole (a3) are a set of relevant parameter values. A combined parametric scan is performed for both. The value of a2 ranges from 3 to 15 mm and the value of a3 ranges from 2 to 6 mm. The analysis results in Figure 6 are shown below.

Firstly, the layering patterns of the four performance figures have similar stratification laws, which are approximately vertically distributed. The results show that the thickness of the support hole has an obvious influence on the performance of the mirror, and its parameter sensitivity is relatively large.

Secondly, the trend of the frequency/mass ratio was opposite to that of natural frequency. When the values of the two variables are smaller, the eigenvalues are larger.

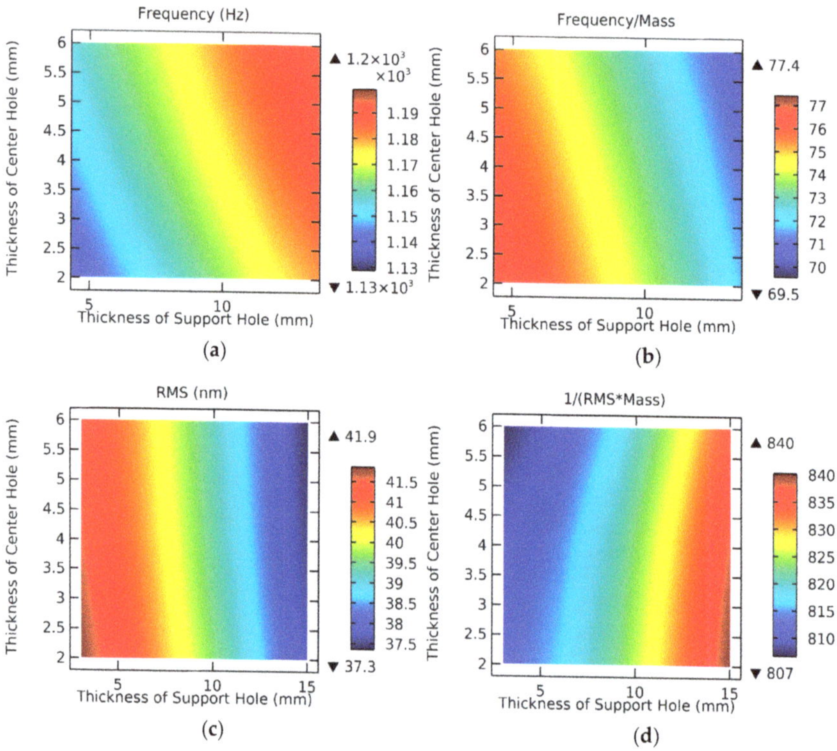

Figure 6. Performance analysis of the thickness of the support hole and center hole under different parameters. (**a**) Frequency; (**b**) Frequency/Mass; (**c**) RMS; (**d**) 1/(RMS*Mass).

Thirdly, Figure 6a–d all show that when the thickness of the support hole is larger, the corresponding performance is better. Therefore, increasing the thickness of the supporting hole is beneficial to improve the comprehensive performance of the structure.

Fourthly, although the distribution pattern is obvious, the range difference between the RMS value and 1/(RMS*Mass) is not large. The range of RMS values in the whole region was only 4.6 nm. Therefore, their values can be approximated to be equal in a wide range. These data need to be taken into account when referring to them for optimization.

Fifthly, according to Figure 6c, when the two parameters are coupled to analyze a certain performance, if the distribution law of the influence on the performance is obvious (almost horizontal or vertical), it indicates that the coupling degree of the two parameters to the performance is not high. On the contrary, if the distribution regularity of the two-dimensional performance figure is not obvious, it indicates that the two parameters have a high influence on each other and have a great impact on the performance change.

3.3. Coupling Analyzing the Properties of the Thickness of the Rib and the Outer Wall

The thickness of the rib (a4) and the outer wall (t1) are important parameters for the lightweight and structural stiffness of the mirror. The two variables were combined for parametric scanning. The value of a4 ranges from 2 to 7 mm and the value of t1 ranges from 1 to 8 mm.

As can be seen in Figure 7, the two parameters have different effects on the performance of the mirror. The analysis of the results is shown as follows.

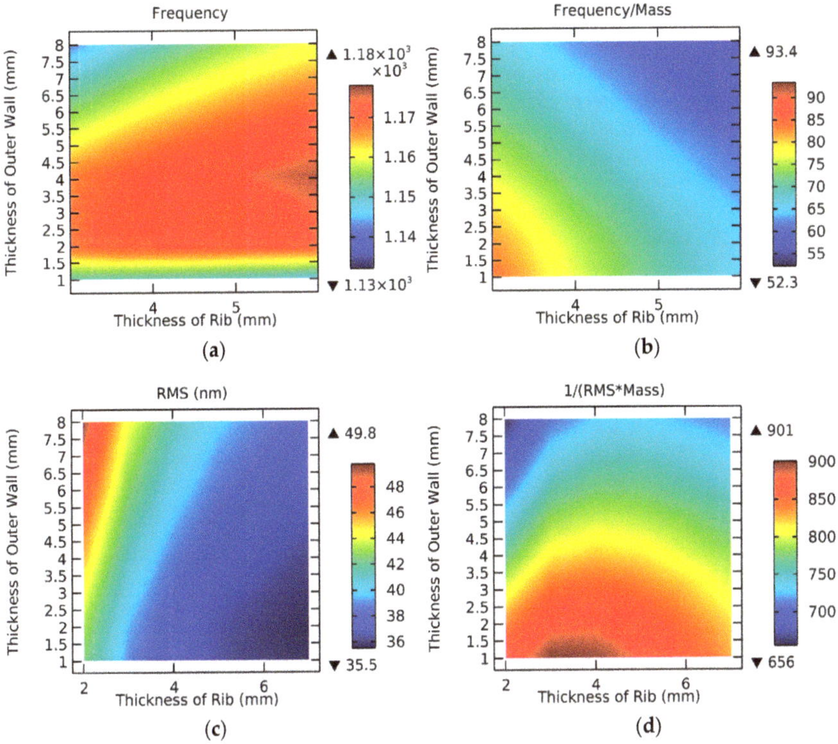

Figure 7. Performance analysis of the thickness of the rib and outer wall under different parameters. (**a**) Frequency; (**b**) Frequency/Mass; (**c**) RMS; (**d**) 1/(RMS*Mass).

First of all, under the mutual influence of the two parameters, the distribution of natural frequency value is not obvious. However, the better value in the red region accounted for more than 50%, and the frequency value in the red region was close to each other without significant change. However, the influence of the outer wall on natural frequency is affected by the first increase and then decrease, so the excessive increase in the thickness of the outer wall will only play a negative role.

Then, the four graphs in Figure 7 show different performances, and their distribution laws are different. The contrast shows that there is a contradiction. For example, increasing the thickness of the rib can improve the natural frequency and RMS value. When the quality factor is taken into account, its cost performance will decline. This indicates that the parameter should not be increased blindly and should be considered comprehensively.

The 1/(RMS*Mass) has a large range. The maximum value is about 901 and the minimum value is about 656. Therefore, it is of great significance to be referenced and should be given priority when optimizing the structure of the mirror.

In addition, the different distribution rules in the four figures in Figure 7 show that the two parameters have different impacts on different performances. It shows that the coupling degree of the two is higher. As shown in Figure 7c, the rib has a higher sensitivity, however, as shown in Figure 7d, the outer wall has a higher sensitivity. When the value of one parameter is different, the gradient of the change of the other parameter is different.

4. Optimization

4.1. Optimized Process

In the analysis of the influence of mirror structure on performance, six variables are combined into three groups. For each group, four two-dimensional graph results about performance evaluation were obtained. To verify that the results of the parametric analysis of the main structure of the mirror can accurately reflect the performance of the structure and have application value for optimization, the above data are extended in this section. Based on the initial mirror model, the influence of the main structure size on the performance is obtained by parametric analysis. Afterward, optimization is guided by laws. Since the optimization of the mirror is not a single objective problem, it involves stiffness, mass, RMS, and other factors. For example, in general, the improvement of mass will improve the stiffness. At the same time, it also increases the influence of gravity on the shape, which makes the image quality worse. They are a set of competing factors. That is to say, when the structural parameter of the mirror is optimized, some performance will be improved, however, other performance may be weakened [33]. Therefore, in order to comprehensively improve the mirror performance, it is necessary to transform the single objective problem into a multiobjective for seeking the optimization problem. The six main structure sizes are defined as function variables and are shown in Table 3:

Table 3. The variables are defined by each parameter.

Parameter	Variate
a5	x_1
b	y_1
a2	x_2
a3	y_2
a4	x_3
t1	y_3

Based on the performance analysis of the parameter coupling pair, three groups of parameter coupling were used to calculate the four performances of the mirror in turn. As shown in Figures 5–7, a total of 12 performance analysis data are obtained as the basis of optimization. Then, the six variables in the three sets of coupling parameters are assigned to vector matrix X, where $X^0 = \left[x_1^0, y_1^0, x_2^0, y_2^0, x_3^0, y_3^0\right]^T$ is the main parameter size of the initial structure. $X = \left[x_1, y_1, x_2, y_2, x_3, y_3\right]^T$ is the main structure size, which changes with the value. The change in the final structure size will affect the performance of the mirror. Although there is no explicit functional relationship between the influence of the variables of the mirror structure size on the performance, the optimization can be carried out according to the two-dimensional graph of the performance results of the variables. Based on the parameterized scan results, the feasible region is planned according to the variables and the values in the figure. The feasible region represents the expected degree of optimization. The mathematical expression is $g_{ij}(X) = f_{ij} - f_{ij}^0(1 + I_{ij}) + C_{ij} \geq 0$. According to each set of parameters, it is expanded as follows:

$$\begin{gathered} \text{Find } (X) = (x_1, y_1, x_2, y_2, x_3, y_3)^T \\ \text{Min F } (1/\text{Frequency, RMS, Mass/Frequency, RMS} * \text{Mass}), \end{gathered} \tag{7}$$

s.t.

$$\begin{aligned} g_{i1}(X) &= f_{i1} - f_{i1}^0(1 + I_{i1}) + C_{i1} \geq 0 \\ g_{i2}(X) &= f_{i2} - f_{i2}^0(1 + I_{i2}) + C_{i2} \geq 0 \\ g_{i3}(X) &= f_{i3} - f_{i3}^0(1 - I_{i3}) - C_{i3} \leq 0 \\ g_{i4}(X) &= f_{i4} - f_{i4}^0(1 + I_{i4}) + C_{i4} \geq 0 \end{aligned} \tag{8}$$

$$\begin{cases} 2 \le x_1 \le 11 \; 4 \le y_1 \le 15 \\ 3 \le x_2 \le 15 \; 2 \le y_2 \le 6 \\ 2 \le x_3 \le 7 \; 1 \le y_3 \le 8 \\ i = 1, 2, 3 \\ j = 1, 2, 3, 4 \end{cases} . \tag{9}$$

In the above formula, Min F is the objective function. The goal of the optimization is to minimize four performance parameters, which are 1/frequency, RMS, mass/frequency, and RMS*Mass. Corner marks ij are used to represent the corresponding two-dimensional diagram, where $i = 1, 2, 3$ represent three groups of variables in turn. Graph numbers a, b, c, and d (in Figures 5–7) in the scan results of each group in turn are $j = 1, 2, 3, 4$, and f_{ij} is the value of the performance at different X. The value of a single performance of the initial mirror is f_{ij}^0, and it is also the initial value. The optimal value of a single performance of the mirror is f_{ij}^*. I_{ij} is the expected improvement rate of a certain performance. C_{ij} is the constant term as the expected improvement of a certain performance. The two factors together determine the expected performance improvement. All of the g_{ij} are jointly planned to obtain the initial feasible region. Equation (9) is the range of values of the variables corresponding to structural parameters. They are selected by reference to the range values selected for structural parameters in parameter coupling analysis.

Based on the above, human intervention is added at the beginning of optimization, and constraints are set according to the requirements and expected improvement. That is, according to Equation (8), the range of the optimization domain is set in advance, and the nonconforming domain is removed from the performance data of the parameter coupling analysis. Then, the optimization is carried out based on the two-parameter coupled data and the setting of the objective function. The intersection domain is found in the retained data region of each two-dimensional graph and the points in the intersection domain are compromise solutions. In other words, the value of the points in the intersection after optimization cannot achieve the optimal performance, however, each key performance has been improved compared with the initial value. Through this process, the relevant variable X that meets the conditions in all parameter coupling groups is found. At this time, the optimization is completed. the structure size of the mirror is improved, and the performance is optimized.

4.2. Results of Optimization

Through the above method, size optimization can be achieved to obtain better mirror performance. For example, the thickness and width of the flanging are analyzed. According to the engineering requirements and the performance improvement degree of human intervention, the feasible region is planned by taking the values shown in Table 4. This plan is shown in Figure 8. The gray area is the part removed during the planning of the feasible region, which is the area that does not meet the requirements. Taking the set of intersection points in the feasible region of the four two-dimensional graphs, the size that can improve the overall performance of the mirror is obtained. A point in the intersection is the optimized point.

Table 4. Planning of feasible regions.

Parameter	Value (j = 1)	Value (j = 2)	Value (j = 3)	Value (j = 4)
f_{1j}^0	1169.7	73.7	39.56	822.71
I_{1j}	10%	5%	5%	5%
C_{1j}	43.33	3.615	1.582	16.15
$f_{1j}^0\left(1 + I_{1j}\right) + C_{1j}$	1330	81	36	880

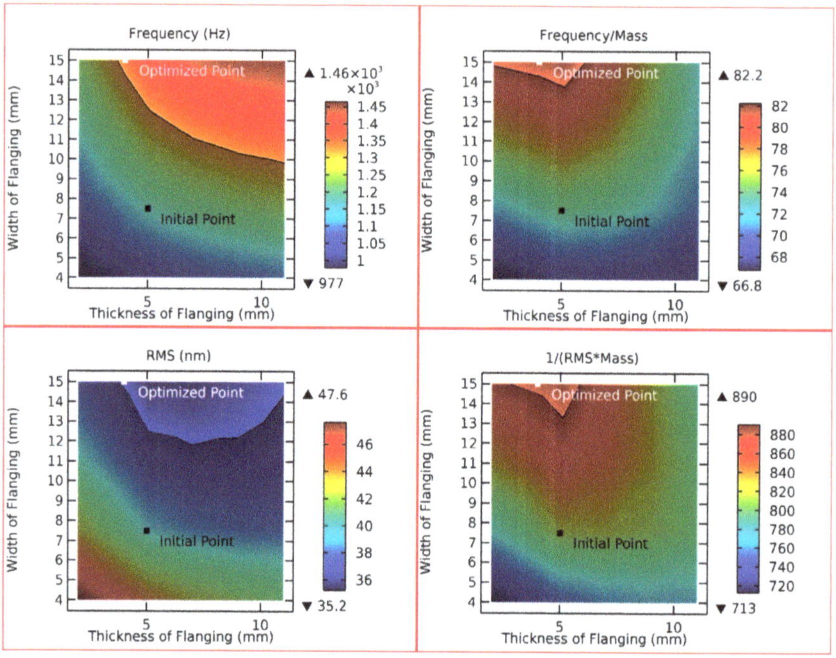

Figure 8. Schematic diagram of an optimization.

Based on 12 performance results obtained by parametric analysis, the proposed method is used to plan and optimize the feasible region. The overall performance of the mirror is improved after the optimization of the main structural parameters. Considering the speed of the optimization, this optimization is only carried out when [X] is taken as an integer. The comparison between the optimized mirror parameters and the initial values is shown in Table 5. The comparison between the initial model and the optimized model is shown in Figure 9.

Table 5. Comparison of parameters between the initial and improved structures (Unit: mm).

Parameter.	Variable	Initial Value	Optimal Value
a5	x1	5	4
b	y1	7.5	15
a2	x2	9	14
a3	y2	4	5
a4	x3	3.5	3
t1	y3	4	2

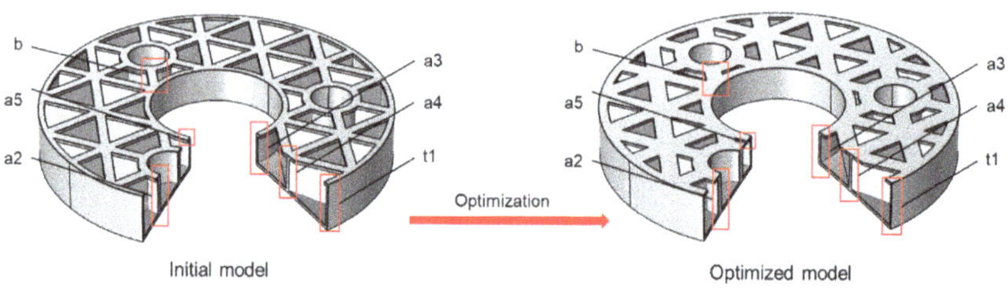

Figure 9. Comparison of the initial mirror and the optimized mirror.

The optimized model was calculated by free mode, mass, optical axis vertical, and the influence of gravity on the face shape RMS value. Through software simulation and data processing, the first natural frequency of the optimized model is 1387.7 Hz. By calculation, the RMS of the model is 33.395 nm and the mass is 15.855 kg. The simulation result is shown in Figure 10. According to the data calculated by the initial model mentioned above, compared with the performance of the initial mirror (shown in Table 6), the performance of the optimized mirror is improved. The weight of the optimized mirror has been reduced from 15.871 kg to 15.855 kg, and it is slightly light (increased by 0.10%). At the same time, the first natural frequency is increased from 1169.7 Hz to 1387.7 Hz, and its stiffness is increased by 18.64%. The frequency/mass value is increased from 73.7 to 87.5, indicating that the cost–performance ratio between stiffness and mass is increased by 18.72%. The RMS value is decreased from 39.56 nm to 33.395 nm, indicating that the shape accuracy is increased by 15.58%. The value of 1/(RMS*Mass) is increased from 822.71 to 975.68, indicating that the cost–performance ratio between shape accuracy and mass is increased by 18.59%. The overall performance is significantly improved, which proves the practicability and reliability of the optimization method.

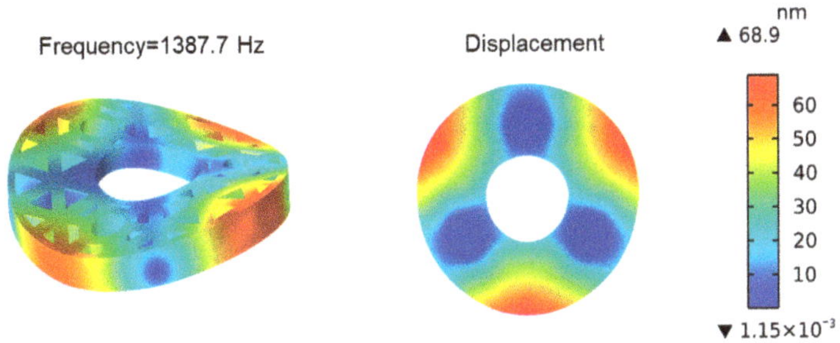

(a) (b)

Figure 10. Analysis of the optimized model. (**a**) The first free mode shape; (**b**) Displacement of surface shape (RMS = 33.395 nm).

Table 6. Comparison of the performance of the initial mirror and the optimized mirror.

Performance	Initial Value	Optimal Value	Increase Rate
Frequency	1169.7	1387.7	18.64%
Mass	15.871	15.855	0.10%
Frequency/Mass	73.7	87.5	18.72%
RMS	39.56	33.395	15.58%
1/(RMS*Mass)	822.71	975.68	18.59%

5. Conclusions

In this paper, a multiobjective optimization method for a SiC mirror based on dual-parameter coupling is proposed. The performance of the mirror is analyzed by using two parameter couplings. The analysis results clearly show the different parameters under the action of coupling to various properties. It improves the deficiency of only considering a single structural change without considering other factors. On this basis, the structural parameters are optimized by adding manual intervention. Compared with the structure designed based on traditional experience, the optimized mirror which uses a SiC material with better overall performance has the following properties. The weight of the optimized mirror is 15.855 kg, and it is slightly light (increased by 0.10%). At the same time, the first natural frequency is 1387.7 Hz, indicating that the stiffness increases by 18.64%. The frequency/mass value is 87.5, indicating that the cost–performance ratio between stiffness and mass increased by 18.72%. The RMS value is 33.395 nm, indicating that the accuracy of surface shape is improved by 15.58%. The value of 1/(RMS*Mass) is 975.68, indicating that the cost–performance ratio between the shape accuracy and the mass has increased by 18.59%. The result shows that this method improves the performance of the mirror and achieves a multiobjective optimization. It provides effective technical means and an engineering reference for the design of a mirror of photoelectric theodolite.

It should be noted that although this method has achieved good results in this work, due to manufacturing errors and actual working conditions in actual engineering, there may be corresponding errors compared with the expected value. In future studies, ray tracing, different optical axis angles, and temperature effects should be added to make up for the deficiency. They will make the results more realistic and universal, and experiments will be added to verify them.

Author Contributions: Conceptualization, Q.D. and Q.W.; methodology, Q.D.; software, Q.D.; validation, Q.D., C.W. and X.W. (Xiaoming Wang); formal analysis, Q.D.; investigation, Q.D.; resources, X.W. (Xiaoming Wang); data curation, X.W. (Xiaoxun Wang); writing—original draft preparation, Q.D.; writing—review and editing, Q.D.; visualization, Y.L.; supervision, C.W.; project administration, X.W. (Xiaoming Wang); funding acquisition, X.W. (Xiaoming Wang). All authors have read and agreed to the published version of the manuscript.

Funding: This research was funded by THE NATIONAL NATURAL SCIENCE FOUNDATION OF CHINA, grant number 6210031610.

Institutional Review Board Statement: The study did not require ethical approval.

Informed Consent Statement: Not applicable. The study did not involve humans.

Data Availability Statement: The data presented in this study are available from the corresponding author upon reasonable request.

Acknowledgments: The authors thank Xiaoming Wang for his theoretical guidance and academic discussions.

Conflicts of Interest: The authors declare no conflict of interest.

References

1. Sun, Q.; Gong, X.F. Optimization design for edge-lateral support of a medium-aperture lightweight primary mirror. *Appl. Opt.* **2020**, *59*, 10498–10505. [CrossRef] [PubMed]
2. Zhu, C.M.; Xu, T.Q.; Liu, S.F.; Yang, B.; Liu, Y.N. Design of primary mirror supporting structure and lightweight of space camera. In Proceedings of the 6th International Symposium on Advanced Optical Manufacturing and Testing Technologies (AOMATT)—Advanced Optical Manufacturing Technologies, Xiamen, China, 26–29 April 2012.
3. Zhang, D.D.; Li, W.Y.; Lv, Q.B.; Liu, Y.Y.; Chen, X.W. Lightweight Design and Finite Element Analysis of Primary Mirror For The Space Telescope. In Proceedings of the Optical Systems Design—Optical Design and Engineering VI, Jena, Germany, 7–10 September 2015.
4. Xin, L.W.; Lin, G.C.; Hui, R.C. Finite Element Analysis of Lightweight, Active Primary Mirror. In Proceedings of the 6th International Symposium on Advanced Optical Manufacturing and Testing Technologies (AOMATT)—Large Mirrors and Telescopes, Xiamen, China, 26–29 April 2010.
5. Li, X.P.; Shi, J.F.; Wang, W.; Wang, Y.J.; Fan, X.W. Review on Splicing Structure Technology of Large Aperture Space Primary Mirror. *Laser Optoelectron. Prog.* **2018**, *55*, 28–40.
6. Kamiya, T.; Yasuda, S.; Okamoto, A.; Sato, S.; Ueno, H.; Mizutani, T. Material selection of 3.5-meter segmented mirror for geostationary Earth observation satellite. In Proceedings of the Symposium on Astronomical Optics—Design, Manufacture, and Test of Space and Ground Systems II held at SPIE Optical Engineering + Applications Conference, San Diego, CA, USA, 12–15 August 2019.
7. Sun, N.; Zhuo, R.S.; Cong, J.F. Optimum Design of the Support System of the SiC primary Mirror with 1m Aperture. In Proceedings of the International Conference on Nanotechnology and Precision Engineering (ICNPE 2012), Guilin, China, 26–27 December 2013.
8. Liu, X.Y.; Zhang, J.X.; Wu, X.X.; Li, J.F.; Guo, P.; An, Q.C. Study on the sensitivity of temperature gradient for large aperture SiC lightweight mirror based on active optics. In Proceedings of the 7th International Symposium on Advanced Optical Manufacturing and Testing Technologies (AOMATT)—Large Mirrors and Telescopes, Harbin, China, 26–29 April 2014.
9. Fox, A.; Hobbs, T.; Edwards, M.; Arnold, M.; Sawyer, K. ULE (R) design considerations for a 3m class light weighted mirror blank for E-ELT M5. In Proceedings of the Conference on Advances in Optical and Mechanical Technologies for Telescopes and Instrumentation II, Edinburgh, UK, 26 June–1 July 2016.

10. Huczko, A.; Dąbrowska, A.; Savchyn, V.; Popov, A.I.; Karbovnyk, I. Silicon carbide nanowires: Synthesis and cathodoluminescence. *Phys. Status Solidi (B)* **2009**, *246*, 2806–2808. [CrossRef]
11. Ning, G.; Zhang, L.; Zhong, W.; Wang, S.; Liu, J.; Zhang, C. Damage and annealing behavior in neutron-irradiated SiC used as a post-irradiation temperature monitor. *Nucl. Instrum. Methods Phys. Res. Sect. B Beam Interact. Mater. At.* **2022**, *512*, 91–95. [CrossRef]
12. Cheng, J. *The Principles of Astronomical Telescope Design*; Springer: New York, NY, USA, 2009; pp. 101–118.
13. Montagnino, L.A. Test And Evaluation Of The Hubble Space Telescope 2.4-meter Primary Mirror. *Proc. SPIE Int. Soc. Opt. Eng.* **1986**, *571*, 182–190.
14. Gehrz, R.D.; Roellig, T.L.; Werner, M.W.; Fazio, G.G.; Houck, J.R.; Low, F.J.; Rieke, G.H.; Soifer, B.T.; Levine, D.A.; Romana, E.A. The NASA Spitzer Space Telescope. *Rev. Sci. Instrum.* **2007**, *78*, 011302. [CrossRef] [PubMed]
15. Zinn, J.W.; Jones, G.W. Kepler primary mirror assembly: FEA surface figure analyses and comparison to metrology. In Proceedings of the Conference on Optical Manufacturing and Testing VII, San Diego, CA, USA, 28–29 August 2007.
16. Werner, M. The Spitzer Space Telescope. *Opt. Eng.* **2012**, *51*, 011008. [CrossRef]
17. Fransen, S.; Doyle, D.; Catanzaro, B. Opto-Mechanical Modeling of the Herschel Space Telescope at ESA/ESTEC. In Proceedings of the Conference on Integrated Modeling of Complex Optomechanical Systems, Kiruna, Sweden, 15–17 August 2011.
18. Minier, V.; Bonnet, R.M.; Bontems, V.; de Graauw, T.; Griffin, M.; Helmich, F.; Pilbratt, G.; Volonte, S.; Minier, V.; Bonnet, R.M.; et al. Silicon Carbide Telescope: Radical Innovation. In *Inventing a Space Mission*; ISSI Scientific Report Series; Springer: Cham, Switzerland, 2017; Volume 14, pp. 139–157.
19. Bougoin, M.; Lavenac, J. From HERSCHEL to GAIA, 3-meter class SiC space optics. In Proceedings of the Conference on Optical Manufacturing and Testing IX, San Diego, CA, USA, 22–24 August 2011.
20. Kihm, H.; Yang, H.S. Design optimization of a 1-m lightweight mirror for a space telescope. *Opt. Eng.* **2013**, *52*, 091806. [CrossRef]
21. Butova, D.V.; Tolstoba, N.D.; Fleysher, A.G.; Orekhova, M.K. Optimization of the parameters of space-based mirrors. In Proceedings of the Conference on Optical Components and Materials XIV, San Francisco, CA, USA, 30 January–1 February 2017.
22. Liu, S.; Hu, R.; Li, Q.; Zhou, P.; Dong, Z.; Kang, R. Topology optimization-based lightweight primary mirror design of a large-aperture space telescope. *Appl. Opt.* **2014**, *53*, 8318–8325. [CrossRef] [PubMed]
23. Shao, M.; Zhang, L.; Jia, X. Optomechanical integrated optimization of a lightweight mirror for space cameras. *Appl. Opt.* **2021**, *60*, 539–546. [CrossRef] [PubMed]
24. Qu, Y.; Jiang, Y.; Feng, L.; Li, X.; Liu, B.; Wang, W. Lightweight Design of Multi-Objective Topology for a Large-Aperture Space Mirror. *Appl. Sci.* **2018**, *8*, 2259. [CrossRef]
25. Chen, Y.-C.; Huang, B.-K.; You, Z.-T.; Chan, C.-Y.; Huang, T.-M. Optimization of lightweight structure and supporting bipod flexure for a space mirror. *Appl. Opt.* **2016**, *55*, 10382–10391. [CrossRef] [PubMed]
26. Xu, H. Structural design of primary mirror subassembly for space telescope. In Proceedings of the 2nd International Conference on Mechatronics and Applied Mechanics (ICMAM2012), Hong Kong, China, 6–7 December 2012. Taipei, China, 8–9 December 2012.
27. Han, Y.Y.; Zhang, Y.M.; Han, J.C.; Zhang, J.H.; Yao, W.; Zhou, Y.F. Design and finite element analysis of lightmass silicon carbide primary mirror. *Trans. Nonferrous Met. Soc. China* **2006**, *16*, S696–S700. [CrossRef]
28. Lee, S.-H.; Lee, Y.-I.; Kim, Y.-W.; Xie, R.-J. Mechanical properties of hot-forged silicon carbide ceramics. *Scr. Mater.* **2005**, *52*, 153–156. [CrossRef]
29. Lamon, J. A micromechanics-based approach to the mechanical behavior of brittle-matrix composites. *Compos. Sci. Technol.* **2001**, *61*, 2259–2272. [CrossRef]
30. Margiotta, J.C.; Zhang, D.; Nagle, D.C. Microstructural evolution during silicon carbide (SiC) formation by liquid silicon infiltration using optical microscopy. *Int. J. Refract. Met. Hard Mater.* **2010**, *28*, 191–197. [CrossRef]
31. Chen, K.; Qiu, R. Topological Optimization Design of Imaging Primary Mirror Based on Root-Mean-Square Error of Surface Figures. *AcOpS* **2022**, *42*, 99–104.
32. Fengchang, L.; Wei, L.; Weiguo, Z.; Haibo, Z.; Guanyu, L.; Xiaodong, W. Topology Optimization Based Parametric Design of Balloon Borne Telescope's Primary Mirror. *Appl. Sci.* **2021**, *11*, 5077.
33. Hu, R.; Chen, W.; Li, Q.; Liu, S.; Zhou, P.; Dong, Z.; Kang, R. Design Optimization Method for Additive Manufacturing of the Primary Mirror of a Large-Aperture Space Telescope. *J. Aerosp. Eng.* **2016**, *30*, 04016093. [CrossRef]

MDPI
St. Alban-Anlage 66
4052 Basel
Switzerland
www.mdpi.com

Photonics Editorial Office
E-mail: photonics@mdpi.com
www.mdpi.com/journal/photonics

www.ingramcontent.com/pod-product-compliance
Lightning Source LLC
Chambersburg PA
CBHW061450210926
51850CB00008B/1669
* 9 7 8 3 7 2 5 8 0 9 6 6 0 *